Time-series analysis

A comprehensive introduction for social scientists

Time-series analysis

A comprehensive introduction for social scientists

JOHN M. GOTTMAN

University of Illinois

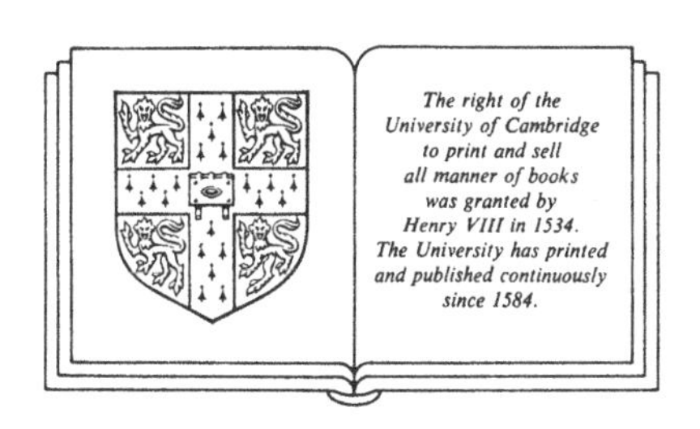

CAMBRIDGE UNIVERSITY PRESS

Cambridge

New York New Rochelle

Melbourne Sydney

CAMBRIDGE UNIVERSITY PRESS
Cambridge, New York, Melbourne, Madrid, Cape Town, Singapore, São Paulo, Delhi

Cambridge University Press
The Edinburgh Building, Cambridge CB2 8RU, UK

Published in the United States of America by Cambridge University Press, New York

www.cambridge.org
Information on this title: www.cambridge.org/9780521103367

First published 1981
Reprinted 1984, 1988
This digitally printed version 2009

A catalogue record for this publication is available from the British Library

Library of Congress Cataloguing in Publication data

Gottman, John Mordechai

Time-series analysis: a comprehensive introduction
for social scientists

Bibliography: p.
Includes index
1. Time-series analysis I. Title
HA 30.3.G67 519.5′5′024301 80-25644

ISBN 978-0-521-23597-6 hardback
ISBN 978-0-521-10336-7 paperback

To Robert E. Bohrer

Oh as I was young and easy in the mercy of his means,
Time held me green and dying
Though I sang in my chains like the sea.

Fern Hill, Dylan Thomas

Contents

Preface

Consider the wind on the waters, the panorama of growth, change, development, evolution, or sudden alteration. We are immersed in a universe of natural events that vary with time. The study of all these things is the subject matter of this book. Despite the ubiquitousness and familiarity of these events, we have to learn how to think of processes that occur in time, because the traditional statistics we have learned ignores the dimension of time.

Time-series analysis is the study of variation across some discrete or continuous dimension, which is usually called "time." There are two major approaches to the study of time-series processes, *time-domain* and *frequency-domain analyses*. The first section of this book consists of six chapters that have a minimum of equations. They are designed to introduce the language and the concepts of *both* domains of time-series analysis. Although time and frequency domains are mathematically equivalent, they are, nonetheless, different metaphors for thinking about data. You will need experience with both of these metaphors for them to become useful. Furthermore, it is valuable to be able to switch from one domain to the other.

The book begins with seven chapters that are a nonmathematical encouragement to learn time-series analysis. Concepts are introduced without the mathematics so some readers who ought to be attracted by the subject matter are not frightened away. The first three chapters discuss cycles, because they are part of the most intuitively natural and yet most difficult metaphor of time-series analysis. Chapter 4 is a discussion of the bivariate time-series case. Chapter 5 is a discussion of autoregressive models and forecasting; this is an introduction to the time domain. Chapter 6 is a discussion of experimentation and levels of causal inference, moving from the time to the frequency domains in the discussion.

Just as in classical statistics, in time-series analysis we are interested in making inferences from a sample of data points to something else; in the case of time-series analysis the inference is to the process that may have generated the sample. This leads to the second section of the book, time-series models. The book thus turns to a discussion of the need for models and the problem of correlated data in Chapter 7; Chapter 8 is a discussion of the basic assumptions of time-series analysis–what is known as stationarity. Chapter 9 takes up the problem of what to do if the data are not stationary, and suggests the use of general models with deterministic and stochastic components.

Part III focuses on stationary time-domain models. Chapter 10 introduces the moving-average models. Chapter 11 begins the discussion of autoregressive models. This discussion continues in Chapter 12, which introduces the second-order autoregressive model, a model that is capable of simulating probabilistic periodicity. In the next two chapters the duality of autoregressive and moving-average models is revealed.

Part IV focuses on stationary frequency-domain models. Chapter 15 begins by discussing the spectral density function and its colorful history; the periodogram is introduced in Chapter 16; the spectral windows are discussed in Chapter 17; and in Chapter 18 an attempt is made to integrate time and frequency domains by a discussion of the spectral densities of moving-average and autoregressive processes, thus explaining the Slutzky effect.

Part V is concerned with estimating the parameters of time-domain models. Chapter 19 is a discussion of the Mann–Wald least-squares autoregressive model-fitting procedure; this procedure is compared to Box–Jenkins models and Wu–Pandit models. Chapter 20 is a more complete discussion of the Box–Jenkins ARIMA models. Most of Part V argues the premise that most users of time-domain methods would do well to use least-squares linear autoregressive models, even if they are not as parsimonious as Box–Jenkins or Wu–Pandit models. Autoregressive models are more familiar models and easier computationally. However, for the sake of thoroughness, all three methods are reviewed. The next two chapters are more applied. Chapter 21 discusses least-squares forecasting, and Chapter 22 presents a worked example of model fitting. This chapter also discusses methods of time-domain and frequency-domain model fitting without the use of a computer.

Throughout the book, primers are included in chapter appendixes that are designed to assist the reader who is unfamiliar with areas such as regression, matrix algebra, the full general linear model, and Fourier series.

Part VI is called "Bivariate Time-Series Analysis." It includes Chapter 23 on bivariate frequency-domain analysis, with a worked example on mother–infant play in Chapter 24. Chapter 25 is about bivariate time-domain analysis; it also includes a worked example, once again on mother–infant play. Chapter 26 is about the interrupted time-series experiment, and Chapter 27 is about multivariate extensions, including time-series regression analysis.

This work is a practical handbook that tries to convey a healthy respect for empirical common sense. An additional goal, however, is to influence the way social scientists think about problems, stressing the facts that much can be gained by searching for patterns across time and the way these patterns themselves change, and that stochastic rather than purely deterministic models may be better at representing data.

Designed to be everyone's first time-series book and to invite use of these techniques, the work will suit many fine scholars who could benefit from a familiarity with time-series analysis but who would be kept away by all the equations that are necessary to discuss the subject. If I could, I would materialize like a genie from the pages and relax the fear away from these scholars whenever it arises, and coax them back to the text. Instead, what I have chosen to do is to make it possible to read the book at several levels and at several speeds. The chapters are usually short, so that only a few concepts are introduced at a time. Although the effect may be a choppy book, completing chapters always gives me a sense of accomplishment as a reader. I have tried to avoid making the reader wade through long presentations that lose the point. There is a lot of prose, and the concepts are presented for both the mathematical and nonmathematical reader. The book should thus serve as an introduction for both kinds of readers.

This work would not have been possible without the release time provided by a National Institute of Mental Health Research Scientist Development Award 1K02MH00257 and, previous to that, by a University of Illinois Faculty Study in a Second Discipline fellowship. The Second Discipline fellowship made it possible for me to spend a year focusing on time-series analysis in the mathematics department at the University of Illinois.

I am indebted to various colleagues and teachers over the last ten years: Robert Bohrer, Richard Devor, Gene Glass, Robert Kuhn, James Ringland, and Takamitsu Sawa. In particular, I wish to acknowledge the superb tutoring I have received from James Ringland over the last two years. James Ringland is a gifted teacher, a talented statistician, and a valuable colleague. I am indebted to Duane Steidinger, who worked in my laboratory from 1977 to 1979 as a computer programmer and research associate. I have also benefited from conversations with my colleagues Robert Bohrer, Steve Porges, and Jim Sackett. I would like to express my appreciation to those who have shared their data with me for use in this book, in particular Professors Gene Glass, Robert Levenson, Steve Porges, and Edward Tronick. In some cases data from Glass, Willson, and Gottman (1975) were used. I wish to acknowledge the feedback I received from the members of a class I taught at the University of Illinois, particularly the editorial comments of Ms. Janet Kruse. Merle Thorne was an extremely helpful, intelligent typist of the last several drafts of the manuscript. Most important, this work has thrived from the support I have received from my department at Illinois. For the magnificent intellectual climate that my colleagues at Illinois provide, I am grateful.

April 1981

J. M. G.

Part I

Overview

The first six chapters of this book are an advertisement for time-series analysis. They are written to be read casually, to introduce some new vocabulary to the reader on primarily an intuitive level, with visual illustrations. Do not read these chapters expecting the technical terms to be defined precisely. Relax. Eventually, the text will cycle back over these ideas and add the necessary precision to this intuitively based introduction.

1
The search for hidden structures

This chapter introduces one concept, the spectral density function, which is capable of detecting the frequencies embedded in the time series even when the data are masked by noise. For the time being, we will discuss the case when there are true cycles in the data. Later, this notion will be generalized to the case when the oscillations are not perfectly cyclic.

Brain waves

When we record changes in electrical potential on the scalp of a subject who is either awake, drowsy, or in different stages of sleep, we obtain electroencephalographic (EEG) brain-wave patterns of the type illustrated in Figure 1.1. When a person is awake and active, the EEG record consists of rapid waves of low amplitude, as in part A. As a person becomes relaxed and eventually drowsy, the EEG is characterized by a wave of 8 to 12 hertz (Hz; cycles per second) called an *alpha wave* (B). As the person falls asleep the waves become slower and have larger amplitude (C to F). During

Figure 1.1. Electroencephalographic patterns characteristic of various states of wakefulness and sleep.

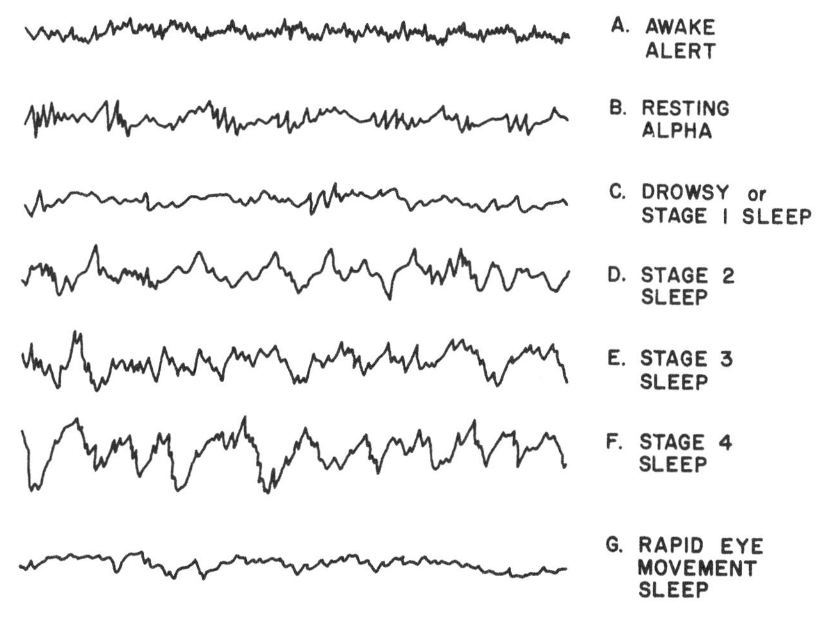

Figure 1.2. Hypothetical spectral density function for stage 2 and alpha EEG patterns.

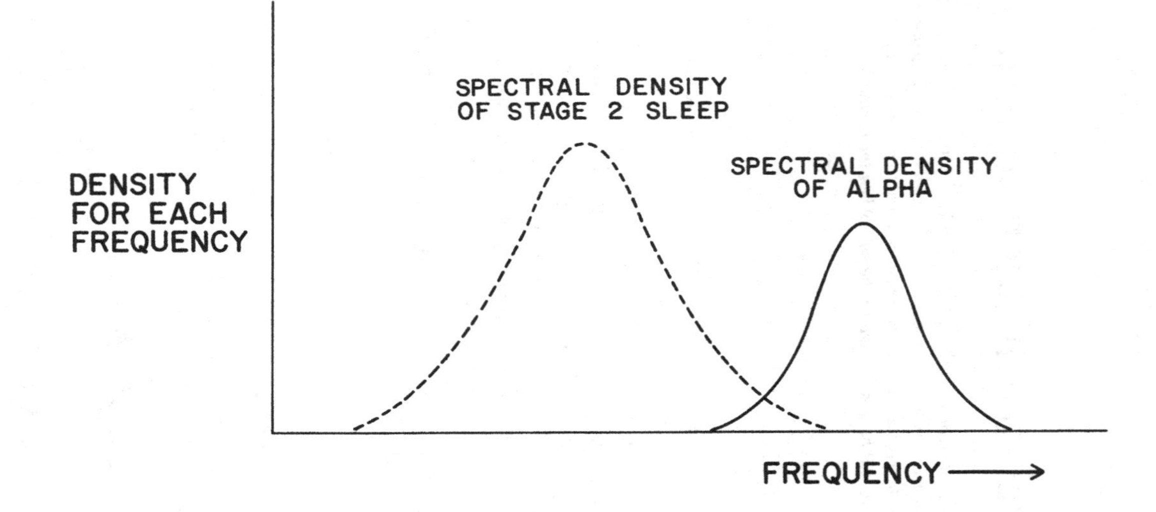

rapid-eye-movement dreaming (G) the EEG pattern is much more like that of wakefulness shown in part A than those of the other sleeping waves.

We can examine these graphs visually, but is there some way of describing them systematically? The waking EEG is usually described as containing "high-frequency, low-amplitude, erratic cycles." What do the terms "frequency," "amplitude," and "cycles" mean? The term "frequency" in time-series analysis does not mean how often an event occurs, as it does in introductory probability and statistics. Instead, *frequency* means how rapidly things repeat themselves. For example, a stagecoach wagon wheel that is rotating slowly is rotating at a slow frequency; a particular spot on the wheel will be on top every full revolution of the wheel. This will occur slowly (slow frequency) if the wheel is rotating slowly, and rapidly (fast frequency) if the wheel is rotating rapidly. In terms of time-series data, if the data wave smoothly and slowly up to a peak and then smoothly and slowly down to a valley, the data are said to have a lower frequency than if they moved back and forth from peak to valley with every successive observation. *Amplitude* refers to how large the magnitude of the change is from peak to valley. In Figure 1.1 the graph for A, awake–alert, has less amplitude than the graph for F, stage 4 sleep. *Cycle* implies *regular* repetition over a fixed time, called the *period*, of this transition from peak to valley and back to peak. A rapid cycle refers to a high-frequency wave that moves from peak to peak in a short time, or short period.

In the EEG of a drowsy person, cycles called *alpha*, the rhythm of relaxed wakefulness, predominate. In the change from wakefulness to drowsiness, the EEG picks up a cycle. To describe this, a function called the *spectral density* would be used. The spectral density would tell us something like how much variance is accounted for by each frequency we can measure. The spectral density of the awake and relaxed state would thus show a peak in those frequency ranges where we expect to find alphas, and would not show a peak elsewhere. By using the spectral density, we could quantitatively assess the state of our subject. Figure 1.2 illustrates this concept of spectral density.

Why is the spectral density necessary? The answer is that the textbook example of Figure 1.1 is not what we usually observe in practice. In the next section we see how a pure cycle can be hidden by noise so that visual inspection alone is inadequate, but the spectral density can easily pick out our pure cycle.

Hidden cycles

Figure 1.3 shows a pure sine wave. No one would have to compute a spectral density or do any statistics at all if our data were this regular. In this case,

Figure 1.3. Example of a pure sine-wave process.

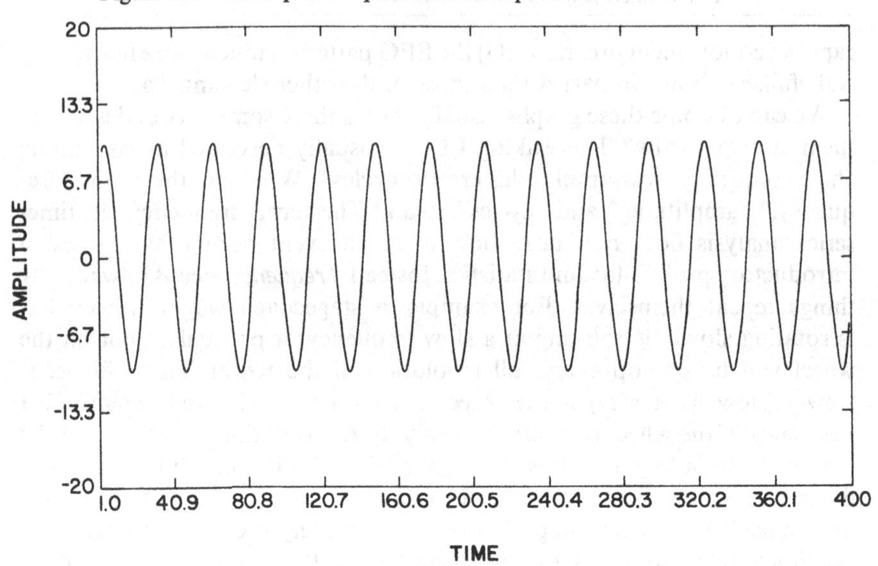

Figure 1.4. Example of a purely random ("white-noise") process.

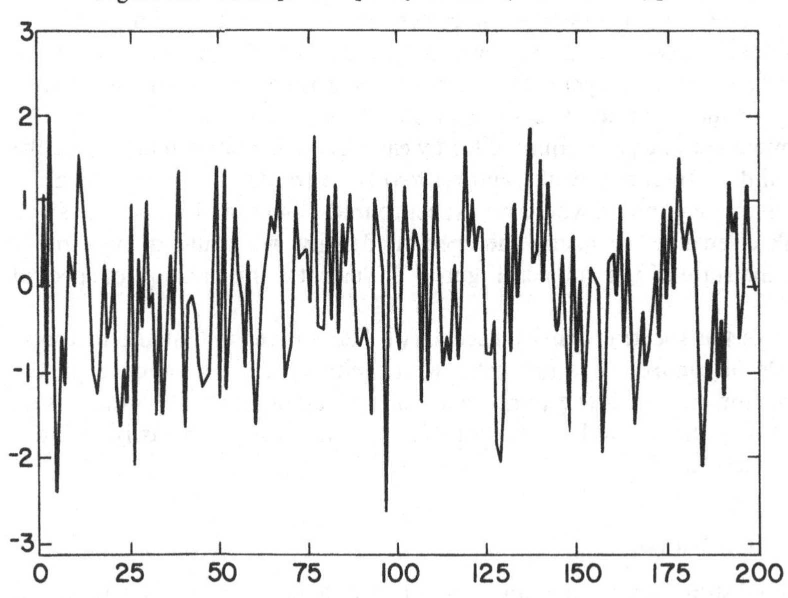

in fact, the spectral density function would be a spike at one frequency and zero everywhere else. Let us examine the opposite state of affairs, a set of completely independent random numbers, called *white noise*, depicted by Figure 1.4. Here there is absolutely no regularity, and all cycles are present with equal intensity, much as white light consists of light of all colors mixed with equal brightness. The spectral density of white noise should theoretically be a straight line with no peaks.

Now consider a pure tone masked by white noise, displayed in Figure 1.5. It is extremely difficult to detect any pure cycle in Figure 1.5a. But the spectral density of the data in Figure 1.5b shows a spike at exactly the right frequency to detect the hidden cycle. This example shows how a spectral analysis of the data can point out structure in the data not visible to the eye.

Detecting hidden changes

In some cases the mean and variance of a time series may not change after some planned experiment, but some other facet of the data, such as its time dependence, does change; however, if we rely only on analysis of variance we might falsely conclude that no change had taken place. In Figure 1.6 we

Figure 1.5. (a) Example of a sine wave masked by white noise. (b) The spectral density of a sine wave masked by white noise reveals a spike at the correct frequency.

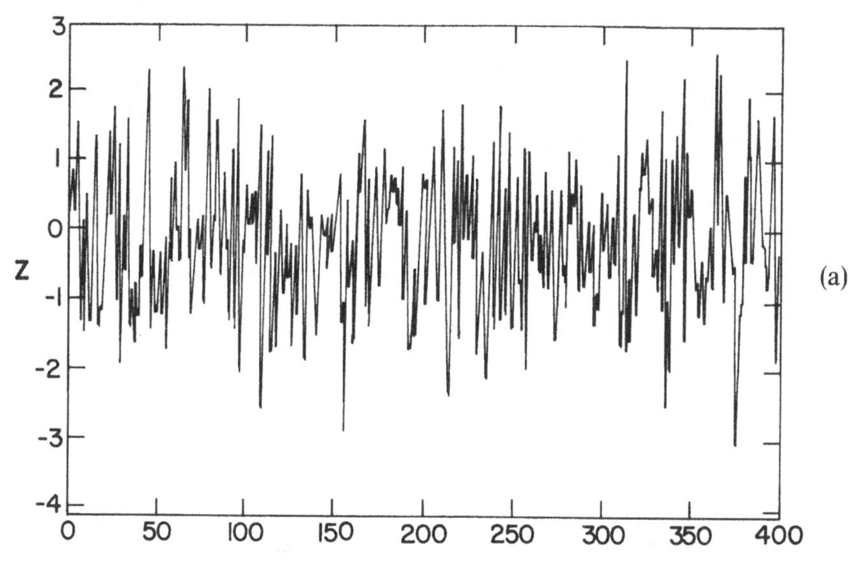

Figure 1.5. (*cont.*)

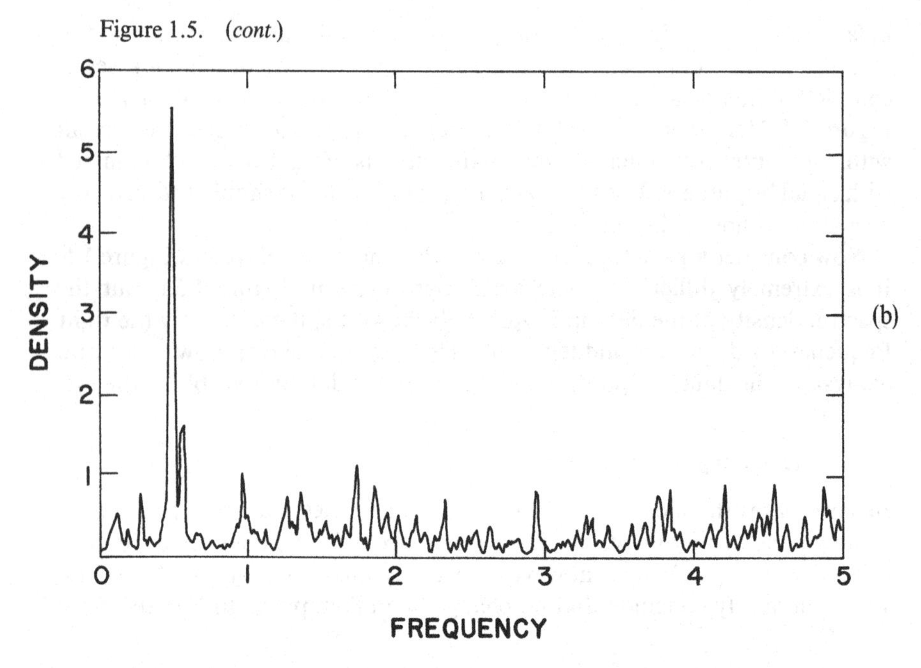

Figure 1.6. Word-association test scores of one schizophrenic patient in two drug conditions.

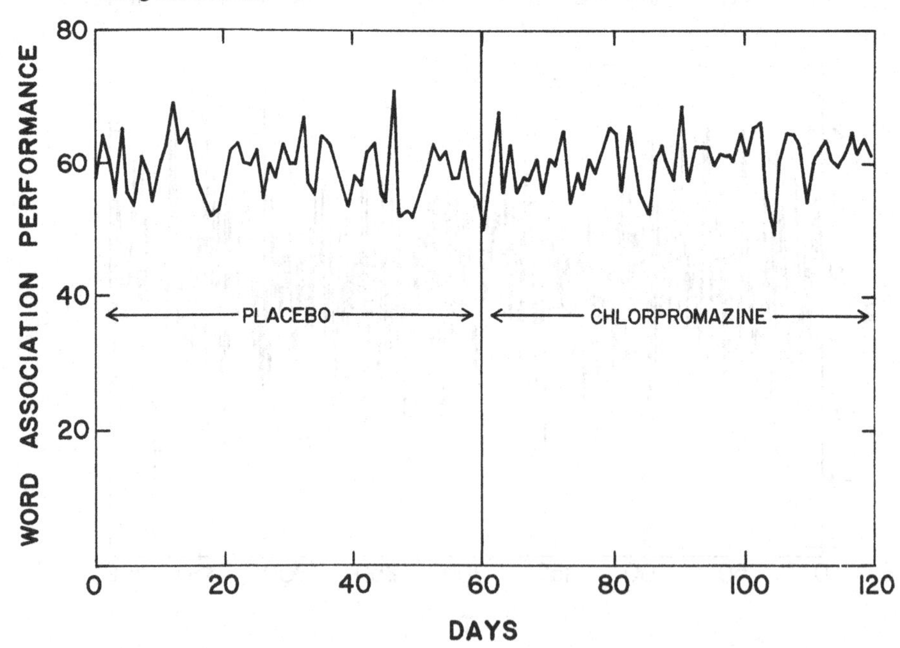

see the daily Word Association Relatedness scores for one schizophrenic patient for 120 days (Holtzman, 1963). This measure presumably assesses the normality of associations or thought processes, with lower scores indicative of more private, or "strange," associations. The means of the two halves of the series are 59.10 for placebo and 60.75 for drug, and the variances are 18.32 and 16.25, respectively. There really has been no change in mean level or variance in the behavior of this person from placebo to drug.

However, if we search for cyclicity in the placebo condition, we find (Figure 1.7) a peak at one cycle, with a frequency of .1. The period of the cycle is the time it takes for the behavior to repeat itself, and it is the reciprocal of the frequency, or $1/.1 = 10$ days. There is a dominant cycle, when the person is on the placebo, of 10 days.

What happens to this cycle on the drug? If we look at the spectral density of the 60 drug days we find that there is no longer a peak at .1, but that the peak has shifted over to the right, and is over a frequency of .2. This is a new period of $1/.2 = 5$ days (Figure 1.8). Therefore, the drug has shifted the cycle to a more rapid frequency.[1] The behavior of the schizophrenic has become more erratic and less smooth. At this point in our research we may not know if this change is "good" or "bad." This example illustrates how thinking in terms of means and variances may miss what could be valuable

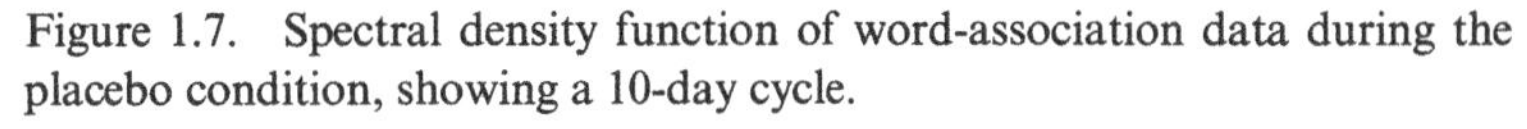
Figure 1.7. Spectral density function of word-association data during the placebo condition, showing a 10-day cycle.

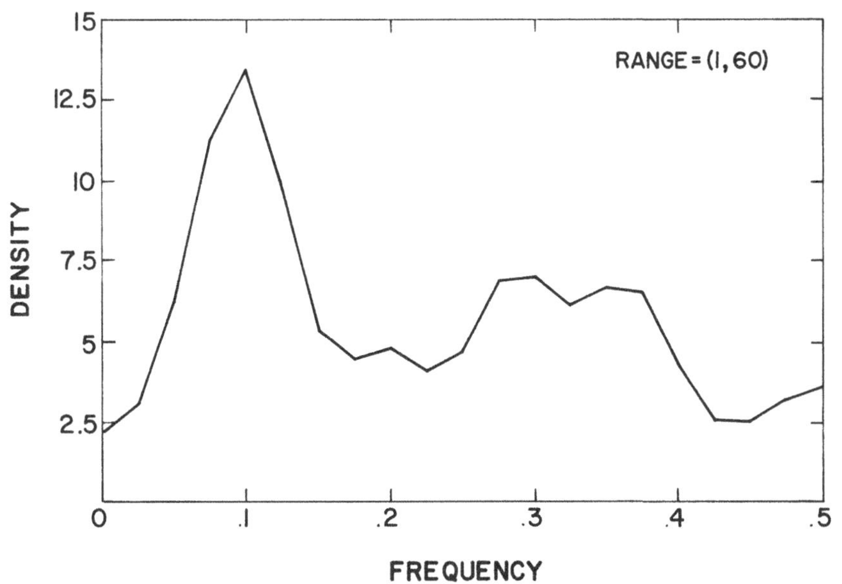

Figure 1.8. Spectral density function of the word-association data during the chloropromazine condition, showing a 5-day cycle.

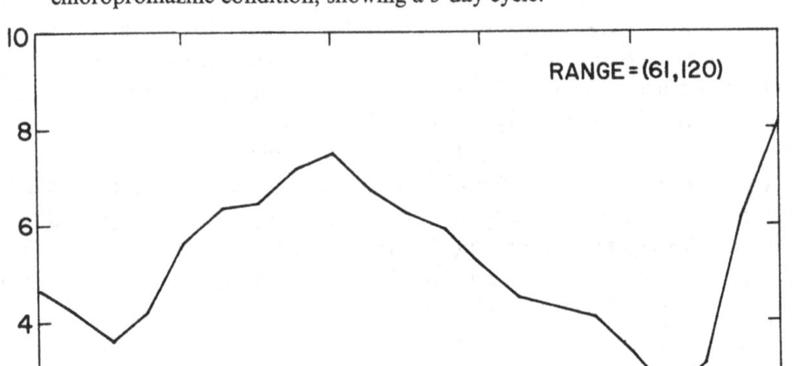

information in the data. These shifts in frequency are statistically significant, and this book will discuss how to perform tests of significance on data such as these.

2
The ubiquitous cycles

This chapter continues the discussion of cyclicity and introduces the idea of spectral decomposition, in which a time series is represented as a sum of independent cycles of different frequency. Traffic fatalities data are analyzed visually using this concept of spectral decomposition. The idea of a sum of cycles as an approximation is discussed, which introduces the notion of a Fourier series.

Most people are familiar with the rainbowlike spectrum into which sunlight is transformed when passed through a prism, and many have also seen the sharp line spectrum that results when monochromatic light is passed through a prism. Mixtures of monochromatic light give rise to lines corresponding to the separate components when passed through a prism, and chemical elements, when properly excited by flame or electric spark or some other means, exhibit characteristic line spectra or band spectra or some combination of the two types. The line spectra are caused by electrons changing energy states in their atoms, but the band spectra are caused by more complicated motions of atoms and molecules. The many frequencies present in sunlight result from the numerous energy transitions going on in the chaos of the thermonuclear reactions. Since any steady-state source of visible light will have a definite spectrum, we can characterize the source by measuring the line and band components. In all cases, broad-band spectra will be associated with more complicated and random atomic and molecular motions, which I will call the "stochastic" component.

A similar situation obtains if we feed an electrical signal composed of a sine wave plus a stochastic component (such as might be produced by vacuum-tube shot noise or an intermittent spark) into a radio receiver. If the radio tunes reasonably sharply and with a fairly flat response over frequencies surrounding that of our sine wave, we will hear the sine wave as a pure tone when we are tuned to its frequency, and we will hear the same tone but with diminishing volume as we tune away from that frequency. Actually, since the ear has a nonlinear response, there will also be a slight tone shift. However, the stochastic component will be heard at about the same volume through a wide tuning range. Anyone who has ever tried to listen to AM radio during a lightning storm and tried to tune out the accompanying static is aware of this phenomenon. Once again we have a case in which a pure tone is characterized by a narrow band of radio frequencies, or "spectral line," and the stochastic component by a broad band of radio frequencies.

In many areas of human activity, measurements of some quantity are made either continuously or at discrete points along the time axis. Furthermore, over long periods of time many of the phenomena being measured have a steady-state quality, or the data can be transformed to exhibit steady-state behavior. (Chapter 8 will explicate this concept of steady-state behavior by introducing the notion of "stationarity.")

In view of the preceding discussion, it is then natural to ask whether there exists a "mathematical prism" or "mathematical tuner" through which we can pass such data and pick out line spectra and band spectra that help characterize and classify the data. Return to the example of brain-wave data measured at various stages of sleep (Figure 1.1). The stochastic component will have a spectral density function that covers a band of frequencies and thus the data will be repetitive, but not as precisely as one would expect if the spectral density function were composed of precise lines. The frequencies of the bands of cycles shift with different stages of sleep. There may also be precise cycles, or line spectra, represented in a set of data, in which case the data would have a spectral density composed of both bands (corresponding to the stochastic component) and lines, corresponding to the pure cycles (called the "deterministic component"). It would be very useful to accurately quantify the cyclical properties, that is, determine the line spectra, but it might also be very useful to profile the band spectra, in order to more fully understand the brain's activity. As might be supposed from the existence of this introduction, the answer to the question posed above is that there is such a technique, called the *spectral analysis of time series*.

Since the spectrum of electromagnetic radiation and the corresponding mathematical techniques applied to other data involve viewing the wave features of the observed quantities after they have been "prismed," we call such an analysis a *frequency-domain analysis*. If we study the data directly as a graph of measurement against time, the analysis is called a *time-domain analysis*. In this chapter we talk in the frequency domain.

To describe those EEG patterns, we have been using the language of frequency-domain time-series analysis. The language is relatively simple until we realize two things. First, most time series are not composed simply of one basic frequency band but are composites of several basic frequency bands. This will make it necessary for us to speak of frequency *components* of a time series. Second, most time series are not *deterministic* sums of sinusoidal components but are, instead, probabilistic. These two facts make frequency-domain time-series analysis somewhat more complex.

To simplify our discussion, it will be helpful if we build some of the foundations of spectral time-series analysis by introducing the concept of frequency decomposition.

Frequency decomposition

The most common time-series problem in the frequency domain is called the *spectral decomposition* problem. It consists of attempting to identify oscillations of major frequencies that explain variance in an observed time series. We will see that the solution of this problem was not as simple (historically) as it seemed at first. However, let us begin by discussing the problem conceptually.

Imagine the electronic arrangement depicted in Figure 2.1. We see a time-series process displayed on an oscilloscope labeled "input." Imagine a "spectral frequency analyzer" as a device that receives the input signal and has electronic resonators that detect major frequencies in the input signal. If there is sufficient energy at a range of particular frequencies the column of lights corresponding to each frequency band will light up; the number of lights in each column is proportional to the variance accounted for, or power, for each frequency band.

The display shows the most energy (number of black dots) at a midrange of frequencies, with some (smaller) contribution at lower and higher frequencies.

The search for a mathematical function like our hypothetical spectral decomposition frequency analyzer is basic to the spectral decomposition problem. The problem is to decompose the observed series into its basic frequencies, much as a prism breaks light into its basic colors. Each color is visible electromagnetic radiation of a particular frequency. The relative brightness of the colors corresponds to the relative heights of the spectral density function.

Introduction to frequency-domain analysis

During the latter part of the last century, the problem of extracting periodic oscillations from observational data began to receive considerable attention. This was particularly true in the area of astronomy, where the improvement and refinement of observational equipment resulted in more noise being detected in the observations. Since even a periodic component that contributes a high proportion of the variance to a signal can sometimes be distorted by noise to the extent that it cannot be detected by "eyeballing" the data, it is imperative that an analytical method of detecting these "hidden periodicities" be available. One of the first analytical techniques developed was *periodogram analysis*, which is based on what is known as the "finite Fourier series fit" to the data points. Although the periodogram itself has certain defects, it continues to occupy an important role in the analysis of stationary time series.

Figure 2.1. Hypothetical electronic spectral density analyzer.

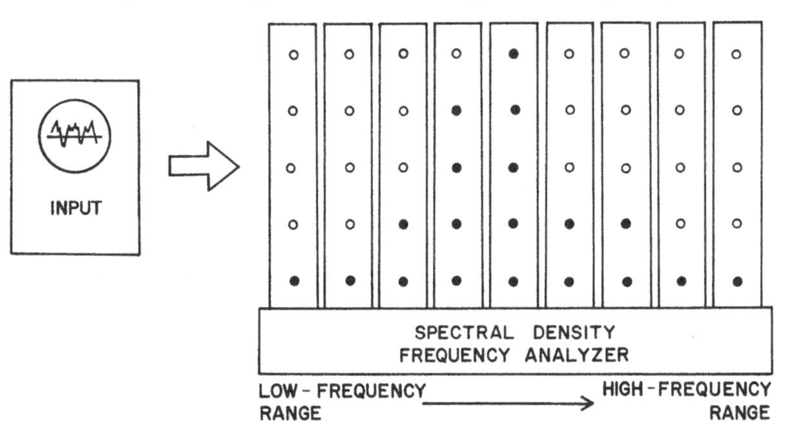

Periodic functions

A function $f(t)$ is called a *periodic function* with period T if, for all t,

$$f(t) = f(T + t)$$

If a function has period T, it also has the period $2T, 3T, \ldots$. Figure 2.2 illustrates a periodic function.

A simple example of a periodic function with frequency f is the *sine wave*:

$$x(t) = A\sin(2\pi ft + \phi)$$

Figure 2.2. Example of a periodic function, with period T.

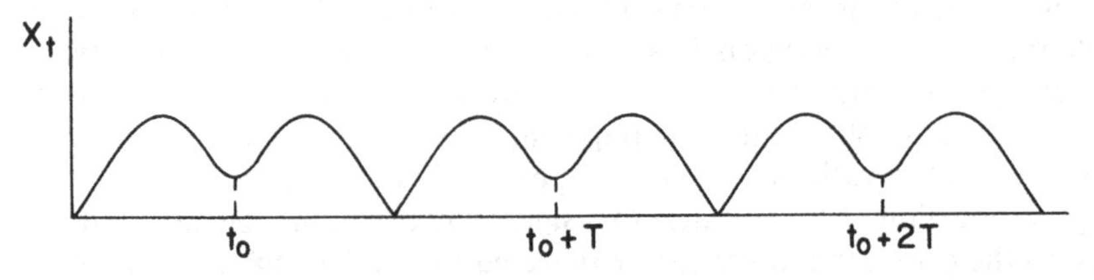

Figure 2.3. Example of a sine wave with amplitude A, frequency f, period T, and phase ϕ.

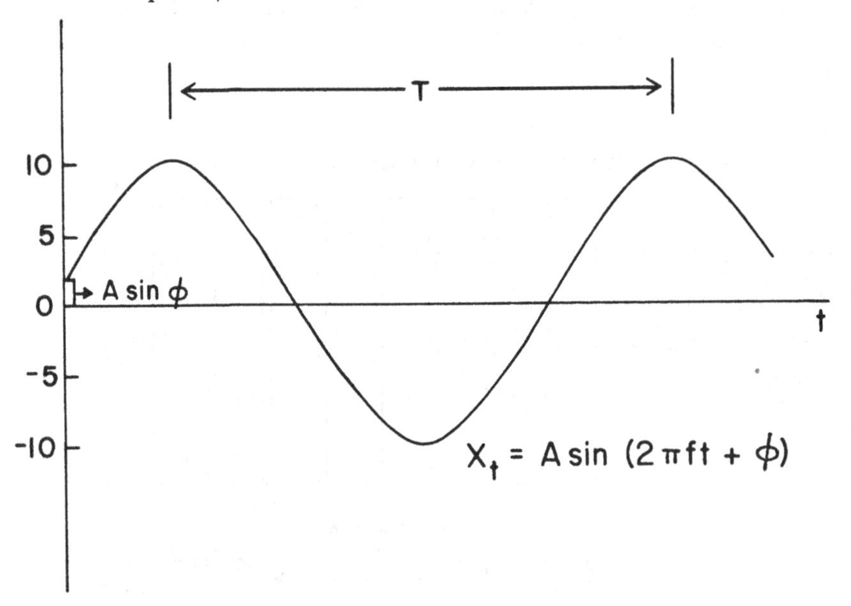

The period of this function is $T = 1/f$ because

$$x\left(t + \frac{1}{f}\right) = A \sin\left[2\pi f\left(t + \frac{1}{f}\right) + \phi\right]$$
$$= A \sin[(2\pi ft + \phi) + 2\pi] = A \sin(2\pi ft + \phi)$$

The period T is the peak-to-peak time in Figure 2.3. The constant A is called the *amplitude*, the constant f is called the *frequency*, and the constant ϕ is called the initial *phase*.

The reader will recall from high school trigonometry that a sine wave is a periodic function. The simplest example of how a sine wave can be generated is from the motion of the pendulum of a grandfather clock. If a pen is attached to the pendulum and a sheet of paper moved under it at a uniform speed, the line drawn by the pen will be a sine wave (see Figure 2.3). If the pendulum moves back and forth faster, the pen will describe a sine wave with faster frequency (see Figure 2.12). The amplitude of the sine wave is how far the pendulum swings to the left and right of its center, and the phase is where in the arc of its swing the pendulum is released. If it started at its center, the phase is zero. The reader who cannot follow the next exercise should probably review some elementary trigonometry.

> *Exercise.* If time is measured in months, what is the period of $\sin(\pi/6)t$?
> Answer: $\sin(\pi/6)t = \sin 2\pi ft$. Solve for f and get $f = \frac{1}{12}$. The period $T = 1/f = 12$ months. This is an annual cycle.

The fastest frequency we can detect

If we have a time series measured at equal intervals as shown in Figure 2.4, the fastest oscillation we can observe is one that goes above the mean on one time point and below the mean on the subsequent time point. The period of this fastest frequency is two time points, shown as 2 days in Figure 2.4, and the fastest frequency is thus .5. This fastest frequency is called the *Nyquist frequency*. The slowest frequency we can detect is zero. This is a frequency that never repeats. Trend lines have zero frequencies and infinite period. *For these reasons the frequencies on our graphs always range between zero and .5.*

In many instances, the data are a continuous time series that is sampled at a constant interval. In these cases the fastest frequency we are capable of detecting in our data may actually be slower than the actual frequencies in the data. Figure 2.4b illustrates the problem. The solid line represents the actual frequency of the data, but because of the sampling rate, an entirely different frequency will be detected. This phenomenon was called *aliasing* by John Tukey. The familiar cinematographic effect of seeing stagecoach or wagon wheels going first forward, then backward, then forward again is an example of this effect of aliasing. This problem is discussed by Blackman and Tukey (1958).

Figure 2.4. (a) The Nyquist frequency, the fastest frequency we can detect, has period $T = 2$ and frequency $f = 0.5$. (b) Aliasing.

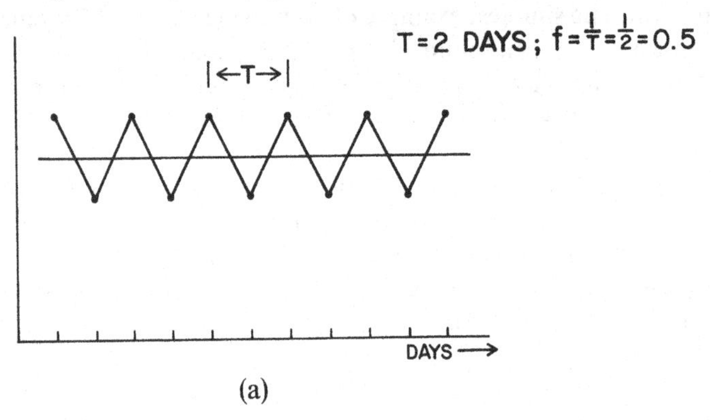

(a)

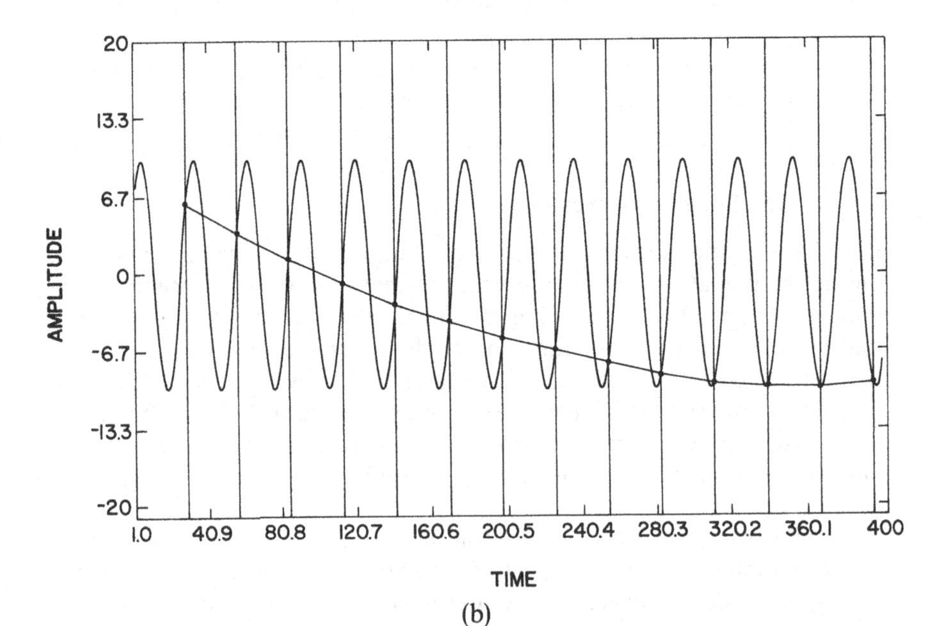

(b)

Case example: traffic fatalities data

On December 23, 1955, the governor of Connecticut initiated a program to crack down on speeding. Glass (1968) undertook the evaluation of the Connecticut intervention, in part by graphing traffic fatalities in four neighboring "control" states (Massachusetts, New Jersey, Rhode Island, and New York). Figure 2.5 is a graph of the unweighted average of these fatalities for the four control states, plotted monthly. Let us examine the cyclicity of the data prior to the Connecticut crackdown, that is, for the first 60 months. The spectral density function is plotted in Figure 2.6. Note that it has one peak near $f = .075$, or a period of $T = 1/f = 13.3$ months, which is close to an annual cycle. This is consistent with our knowledge that traffic fatalities tend to peak in August.

Figure 2.7 is the spectral density function for the postintervention months. This plot does not look very different from the plot for the preintervention months, and the peak occurs at the same frequency.

Parenthetically, there is one difference between Figures 2.6 and 2.7 worth noting – the height of the spectral density at zero frequency. It is much higher before the intervention. In Chapter 9 we again note that this zero-frequency elevation of the spectral density in Figure 2.6 reflects a trend, which in this case is the downward trend in the data that can be noticed by visual inspection before the Connecticut speeding crackdown. The absence of the zero-frequency elevation suggests that this downward trend (or any trend) no longer exists in the postintervention data (i.e., the data appear to have leveled off).

Figure 2.5. Average monthly traffic fatalities in four New England control states, before and after the Connecticut speeding crackdown.

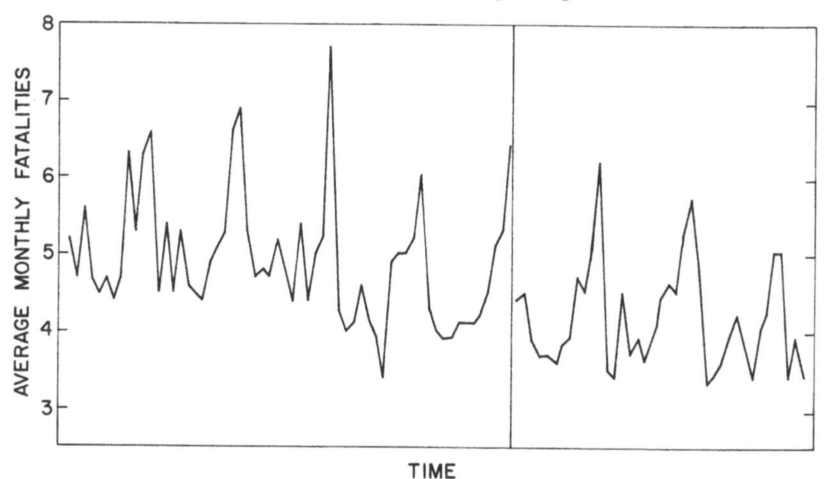

Figure 2.6. Spectral density function of the New England traffic fatalities data prior to the Connecticut speeding crackdown.

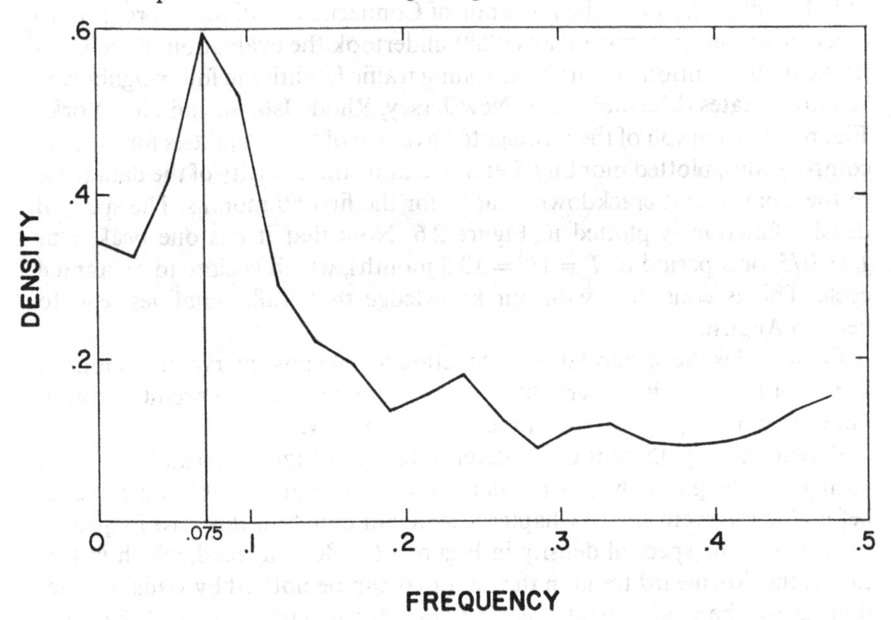

Figure 2.7. Spectral density function of the New England traffic fatalities data after the Connecticut speeding crackdown.

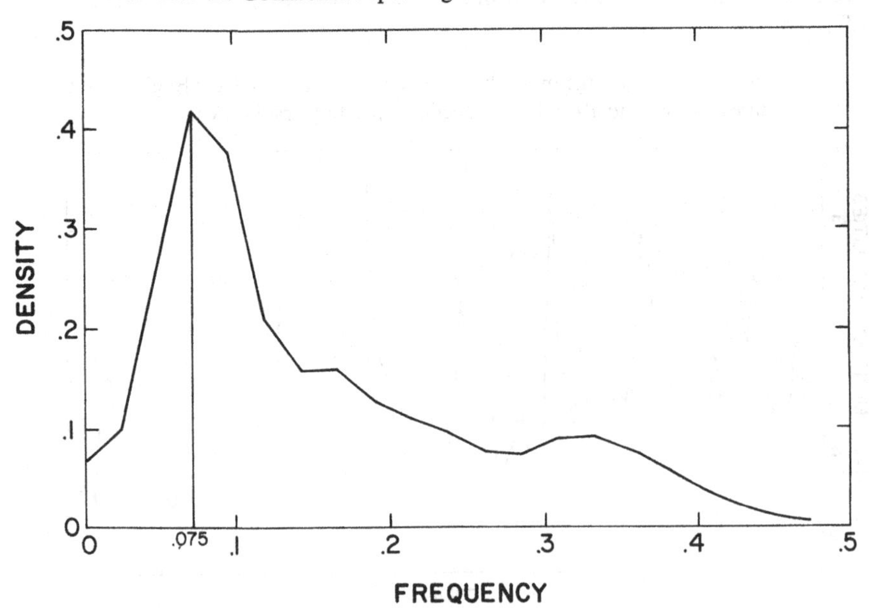

The reader may be wondering if there is any "effect" of the Connecticut crackdown in these control states. If there were, it would bode ill for the conclusion that the Connecticut crackdown had any real effect. Unfortunately, there is a drop in mean level in these fatality data after the Connecticut intervention. From what we know of the behavior of drivers, it is not too likely that this drop in level was the result of the Connecticut governor's program. Hence, the evaluation results in Connecticut must demonstrate an effect greater than this drop in level in the control states, which, sad to say, was not the case (see Glass et al., 1975).

Approximation

One of the ways we can understand the frequency decomposition problem is to try to understand what results when we construct sums of mathematical functions to approximate another function.

In an experiment suppose that we obtain a set of points and plot these points in a graph (see Figure 2.8). We can approximate these points using any of a set of polynomial functions (see Figure 2.9). The cubic seems like the best approximation. Notice, however, that the cubic approximation requires more parameters than does the linear approximation.

More important, notice that we may think of the cubic curve as a weighted sum of four functions, the constant (1), the linear (t), the quadratic (t^2), and the cubic (t^3). In this sense the set of functions $(1, t, t^2, t^3, \ldots)$ may be thought of as the *building blocks* of more complex functions that they can be used to approximate. These building blocks are called a *basis* for the approximation.

Why not simply use polynomial approximation for any set of time-series data? The answer is that, although we would usually obtain a good approximation within our range of data points, beyond this range the polynomial approximation will rapidly move to plus or minus infinity. However, if we believe that our set of data is reasonably representative of what would be

Figure 2.8. Hypothetical time-series data set.

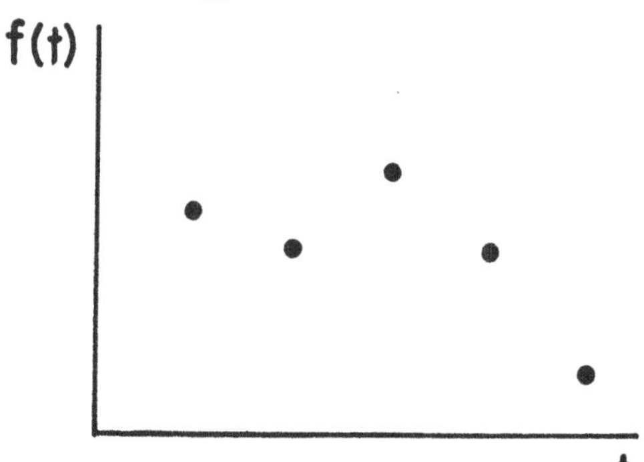

Figure 2.9. Polynomial approximations to the hypothetical time-series data set.

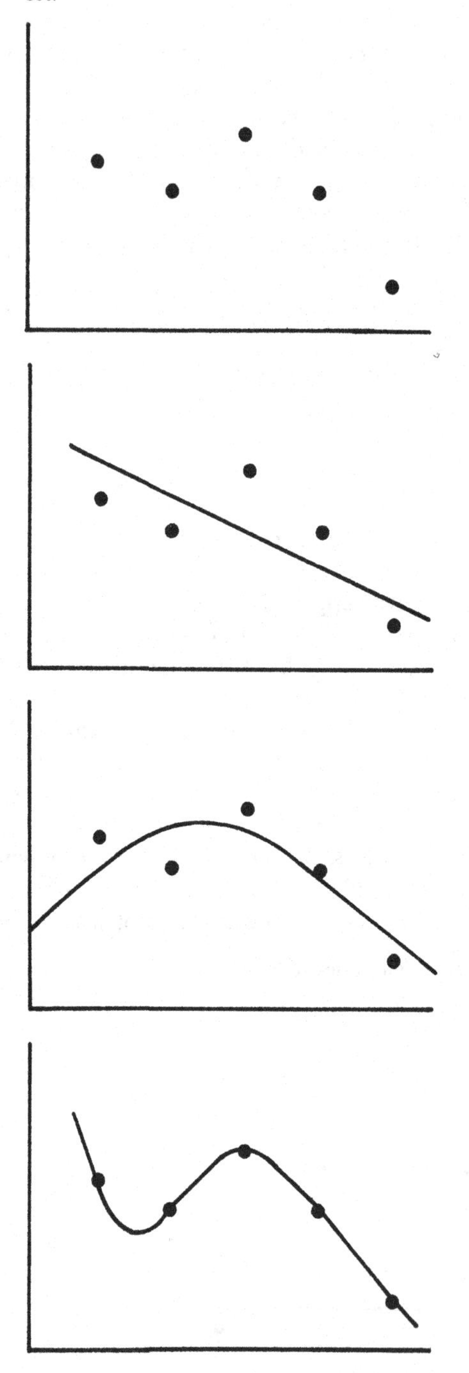

obtained regardless of when we began observing, we expect steady-state or more-or-less repetitive patterns. This kind of behavior cannot be represented by a polynomial approximation. Instead, we need a set of building block functions, a basis that is itself composed of repetitive, or periodic, functions. This alternative to polynomial approximation is a *Fourier approximation.* Here the basic building blocks are the set of functions $(1, \cos t, \sin t, \cos 2t, \sin 2t, \ldots)$. In this book we see how remarkable this particular kind of approximation is. For now, let us consider a visual illustration of Fourier approximation.

Fourier approximation: visual demonstration

Whittaker and Robinson (1924) observed the brightness of a variable star for 600 successive days. Their data are plotted as Figure 2.10. Using Fourier approximation, best estimates were derived of the two component frequencies in their data. The methods for deriving these frequencies are described in Chapter 16 (see also Appendix 16A, which is a primer on Fourier series). The procedure is basically to find the peak frequencies of the spectral density function and then to approximate the phases of these component frequencies. Using these approximations, the data were simulated. The dashed line in Figure 2.10 is the result of the Fourier synthesis of the variable star data.

Figure 2.10. Six hundred consecutive daily observations of the brightness of a variable star (solid line) and a Fourier synthesis of these data (dashed line).

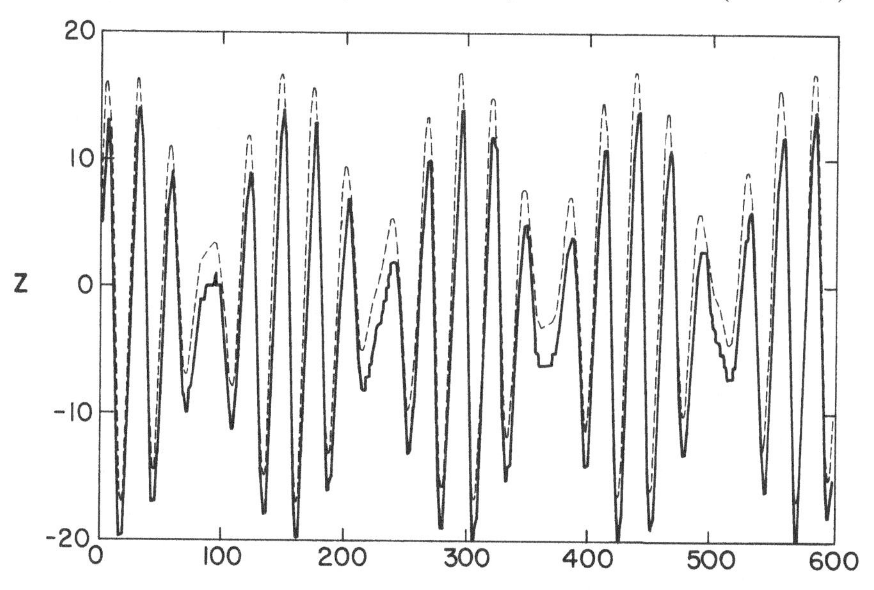

In this approximation process, several criteria can be employed. One criterion is that the least-squares error between the data and the synthesis be a minimum. Another criterion is that the residual be an uncorrelated, random process, called white noise. We see in this book that these two criteria are independent and under what conditions one criterion is preferable. The approximation is quite accurate, but upon closer examination (see Figure 2.11) it becomes clear that the error increases with increasing time. Therefore, the residual plotted in Figure 2.11 is not noise, because it has a clear pattern over time.

Finding hidden periodicities: intuitive overview

When we discussed the problem of approximating a function, first with a linear term, then linear plus quadratic terms, then linear plus quadratic plus cubic terms, and so on, we referred to the basic functions in the approximation, $(1, t, t^2, t^3, \ldots)$, as the basic building blocks of the approximation. An important set of numbers are the weights assigned in a least-squares approximation that uses the sum of these building blocks. Recall that the problem

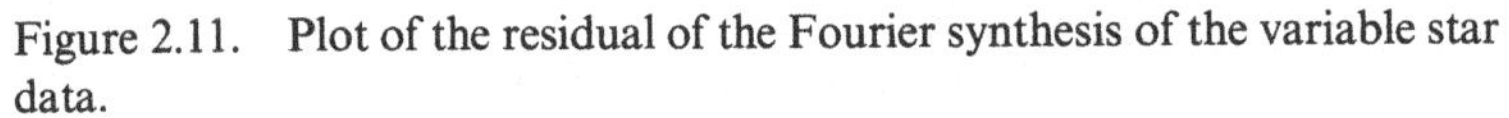

Figure 2.11. Plot of the residual of the Fourier synthesis of the variable star data.

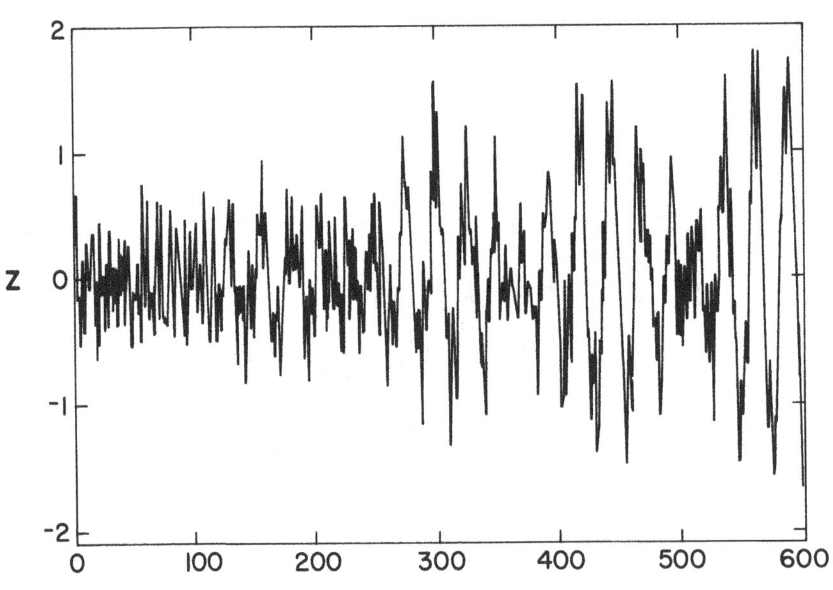

with this polynomial basis for approximation is that although it fits the set of points we have in our range of observations, it wanders off to plus and minus infinity beyond this range. This is very bad if we have a phenomenon that more-or-less repeats a recurring (albeit complex) pattern. Because of this fact, the Fourier approximation provides the best basis for time-series approximation.

When the building blocks are a set of sines and cosines, the least-squares approximation is called a *Fourier series*, and we can always determine the weights of the sine and cosine building block functions directly from the function we are trying to approximate. In practice, a small set of frequencies, called the overtone series, is used to approximate the real frequencies in the data (see also Chapter 16).

It would be handy to have a function (obtainable from an observed time series) that would tell us at a glance the mix of basic frequencies that made up the observed time series. This function is, of course, the spectral density function.

There is one problem with the spectral density function as a decomposition of a time series that must be mentioned. Many different (very different) time series can have the same spectral density function. To see this, consider Figure 2.12, which is an imaginary perfect decomposition of a time series into four sine waves. Note that these time series have fixed *phases;* that is, they cross zero time at given heights. By varying the phases of these components we can construct very different time series – but they will all have the same spectral density function. The spectral density function, although it is very useful, loses information that would be vital in reconstruction. However, this problem is not serious, because, once the component frequencies and their amplitudes are estimated by the sample spectral density function, the phases can be approximated by a least-squares procedure.

To illustrate the use of the spectral density function, three waves of different frequencies, amplitudes, and phases were added together to produce the data displayed as Figure 2.13. The three frequencies were .1, .2, and .3; and three amplitudes were 1.0, 2.0, and .5, respectively; the three phases were -3.0, .0, and -3.0, respectively. Figure 2.14 shows the spectral density function for these data. The peaks occur at component frequencies. Also, note that the heights of the peaks are related to the amplitudes of their respective component frequencies; the height of each peak is actually equal to half the square of the amplitude of each frequency.

Example of detecting hidden cycles

This section is designed to illustrate the fact that visual examination of time-series data is often not adequate. A simple example of detecting patterns will

Figure 2.12. Decomposition of a time series into its components, four sine waves differing in amplitude, frequency, and phase.

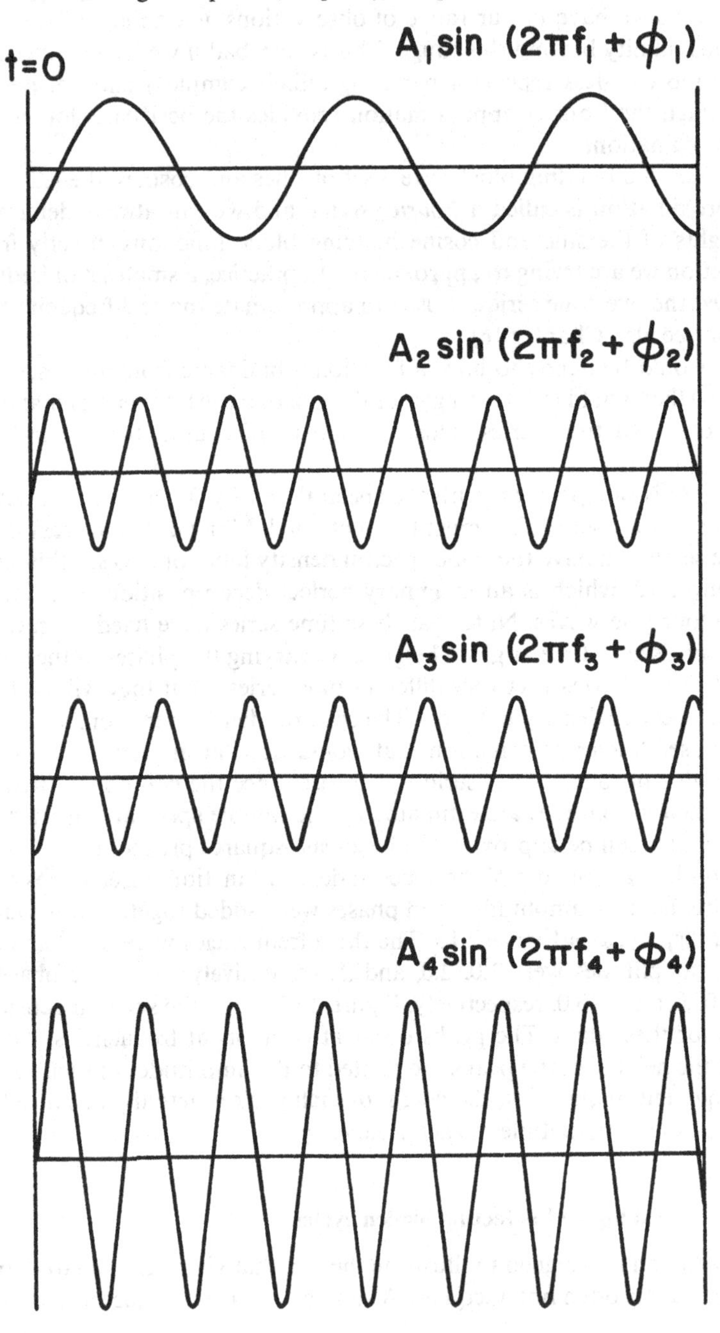

Figure 2.13. Example of a time series generated with three component frequencies.

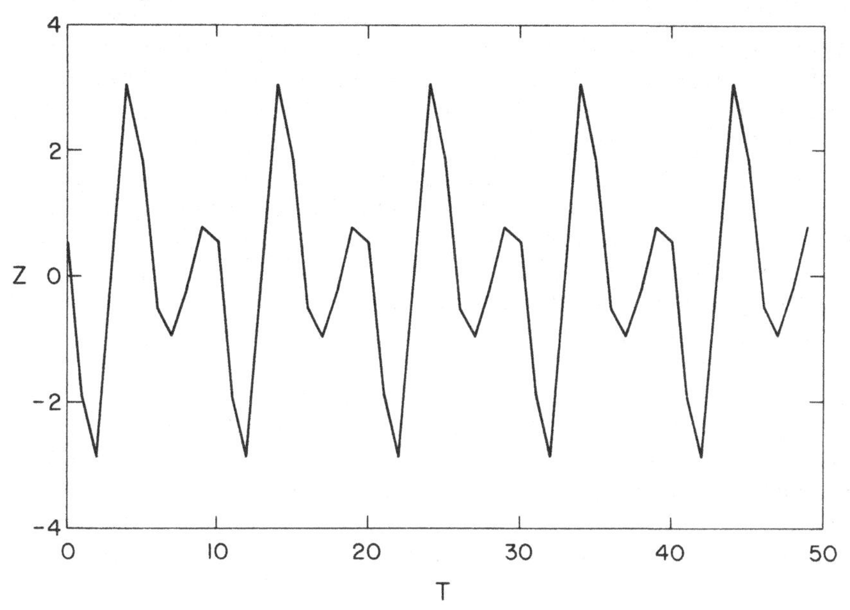

Figure 2.14. Spectral density function of the periodic function displayed in Figure 2.13.

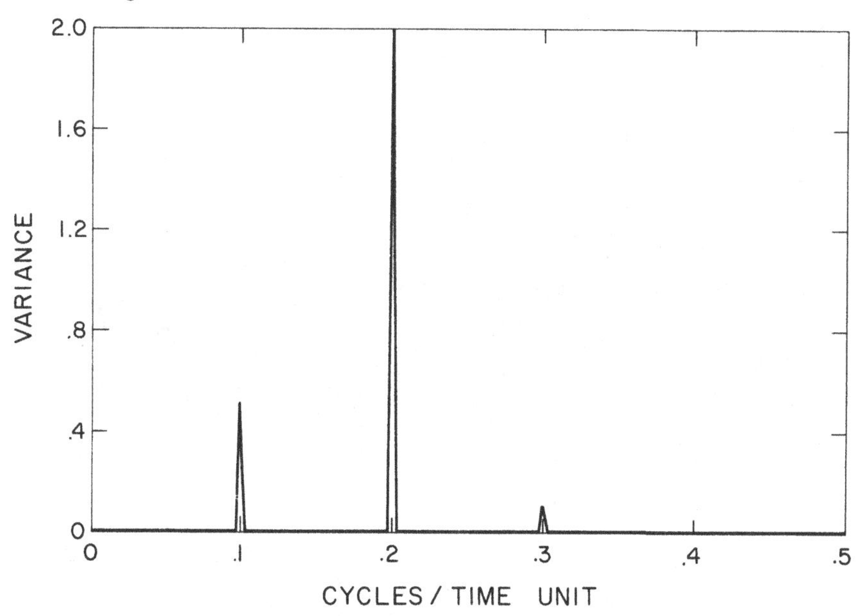

illustrate this point. This section will stay in the frequency domain. The superposition of simple sine waves may generate surprisingly complex patterns. An excellent illustration of the complexity that is possible is provided by the pattern of *beats* obtained simply by adding two sine waves of equal amplitude but different frequency. The data in Figure 2.15 were generated by adding two sine waves, with frequencies $f_1 = .085$ and $f_2 = .045$ and sampling from the resulting sum at equal time intervals. The resulting time series has a repetitive form with frequency $\frac{1}{2}(f_1 + f_2)$ that oscillates with a sinusoidal amplitude of $2 \cos \frac{1}{2}(f_1 - f_2)$ (i.e., within the envelope of this wave). The spectral density of this beat pattern should show two peaks at .085 and .045, which, indeed, is the case (see Figure 2.16).

Suppose now that noise is superimposed on data that are generated by a beat pattern. Suppose further that 1,000 observations are not available or are not practical to obtain. Would the beat pattern still be discernible by visual examination? The answer is "no" (see Figure 2.17). However, if the spectral density of these data is examined, two clean peaks would be suspected (see Figure 2.18). This example provides an illustration of the potential uses of frequency-domain time-series analysis for simple models of

Figure 2.15. Example of a beat pattern.

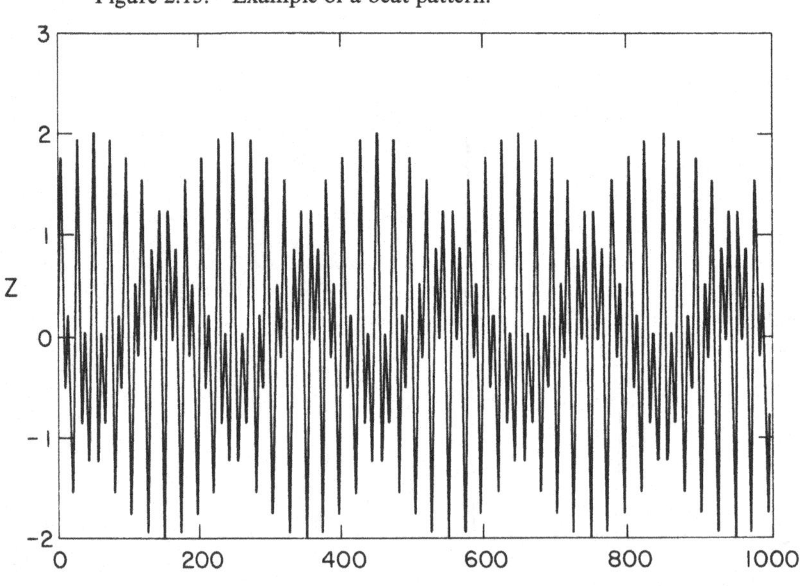

Figure 2.16. Spectral density function of the beat pattern, showing the two component frequencies.

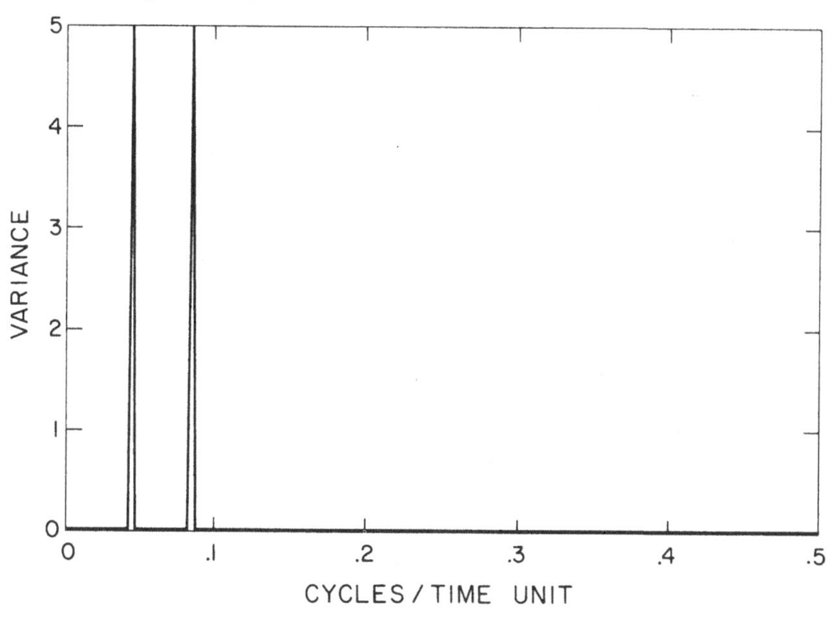

Figure 2.17. Beat pattern masked by noise.

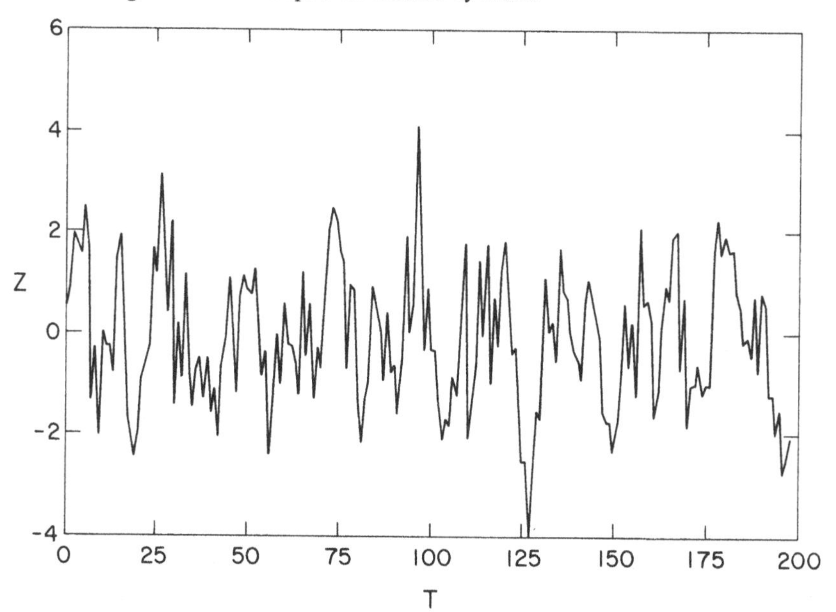

Figure 2.18. Spectral density function of the data in Figure 2.17.

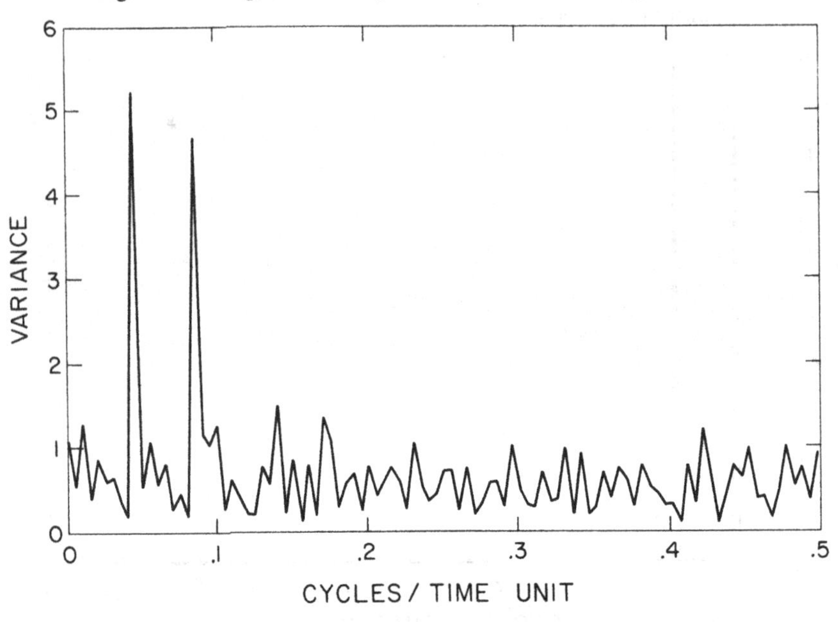

deterministic cycles hidden by noise. Later, it will be seen that this is also the case for cycles whose amplitudes are not fixed but can also be random variables.

3 How Slutzky created order from chaos

This chapter is about time-series models and how they may be based on white noise or how order, in the sense of probabilistic structure, can be constructed out of chaos.

We will have occasion to return to the concept of white noise a number of times. White noise is fascinating because it represents what appears at first glance to be two opposite ideas. The first idea it represents is chaos.

In a uniformly distributed series of random numbers, every number within the range has an equal chance of being selected at any time. Also, the selection of a number at one time is independent of past history. In short, the white-noise series has no memory whatsoever.

On the other hand, white noise is a Pandora's box; just as white light has no color of its own, it also contains all the colors. It only remains for a prism to break white light into its spectrum.

In the same manner a white-noise series is a Pandora's box from which any other series can be constructed. This remarkable fact was discovered by Slutzky, a Russian mathematician, in the 1920s. The effect is called the *Slutzky effect.*

The general idea is simple. An amplifier and filter can selectively amplify any frequency band, and because, theoretically, all frequency bands are present in white noise, the proper amplification system can create any pattern we wish.

One such amplification system is called a *linear filter.* We will demonstrate the Slutzky effect by showing that by suitably filtering white noise we can obtain either a sharp peak in the spectral density or a relatively broad, flat band of frequencies that are amplified. Figure 3.1 illustrates a series of

Figure 3.1. Example of a white-noise time series with zero mean and unit variance.

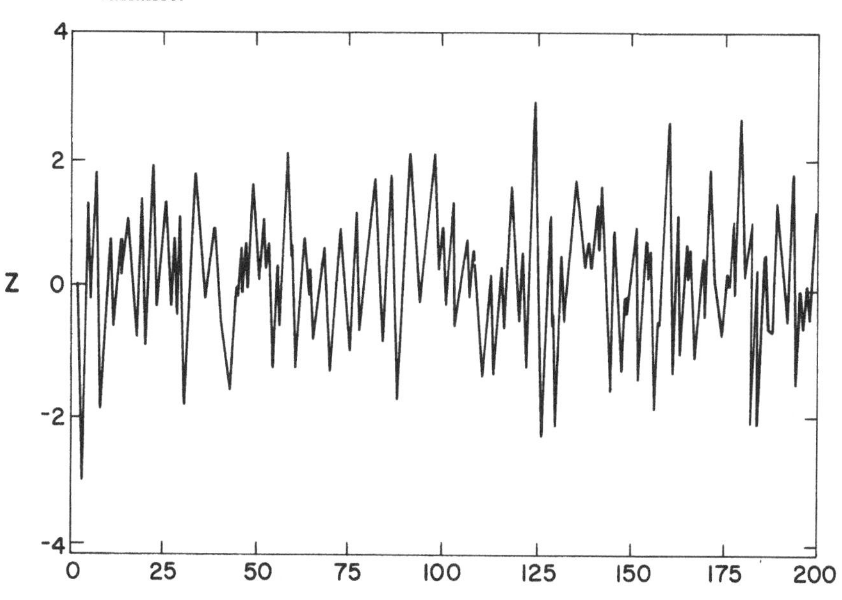

pseudorandom numbers that are generated by a computer to have zero mean and unit variance. Recall that this white-noise series should (theoretically) have a reasonably flat spectrum. Now we filter the white-noise series (in a manner soon to be described) and compute the spectral density of the filtered series. Two different filters of the same white-noise series will yield two different series with two different spectral densities. In this case one has a sharp peak and one has a flat peak. The sharp-peak and flat-peak spectral densities are displayed in Figures 3.2 and 3.3, respectively. How were these two peaks obtained in the spectral densities? Simply by averaging, using two different sets of weights for the average.

Yes, incredible as it seems, by suitable weighting of random numbers we can obtain nearly any pattern we desire. The flat peak was created by a simple four-point weighting of white noise. The weighting is called a *moving average*. This weighting procedure is illustrated in Table 3.1, with the same weights sliding along in time. For example, the new value $-.53$ was created from the first five noise values as $(1)(.46) + (-2)(.14) + (0)(2.46) + (2)(-.32) + (1)(-.07)$. The same weights were used to create the next new value, -4.62, but they were applied to the next five noise numbers: .14, 2.46, $-.32$, $-.07$, and .30. Thus, noise values for $t = 1$ to 5 were used to obtain the first weighted

Figure 3.2. Spectral density of filtered noise, illustrating a sharp peak.

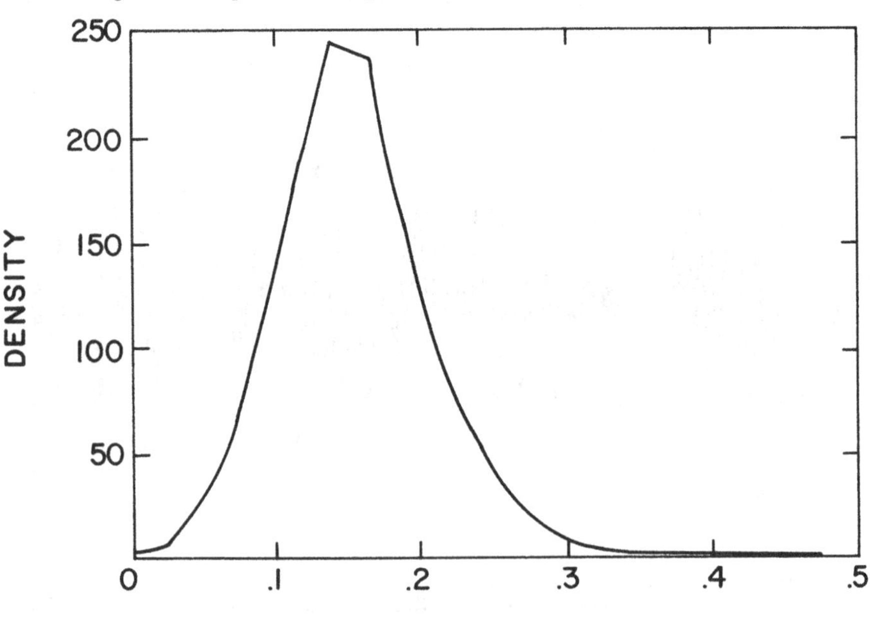

Table 3.1. *Illustration of how a moving average can create the Slutzky effect from a white-noise series*

Time	Noise value	Weight for the average			New value
1	.46	+1			→ −.53
2	.14	−2	1		→ −4.62
3	2.46	0	−2	1	→ 3.41
4	−.32	+2	0	−2	
5	−.07	+1	+2	0	
6	.30		1	+2	
7	−.29			1	
8	1.30				

Figure 3.3. Spectral density of filtered noise, illustrating a flat peak.

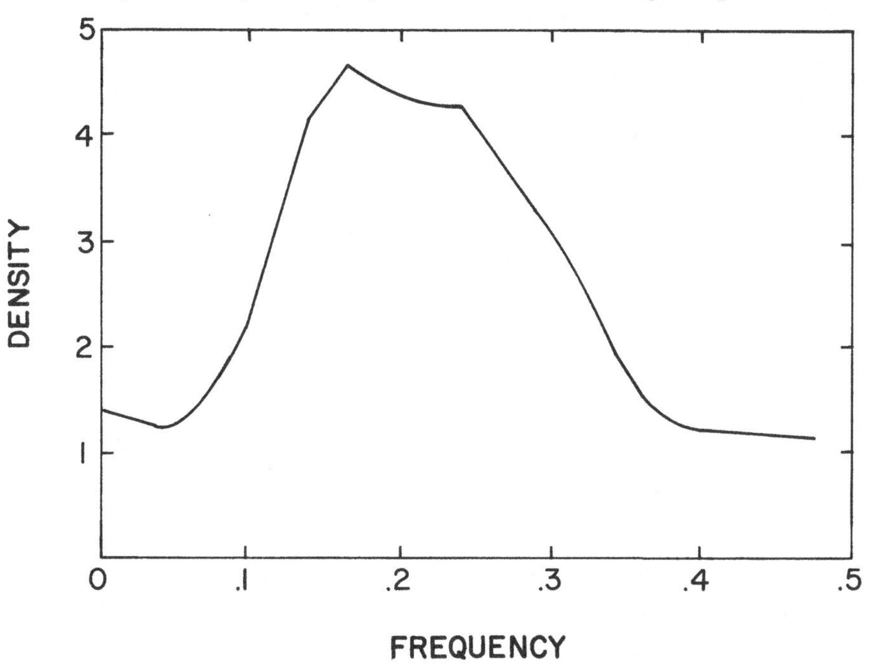

value; noise values for the points $t = 2$ to 6 were used for the second weighted value, and so on. The weights remain constant.

The Slutzky effect was upsetting to Slutzky's contemporaries because they were freely applying all kinds of averages to their data without expecting to *introduce* spurious correlation and cycles into their data. In short, they may have been *inventing* pattern in their data rather than discovering hidden pattern in the data. The Slutzky effect was *initially* horrifying.

Perhaps the reader is asking: How can it be otherwise? The answer is that if we can discover how to take a time series whose structure we wish to understand and convert it, with a suitable linear filter, into white noise, it follows that if we go the other way around, *we can create our series from white noise.* This would give us a *model* for our time series.

The models that result from this process are called, not surprisingly, *moving-average models*, since they are created by averaging noise.

At about the same time that Slutzky was designing moving-average models that create order out of chaos, an American mathematician, G. Udny Yule, discovered the same fact, but by the reverse process. Yule found that he could transform a series into noise by subtracting out its own dependence on its past. The models that result from Yule's discovery are called autoregressive models and they are the dual models for the moving-average models Slutzky discovered. This is not the only occasion in time-series analysis when a Russian and an American, working independently, in different ways, arrived at the same fundamental truth (see Chapter 15 and the Wiener–Khintchine theorem).

This concept is illustrated in Figure 3.4. In the first illustration we take our series and by suitable linear filtering we obtain white noise. This is done by essentially the reverse process that Slutzky used to generate a time series from noise using a moving average. This procedure will be discussed later in the book when autoregressive models are discussed in greater detail. The second illustration shows Slutzky's procedure (the reverse linear filter), which has now become a moving-average model for our observed time series.

Figure 3.4. How the Slutzky effect in reverse can produce time-series models.

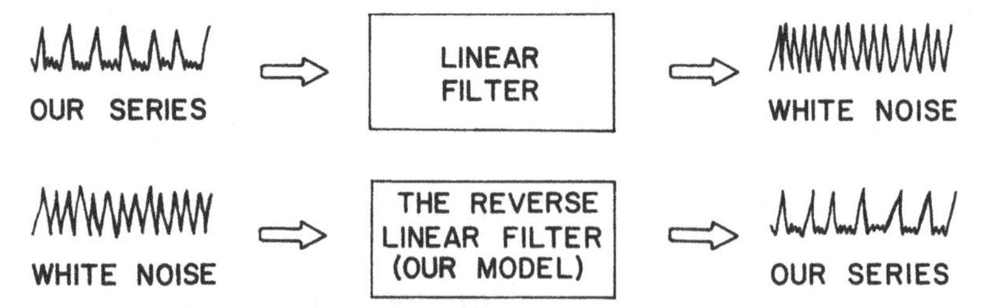

4
Forecasting: Yule's autoregressive models

This chapter begins to introduce some of the assumptions of time-series analysis; in particular, the concept of stationarity is introduced. Models are also alluded to that make it possible to predict the future of a series from its past. The notions of lag and autocorrelation are part of this focus on forecasting. Think of "autocorrelation" as the extent to which the present values of a series are predictable from the past values.

If we think about the problem of trying to predict the future from our knowledge of the past, it will become clear that to solve this problem *we must assume that something does not change very much.* Obviously, the problem is trivial if the data are a horizontal constant; we would simply forecast this value. If the data were just zero-mean white noise superimposed on a constant, we could once again forecast the mean level. If the data were white noise superimposed on an increasing trend line, we would assume that the slope of the trend line remained constant.

In the case of white noise superimposed on a constant, knowledge of the value of the series at any point in time does not make it possible to gain in prediction over the constant value. This is the nature of white noise. But now consider the case of autocorrelated data.

Data are called *autocorrelated* if there is some predictability from the past of a series to its current values. Suppose, for example, that we knew that if a value was one standard deviation above the mean at a particular time, t, it would certainly be one standard deviation *below* the next mean on the next observation, at time $t + 1$. This fact is stated in the following way: "At lag 1 the data have a perfect negative autocorrelation of -1.0." That is true because if we pair observations at time t with observations at time $t + 1$, which means pairing observations one lag apart, and compute a correlation coefficient, we would get -1.0.

The existence of negative (or positive) autocorrelation *does* allow us to improve over the mean as a forecast. Knowledge of the autocorrelations at each lag is called knowledge of the "correlational structure" of the time series, and it is true in general that knowledge of the correlational structure will improve our forecast.

Figure 4.1 is a plot of a migraine-headache patient's daily rating of the severity of his headache pain. Information about the autocorrelational structure of the data was used to compute least-squares forecast values in two ways, either one point at a time, checking the one-step-ahead forecast with the actual data for that point as they become available, or a forecast without this updating procedure, that is, forecasting three steps ahead. The error variance for this example can be computed by calculating the sums of the squares of the deviations divided by the number of forecasted points. This error is obviously lower for the updated than for the nonupdated forecast. This illustrates that the further ahead forecasts are projected, the greater the error likely to be incurred.

The forecasting in this example was accomplished by our knowledge that a sixth-order autoregressive model provided a good fit to the data. If we *know* the model for the time series, we can always derive least-squares estimates of the forecasts. On the other hand, if we *estimate* the model, the error of our estimate enters into the forecast error.

More generally, we can always approximate a wide class of time series (called "stationary" series) by autoregressive models. These models will be introduced in detail in Chapter 11. At this point the reader should consider the following introduction an advertisement of what will come later, and an

Figure 4.1. Migraine-headache patient's ratings of the severity of his headache pain (left). Forecasts (dashed line) and actual data (solid line) without updating the forecast (upper right). Forecasts (dashed line) and actual data (solid line) with updating (lower right).

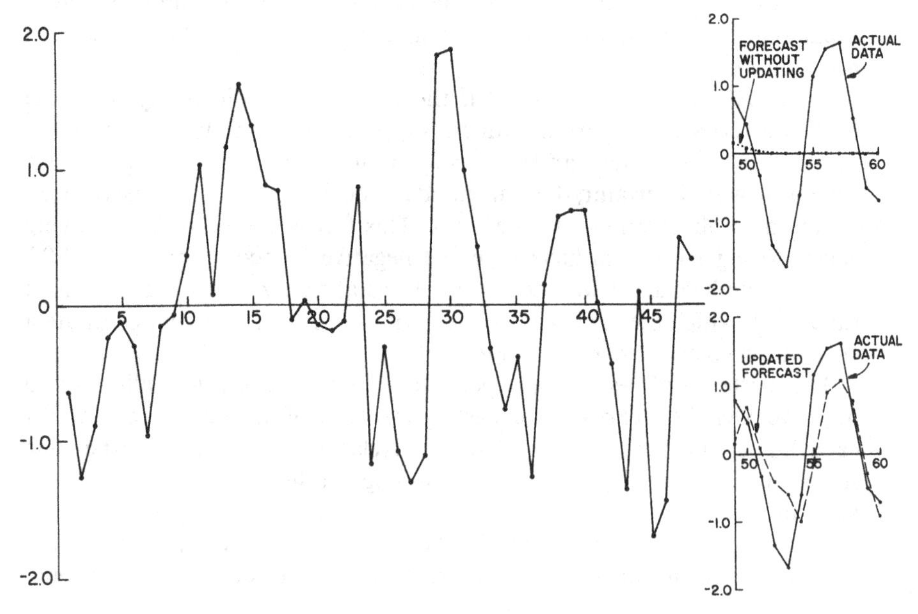

introduction to some new vocabulary. An autoregressive model in which we go back p time units in the regression to predict is called an *autoregressive model of order p*, abbreviated AR(p). In an autoregressive model we attempt to estimate an observation as a weighted sum of the p previous observations. In addition to the mean and the error variance, it has p parameters that we must estimate, and this is accomplished using a theorem derived by Mann and Wald (1943), and a simplified test developed by Quenouille (1947). Once we estimate these autoregressive parameters, we have automatically derived the equation for the least-squares forecasts, under the assumption that the AR(p) model is correct. This procedure is discussed in detail in Chapter 18.

To summarize, we can use the data to construct an AR(p) model and then use this model to forecast, updating our forecast as information becomes available and being careful not to predict very far into the future.

5
Into the black box with white light: time-series models

This chapter is about how to study the relationship between two time series. This is a complicated concept in the time domain, with ideas such as lead–lag and cross-covariance. It is even more complicated in the frequency domain, in which we must discuss how frequency components of each series are related.

If we can find an estimate of a linear filter that transforms an observed time series into white noise, why not try to find a linear filter that transforms any one observed series into any other observed series?

This simple idea is an approach to the solution of an extremely complicated and venerable problem in engineering called the *black-box problem*. One variation of this problem is the input–output problem. The goal is to understand the processes that transform an input into an output.

Why is the problem important? Well, let's assume that we are interested in the relationship between military spending in the United States and military spending in the USSR, or presidential popularity and the cost of living, or the smiling of a mother and the smiling of her infant, or the aggressiveness of a union representative and the compliance of a management representative, and so on. In each case we have two time series and our question is: What is their relationship?

The solution to this problem is conceptually simple. Suppose that our two time series are x_t and y_t. If we can apply one filter, A, to x_t and get noise, and apply another filter, B, to noise and get the series y_t, it follows that the filters in sequence, known as the *convolution* of the two filters, will transform x_t into y_t. This is illustrated in Figure 5.1.

In the frequency domain the spectral density function of one series is transformed into the spectral density function of the other series. The function that relates the spectral density of one time series to the spectral density of another is called a *transfer function.* It is important to note that actually the transfer function is the equivalent in the frequency domain of a filter in the time domain. First let us discuss the time domain.

In the time domain we are really talking about some form of regression of one time series onto the other. That is, we are using the past of one series to predict the present of the other series. This is a reasonably familiar concept.

There are modifications of this time-domain filter that can be very important conceptually. For example, suppose that we are studying the face-to-face social interaction of a mother playing with her infant. We may be interested in predicting a mother's behavior from the past of her infant's behavior. This is one kind of filter. On the other hand, we may argue, what we are interested in is in seeing how much *improvement* we can obtain in predicting the mother's behavior from the infant's past behavior *over and above* the prediction we would get by using the mother's past behavior. This is a different kind of filter. However, in all these filters in the time domain we are relating one *whole* time series to another.

In the frequency domain we take a different view. Because we have performed a spectral decomposition of x_t and a different spectral decompo-

Figure 5.1. How the Slutzky effect can be used to construct linear representations of the relationship between two time series using the concept of *convolution*.

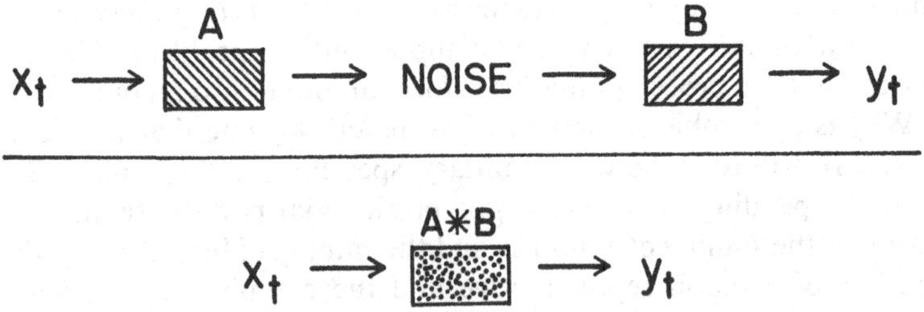

$x_t \longrightarrow$ A $\longrightarrow$ NOISE $\longrightarrow$ B $\longrightarrow y_t$

$x_t \longrightarrow$ A∗B $\longrightarrow y_t$

sition of y_t, we are relating a specific frequency component of x_t with the same specific component of y_t.

To understand this, consider a metaphor. Imagine that x_t is a sound pattern and that we break it up into components of specific phase and amplitude by using the strings in a piano. Now we do the same for y_t, using a second piano. The filter that changes x_t into y_t is a relationship between the vibrations of equivalent strings on the two pianos. Thus, it may be that specific components of x_t "resonate" with parts of y_t, but that this is not the case for other components.

The transfer function in the frequency domain is thus a statement of the amplification necessary at each frequency in the spectral density of x_t to obtain the spectral density of y_t. It is also a statement of what must be changed in the phase of each component. We thus need two pieces of information to define the transfer function at each component frequency: what is to happen to the amplitude and what is to happen to the phase.

A measure of the amplification relationship or correlation is called the *coherence;* the phase relationship, which tells which component (x or y) leads or lags, is simply called the *phase*. Note that each of these two numbers varies as a function of frequency.

In the time domain we also use two sets of numbers, the *cross-covariance* (or cross-correlation) and the *lag*. This is illustrated in Figure 5.2. In this figure we are pairing values of x_t with values of y_{t+2}, so that y lags x in this computation by two time units. We can compute a covariance (or correlation) very similar to those we ordinarily compute between any set of values. Furthermore, we can do this for all lags. Moreover, we can do it going forward or backward. Note that we probably will not get the same covariances (or correlations) with x leading as we would with y leading.

The cross-covariance is essential in calculating the filter and the transfer function; it is important in either the time or the frequency domain. Later, we look at the actual computations of these functions.

Figure 5.2. Two time series are related to one another with a lag of two time points, assessing whether the X series is a lead indicator of the Y series by computing a cross-covariance.

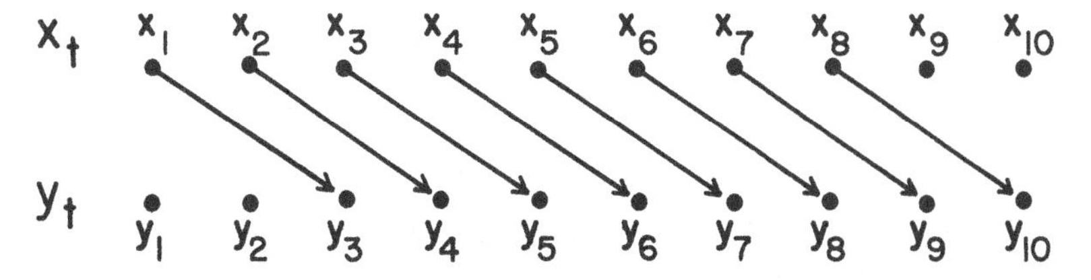

Example: the conversations of young friends

This section presents an example applying the concept of transfer functions to study whether one of two interacting children is dominant, in the sense of one child's behavior being a lead indicator of the other child's behavior. Gottman and Parkhurst (1980) collected audiotape recordings of the conversations of children at home with a best friend or a stranger. This section will discuss the conversations of two young best friends, Matthew, age 2, and Jessica, age 3. The tape recording was done in Matthew's home, so he was the host child and Jessica was the guest.

The tape was transcribed and each statement was coded using a highly reliable, detailed, 22-category coding system. These codes were then scaled so that each statement was given a 1 if it was a statement about the self (e.g., "I'm not finished yet. I'm making a rabbit."); a 2 if it involved negotiation, agreement, disagreement, or exchange of information; and a 3 if it involved the other child (e.g., commenting on the other child's behavior, offering something, helping, or expressing sympathy). This three-point vari-

Figure 5.3. (a) Plot of the self- to other-centered nature of a young child's (Matthew's) statements while playing with a peer (Jessica). (b) Similar plot for Jessica's statements.

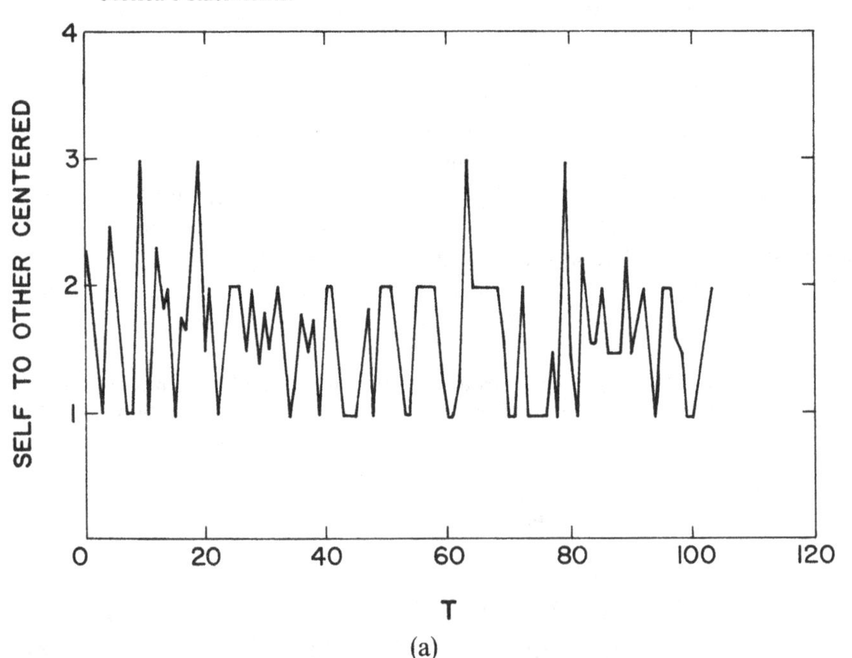

(a)

able is of interest to child psychologists who study social development because young children are assumed by many theorists (e.g., Piaget) to be primarily centered on self rather than on other children. The variable itself could be an index of social development. Within a conversation, the variable could be used to assess social responsiveness during conversation. If one child's conversation moves from self- to other-centeredness, does the other child's data follow suit? To what extent is this influencing symmetrical? If it is asymmetrical, this may suggest a dominance pattern in the dyad.

These number codes were then averaged for all the statements a child made while he or she held the floor of the conversation. Figure 5.3a and b show plots of these data for Matthew and Jessica, respectively. Note that these data are not normally distributed, but time-series analysis may nonetheless proceed. The spectral density functions of Matthew's and Jessica's data were then compared. The spectral densities were not spectacular replicas of one another, but they did have three peaks in roughly the same frequency *ranges;* both spectral densities peaked at about .050 to .075; they also peaked from .175 to .250; and they also peaked from .275 to .325.

Using Matthew's data and a study of the relationship between Jessica's and Matthew's spectral densities, we also constructed a simulated time series

Figure 5.3. (*cont.*)

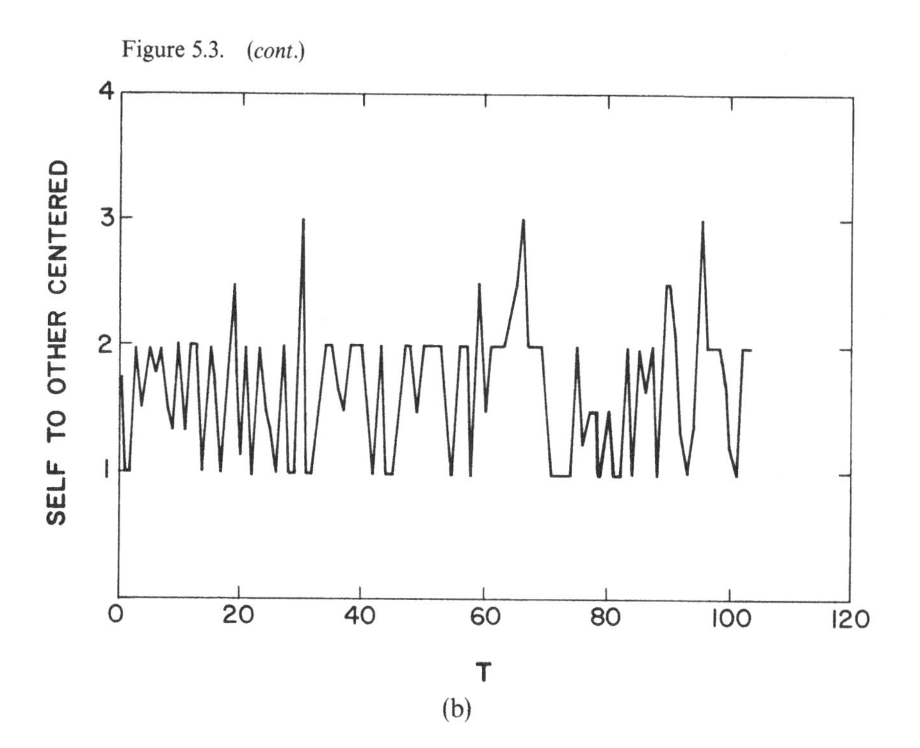

(b)

for Jessica, and this was compared to the actual series. This simulation is displayed in Figure 5.4.

However, the objective of the analysis was not to simulate one data set from the other, but rather to study the relationship between the two children's data. The linear filter assessed this relationship, and in this case, a major advantage of the linear filter derived was its parsimony. Jessica's series at time t could be derived from Matthew's by multiplying .35 times his current value (also at time t) and adding to this .43 times his value six time points (or floor changes) previous to time t. This means that Matthew's self- or other-centeredness was a potent *lead indicator* of Jessica's subsequent self- or other-centeredness. Matthew was thus dominant in this interaction; when Matthew's conversation changed, Jessica's conversation changed soon after, but the converse was not true. In general, Gottman and Parkhurst's data indicate that hosts are dominant for younger children, whereas guests are dominant for older children. This example is presented purely as an advertisement of what can be done in the bivariate case with time-series analysis.

Figure 5.4. Linear simulation of Jessica's data using Matthew's data. This plot is a by-product of a lead–lag analysis that can be used to explore dominance patterns between the children.

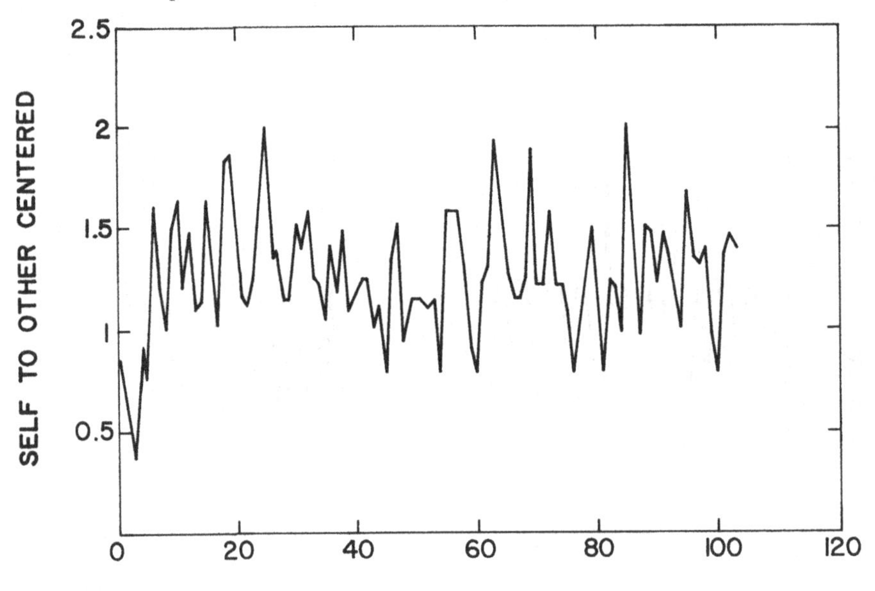

6
Experimentation and change

This chapter is an overview of most of the topics covered in the book. It is a guide to what kinds of things can be done with time-series analysis, the kinds of questions that can be addressed, and the levels of causal inference that are possible. These range from describing variation to explaining variation. The problem of assessing change is introduced.

Historical overview of time-series analysis

Since it is so difficult to disassociate observed events from some sort of idea of occurrence in time, it seems remarkable that most of the body of statistical methodology is devoted to observations for which the temporal sequence is of no importance. Classical statistical analysis requires independence, or at least zero correlations, among observations.

Although intuitively one might think that data in which an observation at time t is in some degree determined by an observation at $t - 1$ should be somewhat smooth and amenable to statistical analysis, quite the opposite has been true, and only relatively recently has progress been made in this area. Physical scientists of the classical period (i.e., pre-relativity and pre-quantum mechanics) succeeded to a large extent in avoiding the problem of correlated data because the variance in their data due to purely deterministic components tended to swamp that due to random observational errors, and/or they were able to control their experiments in such a way that the classical hypothesis of independent errors could be assumed.

It was not until the mid-1920s that a coherent theory of the statistical analysis of observations correlated in time began to take shape in the hands of investigators such as Slutzky, Yule, Birkhoff, and Khintchine. Applied time-series analysis then gained impetus as a full-blown science in World War II. It was during this period, when vigorous attempts were being made to apply the new technology of modern electronics to warfare, that it was discovered that certain problems could not be described by classical "deterministic" mathematical models. For example, it was discovered that the effective tracking by an automatic gunnery control system of a target taking evasive action could not be simulated by conventional mathematical models. Instead, a random, but still somewhat predictable, component had to be incorporated into the equations. Similar problems arose in the extraction of intelligible messages from noisy signals and the distortion-free reception of radar echoes. Indeed, the history of applied time-series analysis during this period has a certain amount of the romance of "cloak and dagger" about it because of the military applications under investigation. Since World War II there have been attempts to apply statistical time-series analysis in a wide variety of fields.

Conceptual overview of time-series analysis

We have been discussing two types of time-series analysis: *time-domain* and *frequency-domain* time-series analyses.

The basic questions asked by the two kinds of analysis are different, although they are really mathematically equivalent, linked by a famous theorem called the Wiener–Khintchine theorem. In the time domain, questions are formulated directly in terms of time, and they can be formulated as an *autoregressive question:* To what extent can I predict the present from the past? This question is formulated mathematically much as any regression equation is written. The present, x_t, is regressed on the past, $(x_{t-1}, x_{t-2}, \ldots)$, with a linear weighted sum,

$$x_t = a_1 x_{t-1} + a_2 x_{t-2} + \cdots \text{ back to infinity} \tag{6.1}$$

This leads to the concept of the *memory* of the series, that is, how far back one must go to be able to continue reducing uncertainty in our ability to

Figure 6.1. Time series generated by a process with a first-order memory.

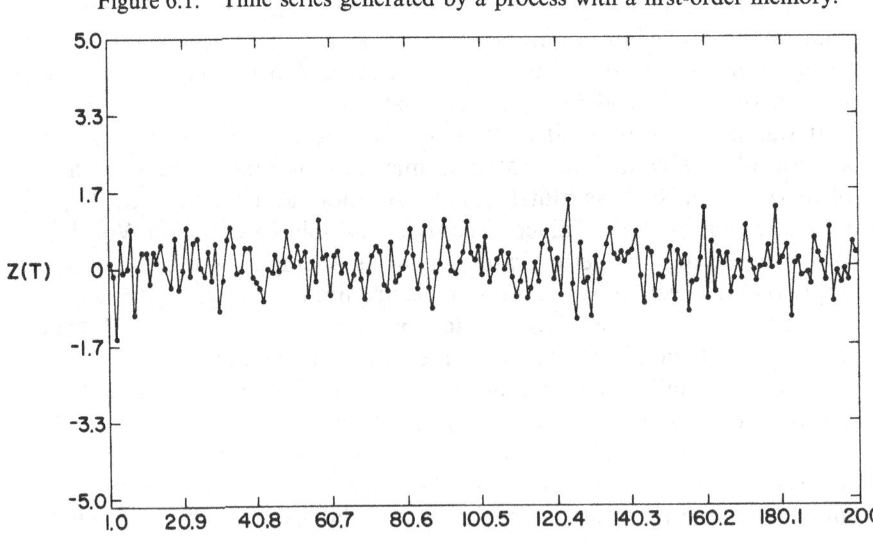

predict. Memory is assessed by a function called the *autocorrelation function*, which I discuss in subsequent chapters.

Figure 6.1 is a computer-generated time series whose memory is first order, and Figure 6.2 is a series whose memory is second order. These figures are presented simply to illustrate the point that it is nearly impossible to discern the degree of memory in a series simply from visual inspection. We need to compute the autocorrelation function.

In *frequency-domain* time-series analysis, the goal is *spectral decomposition*. To return to our analogy, much as a prism breaks sunlight into all the colors of the spectrum (each with different frequencies of electromagnetic oscillation) with different brightnesses, a spectral decomposition breaks a time series into its "basic" frequency components, each with a different amount of energy or variance it accounts for. One goal of frequency-domain time-series analysis we have already discussed is deriving a function called the *spectral density*, which tells us for how much variance each frequency accounts. We will have a great deal more to say about frequency-domain time-series analysis in subsequent chapters.

Some questions that can be asked using time-series analysis

Tukey's (1977) book *Exploratory Data Analysis* dramatized what Richard Vitale (personal communication) has called a Victorian attitude that social

Figure 6.2. Time series generated by a process with a second-order memory.

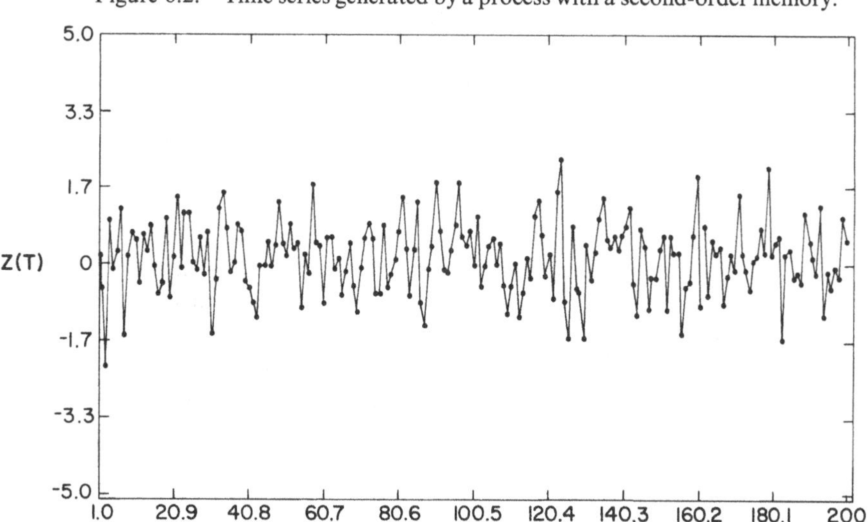

scientists have toward their data. This attitude is that it is not respectable to become too familiar with one's data without the license of a hypothesis. Tukey wrote: "A basic problem about any body of data is to make it more easily and effectively handleable by minds . . ." (p. v). Toward this end, he proposed a set of methods for simplifying data, for presenting data using pictures, and for exploration (hypothesis generating) as well as confirmation (hypothesis testing). He wrote: "The greatest value of a picture is that it *forces* us to notice what we never expected to see" (p. vi).

This chapter is written in the spirit of Tukey's suggestions. Perhaps most valuable to readers at this stage of familiarity with time-series analysis will be a list of some questions currently amenable to time-series analysis. The following list of questions are examples of analyses that can be made using the statistical tools of time-series analysis. The list is by no means exhaustive. In each case assume that the question refers directly to graphs based on data over time.

1. One of a series of observed processes has a long memory, and the other has a short memory. Which has a longer memory? The question is answered using the autocorrelation function.
2. One of two brain-wave patterns characterizes drowsy, and the other, waking states. Are there different cycles in the two? The question is answered using the spectral density function.
3. In an obtained graph of mother–infant interaction, is there a basic cyclicity? Which person is, in some sense, "driving" the interaction? The question is answered using spectral density functions, linear filters, and transfer functions.
4. In a graph of the attention of a child labeled "hyperactive" during a laboratory task, what is the best way of predicting the child's future attention on this task? The question is answered using a time-series model and least-squares forecasts.
5. In a graph of Gallup polls of presidential popularity from Truman to Johnson, what is the best way of predicting the pattern over time? Can this pattern account for Truman's high popularity after Roosevelt's death? The question is answered using time-series regression.
6. What *form* of change of public spending patterns follows an experimental tax rebate of $250 per family? The question is answered using analyses of the interrupted time-series experiment, which is one application of time-series regression.
7. In a record of the Dow Jones average annotated with what are presumed to be "critical incidents," do any of the incidents have an effect on the series? What *kind* of events have an effect on this kind

of series? Do wars have a different effect than natural disasters? Once again, the interrupted time-series experiment can be used, or different time-series models can be fit to different parts of the data.

8. The state of Connecticut institutes a new crackdown on speeding. Does the crackdown affect traffic fatalities? Another interrupted time-series experiment.
9. The social behavior of a withdrawn child is targeted by an intervention designed to teach her to interact more with peers and less with adults. Is there a change in her behavior following the intervention? This is another interrupted time-series experiment.
10. In a graph of the emotional responses of a husband and wife while they are discussing a marital issue, who, if anyone, is dominant in this interaction? The question can be answered using transfer functions.
11. Has the incidence of suicides in a region changed following the establishment of a crisis hotline service? Again, an interrupted time-series experiment is suggested.
12. In a study of the heart rate, respiration, temperature, and wakefulness of a premature and full-term infant, are there cycles and patterns of interconnectedness in the data that discriminate the two infants? Cross-spectral time-series methods can be used to answer this question.
13. In two graphs representing the visual-gaze behavior of two pairs of identical-twin sisters, which pair reports having a more intimate relationship? Cross-spectral time-series methods can be used to study this question.

These are some of the questions that will be asked of time series with the statistical methods introduced in this book. One way of organizing these questions is by examining the process of scientific inquiry itself.

Levels of causal inference in time-series analysis

The kinds of questions that can be asked of time-series data can be organized according to the issues of causal connection. Causal connection is never *demonstrated;* rather, hypotheses are successively eliminated that militate against our confidence in a causal link. Sciences vary greatly in the extent to which it is possible for them to intervene experimentally to exercise the control necessary to make causal inferences. Some sciences, such as astronomy, are limited to observation; some sciences can use simulation techniques. Other sciences study phenomena that admit the possibility of somewhat more controlled investigation. In the social sciences, on the other hand, the entire range of control options is manifest. Fortunately, time-series techniques have been applied in a range of sciences that vary in the amount of control possible, and we can thus benefit from a study of the history of time-series analysis.

In describing variation, the basic objective is to construct a *model* for the process that may have generated our observed data set. This is accomplished using either time-domain models or frequency-domain models. Time-domain models make use of the autocorrelation function, whereas frequency-domain

Figure 6.3. (a) A time-series record annotated with continuous critical events of types A and B. (b) Ten years of the Dow Jones average annotated with critical incidents.

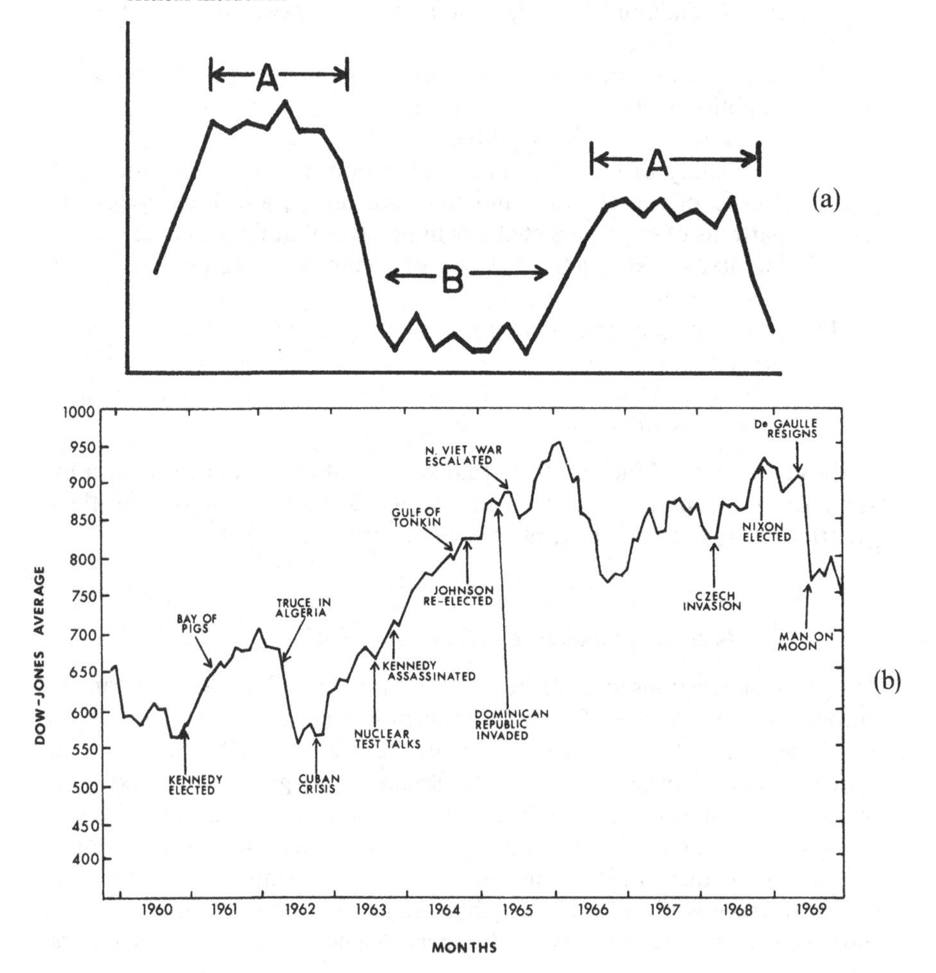

models make use of the spectral density function. *In this book the integrated use of both time- and frequency-domain model building is advocated.*

Once a model has been constructed, it can be used to test our understanding or description of variation. For example, as was discussed earlier, forecasting involves using the past and present of a series $(\ldots, x_1, x_2, \ldots, x_t)$ to predict the future x_{t+L}, where L is called the *lead time.* The equation used to predict x_{t+L} from the past and present defines the *forecast function.* Confidence in the prediction decreases drastically with the lead time, so it usually makes the most sense to predict very close to the present, and *update* the forecast continually as new data come in.

The second level of causal inference involves generating hypotheses about natural covariation. There are two common forms of this level of causal inference. The first employs a time series of interest and critical incidents that are presumed to affect the time series. This is illustrated in Figure 6.3. Figure 6.3a illustrates the general concept, and Figure 6.3b is an example of the Dow Jones average annotated with specific critical incidents. In Figure 6.3a the critical incidents last for specified time periods and we presume they are related to shifts in the time series. In Figure 6.3a the events may be nonrecurring and we will wish to find commonalities among types of events according to the shifts they create in a time series of interest. A fruitful example of this work is Granger and Hatanaka's (1964) book, which presented the results of their analyses of the differential effects on prices of war, strikes, and natural disasters.

A second type of analysis of natural covariation employs two (or more) time series. A famous example is Huntington's (1945) plot of sunspots and thunderstorms in Siberia (see Figure 6.4). Huntington's data show the remarkable covariation of these two processes.

An example of the analysis of covariation between two time series is similar to forecasting, except that one series is used to forecast the other. Using this analogy, more than one time-domain relationship may be interesting. Economists, for example, search for *lead indicators;* for example, wholesale-price increases are lead indicators of retail-price increases. Another relationship of interest is the case in which one series increases prediction of a second series over the second series' own past. This form of the relationship seems logically related to causal connection more than the first form, since causal connection always refers to some pattern of asymmetry in prediction.

In the time domain, *cross-correlation* is used as well as autocorrelation. In the frequency domain two functions are useful, called the *coherence spectrum* (which is like a correlation) and the *phase spectrum* (which specifies lead–lag relationships).

The third level of causal inference involves the use of planned experiments. This book discusses the simplest form of planned experiment, the *interrupted time-series experiment*, in which there is a series of observations both before and subsequent to some planned intervention. The intervention effect itself may have various forms, such as an ephemeral effect on the level of the time series (i.e., a gradual increase and then a return to the preintervention level).

Note that there are many intervention effects that cannot be detected by a *t* test or other analysis of variance. For example, in Chapter 1, I presented a series that had the same preintervention and postintervention mean and variance, but there was a clear effect of intervening in changing the cyclicity of the time series. There are specific methods for detecting other kinds of intervention effects.

Figure 6.4. Sunspots and thunderstorms in Siberia, an illustration of the study of natural covariation.

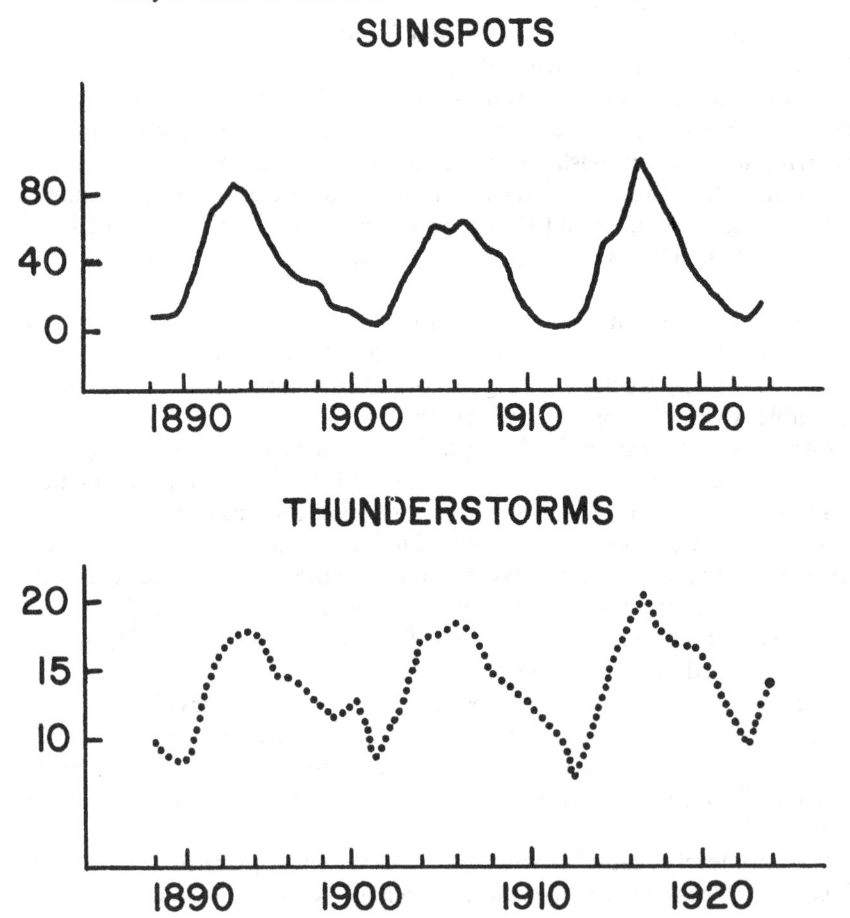

In the interrupted time-series experiment we have some dependent variable that we are studying over time, y_t, and we have some intervention that also has a shape over time, x_t. In some cases we can actually know the shape of the intervention, x_t. In other cases, both x_t and y_t are separate time series. For example, x_t could be the daily caloric intake of a dieter and y_t could be his daily weight. Theoretically, the dieter's caloric intake should simply drop after intervention and remain at a constant lowered level. In this case the intervention effect has been specified in advance of the experiment. However, real-life experimentation is not so deterministic, and, in an actual experiment, caloric intake and weight may both be random variables. In this case we are in a gray area between concomitant variation and experimentation.

In other cases we can hypothesize a reasonable shape for the intervention. In this case the x_t is a series of known constants. For example, Zimring (1975) collected data to evaluate the Gun Control Act of 1968. Zimring assessed the impact of the program on homicide rates. Gottman and Glass (1978) wrote that Zimring:

> assumed that the intervention should have an accretive effect on the scarcity of gun supplies since enforcement officers were added gradually. The design matrix was altered so that the full impact of the intervention would be felt in the following proportions: January 0, February 1/6, March 2/6, April 3/6, May 4/6, June 5/6, July 6/6. A dropoff of 1/6 effectiveness every two months after July was also predicted since enforcement measures were gradually phased out. [Pp. 228–9]

Glass et al. (1975) discussed 10 different intervention effect curves, and these are reproduced in Figure 6.5. The evolutionary operations curve may be unfamiliar to the reader. It represents a temporary decrement (or exponential decay) followed by an exponential growth to a new level. This type of curve might be obtained when an old response must be unlearned before a new response is acquired; for example, the dependent measure could be reading comprehension, and a decay in comprehension might take place until new reading techniques were mastered.

Example

Figure 6.6 presents data from the same schizophrenic we discussed in Chapter 1, except that the dependent measure is a perceptual speed task. Once again we are assessing the effect of the drug. Glass et al. (1975, pp. 137–40) analyzed these data by fitting a time-series model to the data and testing for a change in level after intervention. The reasons for these reasonably complicated procedures will be discussed in Part II. Glass et al. found that there

Figure 6.5. Ten different forms of intervention effects.

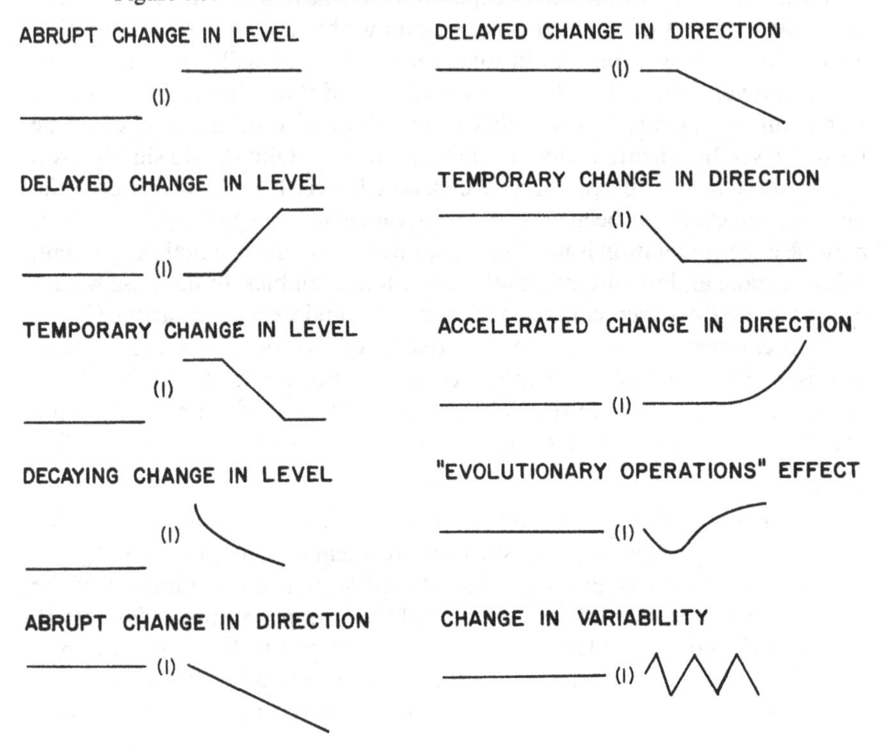

Figure 6.6. Perceptual speed of a schizophrenic patient on a placebo (first 60 days) and on chloropromazine (second 60 days).

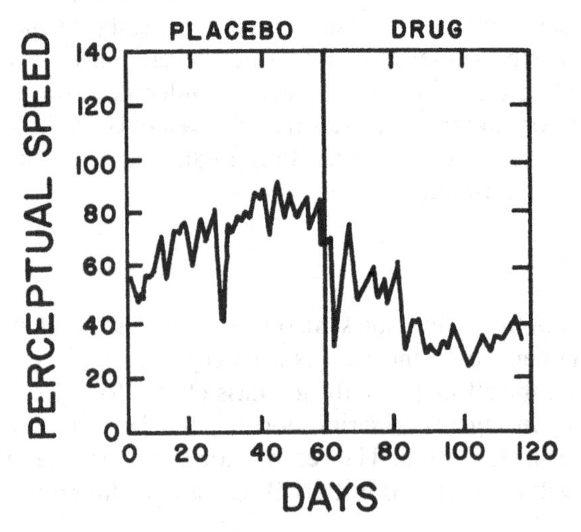

was a highly significant effect (decrement) in performance following the introduction of the tranquilizer. But they noted:

> A "naked eye" examination of Figure [6.6] would have seemed to have indicated such a shift, although the level of the series was dropping for about ten days prior to the intervention, and it is not obvious that the drop in the series after intervention is quite unexpected from {what one would expect given the pre-intervention time-series model}. However, the statistical analysis demonstrates emphatically that the post-intervention level of the series is not the normal progression of the pre-intervention process. [Pp. 139–40]

Thus, across the wide range of experimental control options available to social scientists, a variety of time-series methods exist that could detect patterns in data collected over time.

Part II

Time-series models

Up to now this book has introduced very few equations. Instead, the first six chapters have been an advertisement for time-series analysis. That is about to change.

To make it possible for the reader to avoid getting bogged down in equations, this section presents a tour of the bare bones of this book. Why was this section written? After all, time-series analysis is not easy, and perhaps the "lazy" reader should be discouraged. On the contrary: This book was written to encourage the busy researcher to try time-series techniques; this book was written to coax, to urge, to plead the case with our best investigators that time-series techniques are useful and not necessarily impenetrable. I am therefore afraid that a quick, impatient flip through the rest of this book will discourage some readers. This section is addressed to them.

Recall that there are two major kinds of time-series analyses, time-domain and frequency-domain. The reader must learn to be comfortable in both domains and to move back and forth between them. Chapters 7 through 14 are about time-domain methods.

Two of these chapters (Chapters 8 and 9) address the assumptions that have to be made in time-series analysis, that is, the models that have to be built. These models have two components, a "nonstationary" and a "stationary" component. The next discussion is about general types of stationary models, moving (MA), autoregressive (AR), and autoregressive moving-average (ARMA) models. This takes us from Chapter 10 to Chapter 14. In Chapters 19 to 22 the discussion returns to the time domain to discuss methods for estimating the parameters of these models and forecasting.

Chapters 15 to 18 introduce the frequency domain. In these chapters the reader will learn how to detect cyclicity in a time series.

Chapters 23 to 25 introduce the bivariate case for studying connections between two time series; Chapters 23 and 24 are in the frequency domain and Chapter 25 is in the time domain.

Chapter 26 discusses the assessment of intervention effects (i.e., the interrupted time-series experiment). Chapter 27 is a review of other multivariate approaches, including time-series regression.

7 Models and the problem of correlated data

In this chapter the need for time-series models is discussed. Two practical issues are also mentioned, the inadequacy of "eyeballing" the data and the issue of how many points are needed for time-series analysis.

If all the time series we encountered in research had no autocorrelation, there would be little need for time-series analysis. Nor would there be much advantage to its application. This point will be illustrated by a reexamination of the interrupted time-series experiment, assuming a world in which no autocorrelation existed.

The interrupted time-series experiment revisited

An interrupted time-series experiment consists of a series of observations before and after some intervention. The question of the experiment is: Has the series changed after the intervention?

The question is easily answered if there is no autocorrelation in the preintervention series. Suppose that the observations, x_t, are normally independently distributed with mean L and variance σ^2. A simple test for intervention effects on the mean is provided by determining if the postintervention mean, $\bar{x}$, is not in $L \pm 2\sigma/\sqrt{N}$, where N is the number of observations after intervention. A crude approximation to this test is used in industrial quality control. If consecutive observations after intervention drift outside the interval $L \pm 2\sigma$, there is a statistically significant effect at $p < .05$. In practice the data would be graphed successively in a chart called a Shewart chart (Shewart, 1931). These charts are used in industrial quality control. For example, the diameter of ball bearings produced by a machine might be plotted and when the data drift outside the $L \pm 2\sigma$ interval, the machine may require repair. For a small change in the mean level, the Shewart chart is crude and may be ambiguous, falling out of the band once in a while. Nonetheless, it is a useful graphical technique.

An example of actual data that would be appropriate for a Shewart chart analysis was difficult to find. As a close approximation I selected a study by Hall et al. (1971), discussed by Glass et al. (1975). These data represent the number of times children disruptively "talked out" in a second-grade classroom in a poverty area.

To test whether the data are *not* autocorrelated, correlations were calculated in a manner that will be discussed in Chapter 8. They were not significantly different from zero.[1] The standard deviation of the baseline data is 2.11 and the mean is 19.35. The data in Figure 7.1 clearly show that the data points after interventions drift beyond the 2SD band. The mean after intervention was 4.85 and the standard deviation was 2.39. A t test for these means would yield $t(38) = 19.86$, which is a highly significant effect.

Sophisticated examinations of the effect of even moderate autocorrelation in the data show that estimates of effect size that assume no autocorrelation may be badly biased,[2] and the actual confidence intervals may differ markedly from what we would assume them to be (for reviews, see Glass et al., 1975; Gottman and Glass, 1978; Hibbs, 1974). To see why this is the case, this chapter will briefly examine the interrupted time-series experiment and the problem of estimating the variance of the mean.

What if the observations were strongly autocorrelated? Well, then the problem is slightly more difficult than the use of Shewart charts. It is solvable if we knew that the time series could be represented in some form. For example, suppose that we had a formula for computing the next observation from its past. Let us say that the observation at time t is half the previous

Figure 7.1. When the data are not autocorrelated, a Shewart chart can be used as a graphical technique to assess the effectiveness of an intervention.

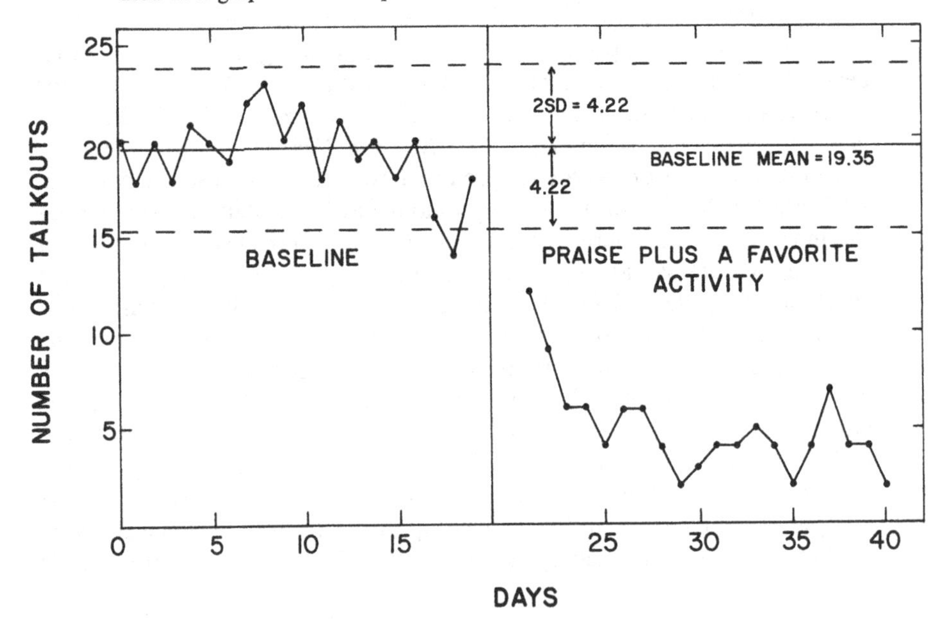

observation, plus some white noise. This is an autoregressive model, going back one step into the past. Call the series x_t, and assume that it has zero mean. This notation means that at time 1 we observe x_1, at time 2 we observe x_2, and so on. Call the white noise w_t. Then our formula is

$$x_t = .5x_{t-1} + w_t \tag{7.1}$$

where the w_t is what we need to be able to use the Shewart chart. The solution is simple in this case. We simply transform the old series to a new one, w_t, as follows:

$$w_t = x_t - .5x_{t-1} \tag{7.2}$$

and see if successive observations *of the new series* after the interventions drift outside the new Shewart band for w.

But the problem is: How can we know the appropriate transformation $x_t - .5x_{t-1}$ ahead of time? The answer is that we need a way of *estimating* this transformation from the data. Any simpler descriptions of a complicated process could be called a model; the transformation is an example of a *model*, and the fitting process is called *estimation*.

Is all of this necessary? How good is the assumption that there is no autocorrelation in most time series we would encounter? Not very good at all. We will see that even though our estimates of model parameters, such as the mean, are unbiased, estimates of their variances that ignore autocorrelation will be biased. Even if we ignored the cases of autocorrelation induced by linear and quadratic trend, out of 95 cases of time series taken (albeit not randomly) from a wide range of applications in the social sciences, Glass et al. (1975) found only 28% that had no autocorrelation. Autocorrelation is clearly something we must learn to understand.

Autocorrelation affects even the simplest statistical tools in a profound way. For example, linear regression assumes that the residuals from the regression are a set of uncorrelated numbers. Once again, what if this is not the case? It turns out that the consequences of violating this assumption about the residuals can be severe.

However, the concept of autocorrelation is not just a statistical warning or a prohibition. It is also an opportunity. We discussed how it could be used to improve on forecasts. Taking autocorrelation in the residuals into account can often improve prediction between variables. Hibbs (1974) demonstrated this for presidential popularity (see Chapter 27).

Beware the interocular test

To return for a moment to the prohibitive side of the matter, one argument that has been made in the literature of applied behavior analysis is that it ought to be possible to *see* intervention effects with the eye (sometimes called the "interocular test") rather than with statistics. This is a misunderstanding of what statistics tries to accomplish. The logical errors of this argument were discussed by Gottman and Glass (1978). However, there is also an ethical argument, which is that if we have a difficult and socially important problem to solve, we may wish to be able to detect small but significant effects. In settings where we have less control, there will be a lot of noise in the data. The question time-series analysis asks is: Have the data before and after intervention been generated by the same process? The attempt is to see a change in the "signal" over and above the noise present in the data.

Another argument that has been leveled against time-series methods in applied behavior analysis is that time-series analysis requires a lot of data for fitting models. This is indeed an issue, but it is a practical issue of power. If a graduate student were to propose a dissertation employing a two-group analysis of variance but said that he or she planned to use only three subjects in each cell, the dissertation committee would probably inform the student that there had better be a good reason for using so few subjects, because such a study will provide little power to detect differences between means. Of course, there are those rare cases when, in fact, this is a necessary course of action. However, in general, it is not a wise course.

The complaint that a lot of points are necessary in time-series analysis is equivalent to the case of analysis of variance. Indeed, more points usually permit one to identify more sophisticated models that provide a better fit to the data and make it easier to detect smaller departures from the process after intervention. Perhaps the reader will think that time-series analysis is possible with few points if one is willing to make some assumptions about the data. But consider the data in Figure 7.2. There are five points before and five points

Figure 7.2. Hypothetical data set illustrating some difficulties with time-series analysis with few observations.

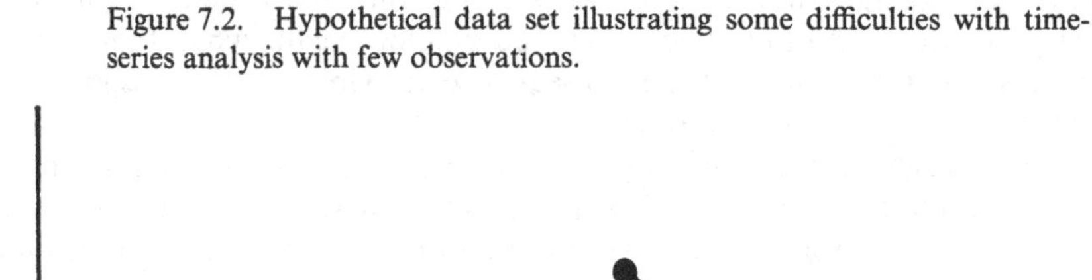

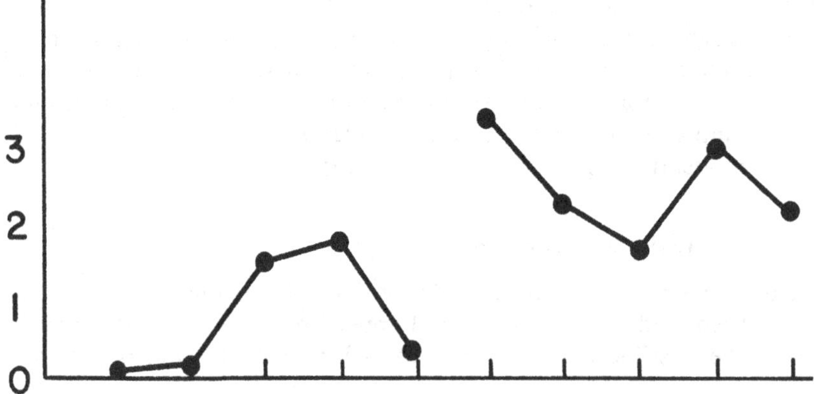

after intervention. What assumptions are we willing to make? If we assume that the data were generated by a white-noise process, we could simply do a t test on the pre- and postintervention means. We would find that $t(8) = 1.99$, and the two-tailed critical t is 2.306. This would lead us to conclude that the intervention was not significant. Everyone hates nonsignificant effects in most cases, so the reader may think that there must be *something* we can do.

Perhaps you will ask: Why not *test* for autocorrelation in the data? The answer is that the confidence interval is enormous when there are only five points; we would find that the sample first-order autocorrelation (the correlation between points only one time unit apart) for the data before intervention is .326, but that we cannot discriminate this from no autocorrelation[3] unless the autocorrelation were bigger than .894.

However, if we *knew* that the data were generated by a first-order autoregressive process, and that the effect of the intervention was to change the mean level of the series, we could find least-squares estimates for both the autoregressive model parameter and the change in level. We could also do a t test to assess whether the change in level is significant. In this case we would find that $t(8) = 4.16$, which is significant. So time-series analysis is possible with few points, but it is an extremely risky business. By the way, these results were calculated without a computer. The methods for these computations are described in Chapter 26.

Whereas in the analysis of variance case a lot of external structure is assumed, time series usually have internal structure, which cannot usually be assumed but must first be found. Perhaps you are asking the question: But you can *see* that the level has changed, so why not use the "interocular test"? The answer is that it has been shown (Glass et al., 1975; Jones et al., 1977) that the result of this test is that both many significant results are judged insignificant and that many nonsignificant results are judged significant. These errors also do not seem affected by the experience of the judge with time-series data or time-series statistics. For *most* data in social science literature, we cannot rely on our eye. The following section illustrates that even in something as basic as establishing confidence intervals for the mean, autocorrelation cannot be ignored.

How autocorrelation affects confidence intervals for the mean

This section considers an extremely basic situation. Suppose that we are sampling observations on some normally distributed variable x, with standard deviation σ and mean $\bar{x}$, and we know that the observations we sample are uncorrelated. The large sample 95 per cent confidence interval for the mean is then

$$\bar{x} \pm 1.96\left(\frac{\sigma}{\sqrt{N}}\right)$$

The figure $\sigma/\sqrt{N}$ is the standard deviation of the mean across samples of size N.

Now, what is the effect of this independence assumption, that is, the assumption that the normally distributed observations are uncorrelated? The answer is that if the data have positive autocorrelation, and if they are above the mean they are likely to continue above the mean for a time, and the same statement can be made for observations below the mean a particular time.[4] Thus, we have *runs* of consecutive observations away from the mean, and this will spread out the distribution of estimates of the mean around its expected value, $\bar{x}$. Thus, for *positive* autocorrelation, $\sigma/\sqrt{N}$ will *underestimate* the variability in $\bar{x}$.

For negative autocorrelation, quite the opposite situation is true. An observation displaced above the mean at one time will be below the mean the next time it is sampled, so for samples of N consecutive points, $\sigma/\sqrt{N}$ will *overestimate* the variability in $\bar{x}$.

To summarize, both as a statistical prohibition and as an opportunity for increased precision (e.g., in the case of forecasting), it is wise to consider the case of autocorrelation in data collected across time. Chapter 8 considers the basic assumptions of time-series analysis.

8
An introduction to time-series models: stationarity

This is an important chapter because it begins to explore the assumptions that underlie time-series analysis. The discussion proceeds with a lot of detail and precision. The development is purposely slow and careful, and the chapter is written at an elementary level. In this chapter the concepts of stationarity, autocovariance, autocorrelation, and the correlogram are introduced. The Bartlett test for the significance of the autocorrelation and a rule of thumb for assessing stationarity are discussed.

Stationarity

A *model* is a set of assumptions made about the mathematical process that may have generated data. We try to search for a simple and reasonable model that is able to simulate the data or some interesting structural property of the data. In this chapter we discuss the kinds of assumptions we need in order to do any kind of statistical analysis of time-series data.

To do this we need to distinguish between a particular observed time series, called a *realization*, and the *process* that is presumed to have generated the realization. Just as in regular statistics we try to make inferences from a sample to a population, in time-series analysis we try to make inferences from a realization to the process that generated it.

We will have to make assumptions about the underlying process that generated the observed series – assumptions that something about the process remains unchanged throughout historical time. Let us consider why this is important. The job of anyone who tries to predict, or forecast, the future is to pick something about the past that is likely to remain fairly stable. It need not be the time series itself, but it must be some function of the time series that remains unchanged with time. For example, we might assume that an observed trend will stay the same; or that the rate of change of the trend will stay the same. Without some kind of assumption of stability we can do nothing logical about making a reasonable forecast.

These stability assumptions are called *stationarity* conditions. In probability and statistics we have a process of some type such that at each point in time, the observation can be considered to be a random variable, each with a *probability density function* (pdf) describing the relative likelihood of the values it might generate. For example, at time $t = 5$, there is a specific probability $P(M)$ that the series will be less than or equal to M. This probability density function (pdf) is usually described by its various moments, such as the expected value (mean) and variance. If the random variables are assumed to be normal, the pdf is completely characterized by its mean and variance. A time-series process is thus a collection of random variables, and the pdf's describe something about its structure.

Since the kind of stationarity assumptions we want to make is that the underlying process is stable over time in some statistical sense, an obvious first requirement is that some aspects of the random variables at each time point remain the same. This is our first condition of stationarity:

> *Condition 1* A stationary process is characterized in part by the fact that its (finite) mean and (finite) variance do not change with historical time.

However, we will need an additional stability condition. Because we have a pdf at each point in time, $t = 0, 1, 2, \ldots$, we can compute how well we can predict the value from one time point t to neighboring time points, k lags away. For example, our process may be such that prediction is *only* possible between adjacent time points.

In general, we must require that the *form* of this prediction not be a function of historical time but only of the amount of lag between time points. This means, for example, that we can predict as well between points 1 and 2 as between points 48 and 49, or 57 and 58, and so on.

How is this stability in prediction to be described mathematically? By the covariance between the random variables at two time points. If the time points are t and $t + k$, the covariance must be a function only of lag, k, and not the starting point, t. This leads us to:

> *Condition 2* A stationary process is characterized in part by the fact that the covariance between two random variables at t and $t + k$ is a function only of their relative lag, k, not of the starting point, t. In other words, the covariance of this process is independent of historical time.

Example

Suppose that we have decided to test whether a normal random number generator is really working as it should. We expect the random number generator to be a process, and this process will have any number of possible realizations. In Table 8.1 this is illustrated symbolically. In Table 8.2 this

Table 8.1. *Family of realizations of a random process: symbolic representation*

		Time			
Realization number	Symbol for the series	1	2	3	. . .
1	$x_1(t)$	$x_1(1)$	$x_1(2)$	$x_1(3)$	. . .
2	$x_2(t)$	$x_2(1)$	$x_2(2)$	$x_2(3)$	. . .
3	$x_3(t)$	$x_3(1)$	$x_3(2)$	$x_3(3)$	. . .
.	.	.	.	.	. . .
.	.	.	.	.	. . .
.	.	.	.	.	. . .
N	$x_N(t)$	$x_N(1)$	$x_N(2)$	$x_N(3)$	. . .
.	.	.	.	.	. . .
.	.	.	.	.	. . .
.	.	.	.	.	. . .

Table 8.2. *Family of realizations of a random process: numbers*

Realization number	Symbol for the series	Time								
		1	2	3	4	5	6	7	8	9
1	$x_1(t)$	.46	.14	2.46	−.32	−.07	.30	−.29	1.30	−1.90
2	$x_2(t)$	.06	−2.53	−.53	−.19	.54	−1.56	.19	−1.19	.02
3	$x_3(t)$	1.49	−.35	−.63	.70	.93	1.38	.79	−.96	−.85
4	$x_4(t)$	1.39	−.56	.05	.32	2.95	1.97	.41	.44	−.04
5	$x_5(t)$	−.53	.60	.88	−.93	1.58	.16	−1.89	.37	.37
6	$x_6(t)$	.97	.71	1.09	−.63	−.26	−.70	−1.50	−.49	−.16
7	$x_7(t)$	.45	.75	−.42	−.43	−.69	.76	−1.62	−.35	−.51
8	$x_8(t)$	−.48	1.68	−.06	−1.23	−.49	.86	−.49	−1.98	−2.38
9	$x_9(t)$	1.36	−.56	−.26	−.21	.22	.78	.95	−.87	−1.01
.	.	.	.	.	.	.	.	.	.	.
.	.	.	.	.	.	.	.	.	.	.
.	.	.	.	.	.	.	.	.	.	.
Estimated mean[a]		.58	−.01	.29	−.32	.52	.44	−.38	−.41	−.71
Estimated variance		.78	1.19	1.01	.59	1.16	1.06	1.07	.99	.92

[a] The average of these nine means is zero.

Table 8.3. *Test of condition 1 for stationarity from a pseudo-random-normal stationary process* $[N(0,1)]$

	Time									
Condition 1	1	2	3	4	5	6	7	8	9	10
Mean	.05	.09	−.06	.08	−.09	−.10	−.67	.29	−.02	.05
Standard deviation	.91	.83	1.20	1.08	.81	1.29	.82	.76	1.02	.83

Table 8.4. *Test of condition 2 for stationarity for a pseudo-random-normal stationary process* $[N(0,1)]$

	Lag				
Starting point	1	2	3	4	5
$t = 1$	−.24	.22	−.11	−.46[a]	−.34
$t = 3$	−.37	−.09	−.15	.08	.10
$t = 5$	.21	.07	.00	.11	.15

[a] $p < .05$.

is illustrated numerically. These numbers were derived from a random numbers table; they are generated by a normal process with zero mean and unit standard deviation [i.e., $N(0,1)$]. The first thing to notice about these tables is that at each time point there is a distribution. For example, at $t = 1$ in Table 8.1 the values are represented symbolically as $[x_1(1), x_2(1), x_3(1), \ldots, x_N(1)]$. In Table 8.2 the values are $(.46, .06, 1.49, 1.39, -.53, .97, .45, -.48, 1.36, \ldots)$. Thus, at each time point we can compute a mean and a variance.

An $N(0,1)$ computer pseudo-random number generator was used to create a matrix such as the one in Table 8.2, with 20 realizations ($N = 20$) and 10 time points. Only 9 of these realizations and 9 time points are shown in Table 8.2. Table 8.3 presents the means and standard deviations for each time point for all 20 realizations. Note that they are quite close to zero mean and unit standard deviation.

Next, the correlations were computed across time points. Each correlation is an ordinary correlation of two columns in the table, correlated as if they were two variables. For condition 2 of stationarity we could see whether these correlations change, depending on where we begin computing them. For example, in Table 8.2 the correlation between time points 1 and 2 pairs .46 with .14, .06 with -2.53, 1.49 with $-.35$, and so on. Thus, the first row of Table 8.4 consists of the correlations between time points (1, 2), (1, 3), (1, 4), etc.; the second row consists of correlations between time points (3, 4), (3, 5), etc.; and the third row consists of correlations between time points (5, 6), (5, 7), etc. Under the null hypothesis of independence, only one of the correlations is significant at $p < .05$, which is to be expected in making 15 tests, each with a 5% chance of being accidentally significant. Thus, the correlations are pretty much all zero, as we would expect from random numbers, and this does not change much as we shift the starting time for computing the correlations. This is consistent with condition 2 of stationarity.

The Wiener definition of stationarity: a problem with our definition

In most applications we have only one time series and not an ensemble. Thus, it will not be possible to test our two assumptions about stationarity. How can we proceed? *We have to break the series into separate pieces.*

Conceptually, we can proceed by altering our definition of stationarity to read that the mean, variance, and covariances do not change if we break the series into successive long chunks. In this definition of stationarity we are making an assumption that a theorem called the *ergodic theorem* applies. An observed time series is presumed to be one *realization of a process;* that is, the time series is assumed to be one member of an *ensemble* (or population) of series generated by the same stochastic (probabilistic) process. Stated mathematically, the time series $x_i(t)$ is assumed to be one realization of a process $x(t)$ and part of an *ensemble* $[x_1(t), x_2(t), \ldots, x_i(t), \ldots]$. If we assume that each realization is infinitely long (in time), *the ergodic theorem states that for stationary processes, the ensemble average (of sample means, variances, moments, cross-moments, etc.) at any point equals the corresponding time average for any one series.*

Perhaps this ergodic property of stationary time series is fine for condition 1 of stationarity, but what do we do about condition 2, the stability of the covariances? How can we even *define* the covariance between two time points? The answer is that we need to define a function called the *autocovariance* and a closely related function called the *autocorrelation.* Theoretically, these functions must have something to do with stable predictability between time-series points as a function of lag only, not of historical time. We now turn to a discussion of these concepts.

Table 8.5. *Average of several daily ratings of headache severity by a migraine-headache patient (5 = most severe pain)*

Day	Average rating	Day	Average rating
1	1.61	21	2.27
2	1.00	22	2.35
3	1.38	23	3.33
4	2.05	24	1.33
5	2.21	25	2.19
6	2.00	26	1.44
7	1.35	27	1.24
8	2.17	28	1.44
9	2.25	29	4.40
10	2.71	30	4.50
11	3.39	31	3.56
12	2.50	32	3.00
13	3.56	33	2.29
14	4.00	34	1.83
15	3.76	35	2.24
16	3.29	36	1.39
17	3.28	37	2.82
18	2.31	38	3.29
19	2.50	39	3.33
20	2.33	40	3.33

Autocovariance and autocorrelation

Suppose that we have one set of data over time. The following example is a series of ratings of the severity of headache pain by a migraine-headache patient. The patient made several ratings on a five-point scale each day. We will analyze this data set several times throughout this book. The data set was used in Chapter 4 to illustrate the concept of forecasting. Table 8.5 lists the first 40 of the 60 points of this series.

Throughout this book, symbols such as x_t, y_t, z_t, and w_t will be used to stand for a time-series process or for the realization of a time-series process. Suppose that we wish to determine how predictable the migraine-headache patient's headache pain is. We can draw a set of scatter plots, first pairing days 1 and 2, days 2 and 3, and so on (Figure 8.1). The first scatter plot always pairs points one lag apart, (x_t, x_{t+1}) for $t = 1, 2, \ldots, 39$. The second scatter plot always pairs points two lags apart, (x_t, x_{t+2}) for $t = 1, 2, \ldots, 38$.

You can see from the scatter plots that there is a high positive correlation at lag 1 and a slightly positive one at lag 2. The lag 1 correlation is called the *first-order autocorrelation coefficient*, written r_1. It is calculated as

$$r_1 = \frac{\sum_{t=1}^{N-1} (x_t - \bar{x})(x_{t+1} - \bar{x})}{\sum_{t=1}^{N} (x_t - \bar{x})^2} \tag{8.1}$$

where $\bar{x}$ is the mean of the observed series.

The formula for the lag 2 autocorrelation is computed in a similar fashion:

$$r_2 = \frac{\sum_{t=1}^{N-2} (x_t - \bar{x})(x_{t+2} - \bar{x})}{\sum_{t=1}^{N} (x_t - \bar{x})^2} \tag{8.2}$$

In both equations (8.1) and (8.2), the numerator of the ratio times $1/N$ is an estimate of the *autocovariance*, at lags 1 and 2, respectively. The autocovariance will eventually become as familiar a function of lag as the autocorrelation. Note that the denominators are the same in equations (8.1) and (8.2). The denominator times $1/N$ is an estimate of the variance of the series.

An approximate standard deviation for the autocorrelation estimate is given by $1/\sqrt{N}$, under the null hypothesis that there is no autocorrelation in the process of which this series is an approximation (Bartlett, 1946). For the headache data (all 60 points) the lag 1 autocorrelation is .564 and the lag 2 autocorrelation is .176. To be significantly different from zero (at alpha = .05), each of these autocorrelations must exceed $2/\sqrt{N} = 2/\sqrt{60} = 0.258$. We might conclude that the lag 2 autocorrelation is thus not signif-

Figure 8.1. Lag 1 and Lag 2 scatter plots of the migraine-headache data plotted in Figure 4.1.

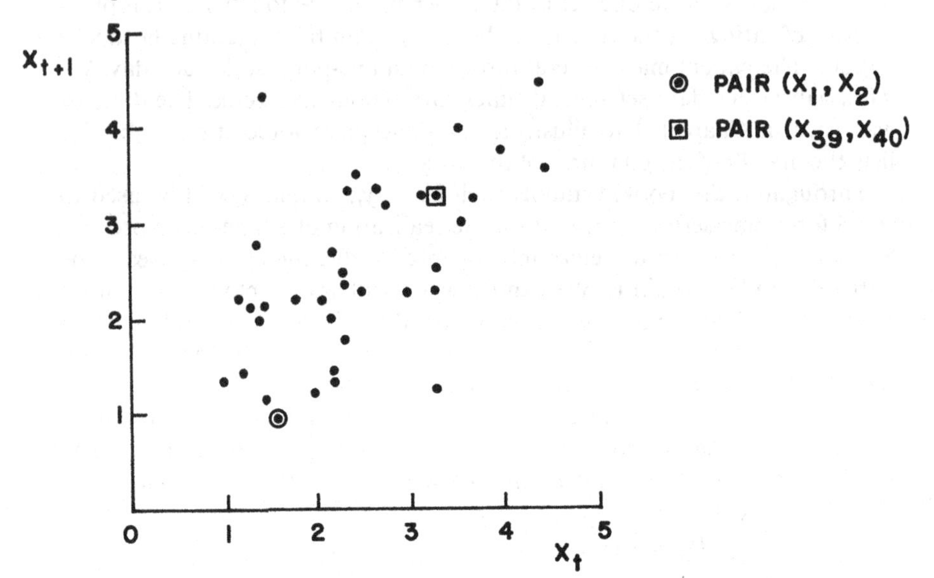

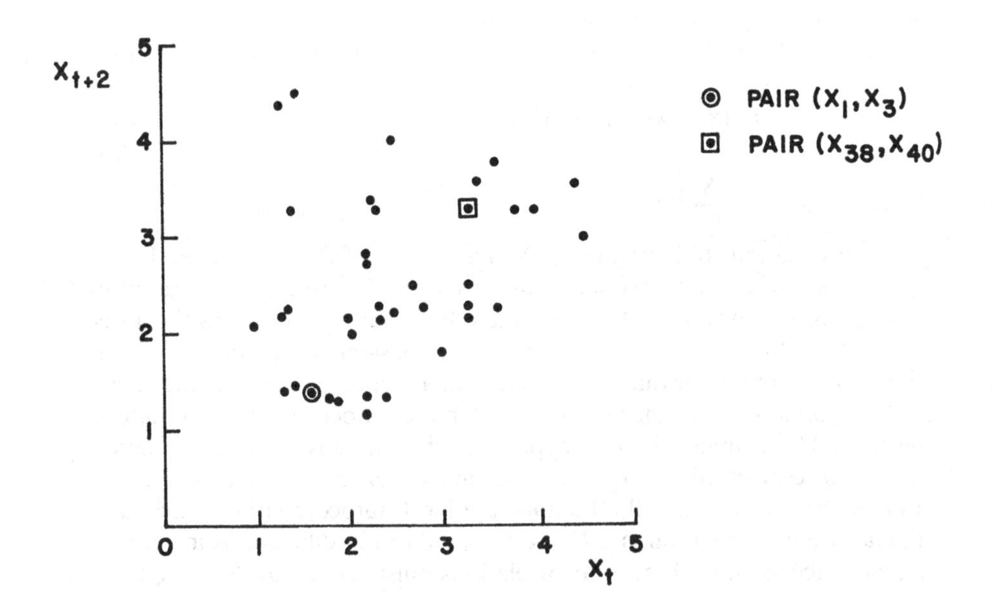

icant. However, later in the book we discover how mistaken this decision would be.

We can generalize this formula and define the kth-order autocorrelation coefficient r_k:

$$r_k = \frac{\sum_{t=1}^{N-k} (x_t - \bar{x})(x_{t+k} - \bar{x})}{\sum_{t=1}^{N} (x_t - \bar{x})^2} \tag{8.3}$$

Furthermore, we can plot r_k as a function of k.

A plot of r_k by k is called the *correlogram* of the time series. A plot of the numerator of r_k divided by N (the autocovariance), as a function of k, is called the *covariogram*.

Example

Calculation of r_1:

t	x_t	$x_t - \bar{x}$
1	47	−4
2	64	13
3	23	−28
4	71	20
5	38	−13
6	64	13
7	55	4
8	41	−10
9	59	8
10	48	−3

$$\sum_{t=1}^{9} (x_t - \bar{x})(x_{t+1} - \bar{x}) = (-4)(13) + (13)(-28) + \cdots + (8)(-3)$$

$$= -1{,}497$$

$$\sum_{t=1}^{10} (x_t - \bar{x})^2 = (-4)^2 + (13)^2 + \cdots + (-3)^2 = 1{,}896$$

$$r_1 = \frac{-1{,}497}{1{,}896} = -.79$$

Note again that the sample autocorrelation is the sample autocovariance ($1/N$ times the numerator) divided by the sample variance ($1/N$ times the denominator).

Condition 2 revisited

We have now defined the autocovariance and autocorrelation functions of a time series. Condition 2 for one series is now: The autocovariance function, which is a function of lag, should have the same shape for different long chunks of the series. This means that the correlograms and covariograms should have the same shape, independent of the starting point in historical time where we begin calculating.

Example

Four hundred observations from one realization of a stationary time-series process were computed, using a noise process from our pseudo-random-number generator. The time series was broken into four chunks of 100 points each. For each of these chunks the autocorrelation function for that chunk, the mean of that chunk, and its variance were calculated (see Table 8.6). Note that this series appears to satisfy our new stationarity conditions. Not only are the mean and variance stable (Condition 1), but the autocovariance function has the same shape throughout. This is the shape characteristic of a random process; that is, there is no significant autocorrelation. In other words, no autocorrelation falls outside the 2SD band (all are less than .20 in absolute value); of course, we can expect about 5% of the autocorrelations to be significant by chance alone.

This concept of stationarity applied to one realization of a time-series process is equivalent to a belief in the stability of natural processes; for

Table 8.6. *Means, variances, and autocorrelations for four chunks of one realization of a pseudorandom process (N = 100 for each chunk)*

			Lag					
Chunk	Mean	Variance	1	2	3	4	5	6
1	.17	.91	−.02	−.15	.09	−.04	.14	.05
2	.05	1.02	−.10	−.03	−.05	.03	−.09	−.02
3	.07	.65	−.16	.04	.00	.09	.09	−.16
4	−.02	.93	−.13	−.17	.02	.08	.06	.05

Note: $2SD = 2/\sqrt{N} = .20$

example, we would expect Whittaker and Robinson's (1924) conclusions about the star brightness data to be independent of when they started taking their daily observations. If they had begun a week or two earlier or later, we would have expected them to have obtained similar results.

Examples of nonstationary processes

In this section we consider computer-generated realizations of various *nonstationary* processes. In each case, of course, I will know what the process is but pretend that I do not know. I will generate 400 points of the series and examine, for each of four equal 100-point segments, the following sample estimates: (1) the mean, (2) the variance, (3) the autocorrelations for eight lags, and (4) the spectral density function. Recall that we discussed the spectral density function; Condition 2 could have been stated in terms of stability of the spectral density function instead of the autocovariance function. It will be helpful for the reader to begin thinking in terms of all these functions, which means being able to think in both the time and the frequency domains.

Case 1: Nonstationarity in the mean

Table 8.7 presents the statistics for the first case. We first examine Condition 1 and would conclude that although the variances appear fairly stable, the means are definitely increasing, and thus this series has an upward trend. It is thus nonstationary in the mean. Note, however, that the autocorrelations have the same shape from segment to segment. That shape is that the autocorrelations are first negative, then positive, and so on, decreasing rapidly in size so that they are within the Bartlett band ($2/\sqrt{N} = .2$) by lag 5.

Note also that the spectral density functions are consistent. They are highest at the high frequencies, and second highest at the lowest frequency (near zero). The first-order autocorrelation was negative, which means that neighboring observations tend to alternate, first above and then below the mean. This is represented by a high-frequency peak in the spectral density. The low-frequency peak represents a trend (in this case). That a trend is equivalent to a low-frequency oscillation can be seen from Figure 8.2, which shows that a trend is indistinguishable from a low-frequency wave that extends beyond the observation period, or one whose period is interminably long.

The case was generated by using e_t, a normal white-noise generator, to obtain a realization of the process:

$$y_t = x_t + z_t \tag{8.4}$$

Table 8.7. *Case 1 of a nonstationary process*

Autocorrelations:			Lag							
Segment	Mean	Variance	1	2	3	4	5	6	7	8
1	.61	1.45	−.45	.32	−.13	.14	.03	.17	−.14	.02
2	1.53	1.91	−.59	.44	−.24	.25	−.11	.10	.05	−.08
3	2.55	1.24	−.59	.42	−.18	.18	.04	−.02	.03	.06
4	3.50	1.60	−.51	.33	−.10	.20	.02	.10	.06	.06
Spectral density:	Selected frequencies									
Segment	.025	.075	.125	.175	.225	.275	.325	.375	.425	.475
1	.36	.16	.12	.23	.32	.15	.47	.61	.84	1.37
2	.33	.14	.13	.18	.22	.32	.48	.58	1.40	2.28
3	.24	.09	.05	.10	.19	.22	.23	.54	.98	1.31
4	.43	.08	.09	.18	.27	.35	.48	.60	1.15	1.48

where

$$x_t = (.01)t \qquad \text{and} \qquad z_t = -.6z_{t-1} + e_t$$

In subsequent chapters we will see that z_t is an example of a "first-order autoregressive" process.

Case 2: Nonstationarity in the mean

Table 8.8 presents statistics for the second case. Once again we see nonstationarity in the mean, but not in the variance. Except for the fourth segment, the autocorrelation functions have the same shape. They do not seem to die down rapidly in the fourth segment, although this is probably a random event, since by lag 11 the autocorrelation is .13, and it continues to decrease rapidly thereafter. Table 8.8 presents only the first eight lags to conserve space. The spectral density for all segments has the same shape. The peak is in the low-frequency ranges, and this is consistent with the high positive autocorrelation of neighboring points. The high positive autocorrelation means that if the series is above (or below) the mean, it will continue to be above (or below) the mean for a while; in other words, it cycles smoothly and slowly, or, in frequency-domain language, it has a peak in the low-frequency range. This case was generated by the process

$$y_t = x_t + z_t \tag{8.5}$$

where

$$x_t = (.01)t \qquad \text{and} \qquad z_t = .6z_{t-1} + e_t$$

Figure 8.2. Trend is not distinguishable from a very low frequency peak.

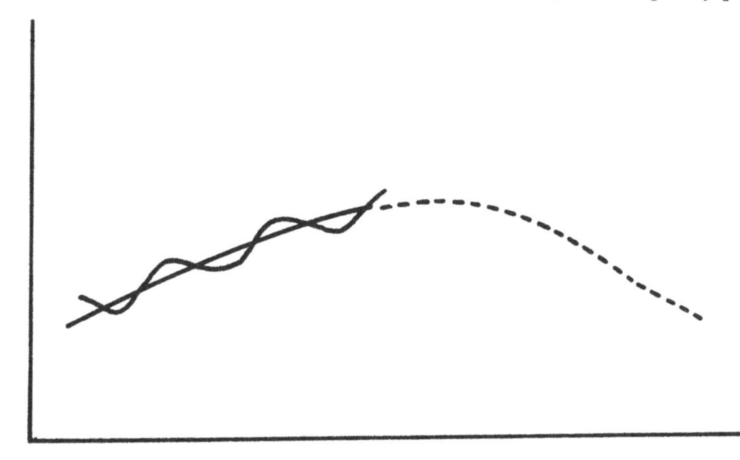

with e_t being normal white noise, as before. Note that the first-order autocorrelation is a good estimate of the .6 coefficient of the z_t equation. In Table 8.7 it was a reasonable estimate of the $-.6$ coefficient of the z_t equation. In subsequent chapters we will see that this is no coincidence.

Table 8.8. *Case 2 of a nonstationary process*

Autocorrelations:			Lag							
Segment	Mean	Variance	1	2	3	4	5	6	7	8
1	1.00	1.59	.63	.34	.18	.11	.10	−.01	−.16	−.09
2	1.70	1.62	.61	.36	.20	.09	−.03	−.11	−.19	−.26
3	2.76	.97	.59	.40	.31	.26	.15	−.03	−.06	−.02
4	3.46	1.62	.64	.46	.45	.45	.43	.38	.32	.26
Spectral density:	Selected frequencies									
Segment	.025	.075	.125	.175	.225	.275	.325	.375	.425	.475
1	1.62	1.33	.61	.47	.42	.11	.15	.15	.10	.10
2	1.54	1.51	.60	.41	.28	.22	.18	.12	.13	.14
3	1.14	.79	.23	.18	.23	.15	.09	.10	.10	.08
4	2.67	.64	.37	.37	.34	.23	.19	.12	.13	.10

Table 8.9. *Case 3 of a nonstationary process: random walk*

Autocorrelations:			Lag							
Segment	Mean	Variance	1	2	3	4	5	6	7	8
1	6.08	22.69	.96	.92	.87	.84	.81	.78	.75	.72
2	18.27	4.83	.89	.81	.73	.66	.59	.53	.47	.41
3	25.70	19.06	.98	.97	.97	.96	.95	.93	.91	.89
4	24.48	7.67	.94	.87	.85	.82	.77	.72	.67	.62
Spectral density:	Selected frequencies									
Segment	.025	.075	.125	.175	.225	.275	.325	.375	.425	.475
1	61.64	6.27	1.74	.90	.49	.26	.23	.26	.24	.21
2	11.16	2.36	.59	.34	.23	.18	.16	.10	.11	.12
3	56.38	2.71	.44	.28	.28	.17	.12	.13	.09	.06
4	20.42	2.19	.54	.34	.30	.21	.15	.09	.10	.07

Case 3: Nonstationarity in both mean and variance

Table 8.9 presents statistics for the third case. Here it appears that the process is nonstationary in both the mean and the variance. Note that the autocorrelations are similar for each segment; they remain close to 1.0 throughout, declining very gradually compared to our first two cases. The spectral density has the same shape in each segment, peaking in the low-frequency range, as we would expect of data with such high positive autocorrelations and increasing mean (see Figure 8.2).

This series is a realization of what is called the *random walk process*,

$$z_t = z_{t-1} + e_t \tag{8.6}$$

where e_t is normal white noise. The random walk process is the path described by a drunkard who starts at a lamp post at time $t = 1$, and is then likely to step in either of two directions on a street in the next time points. These steps form an additive drift. The path is also a good model for the Brownian motion of a molecule in a gas (see Figure 8.3).

Case 4: Shifting covariance structure

Table 8.10 presents the results of case 4. Note that the mean and variance seems fairly stable, and indeed, the case was designed so that the mean and

Figure 8.3. Realization of a random-walk process.

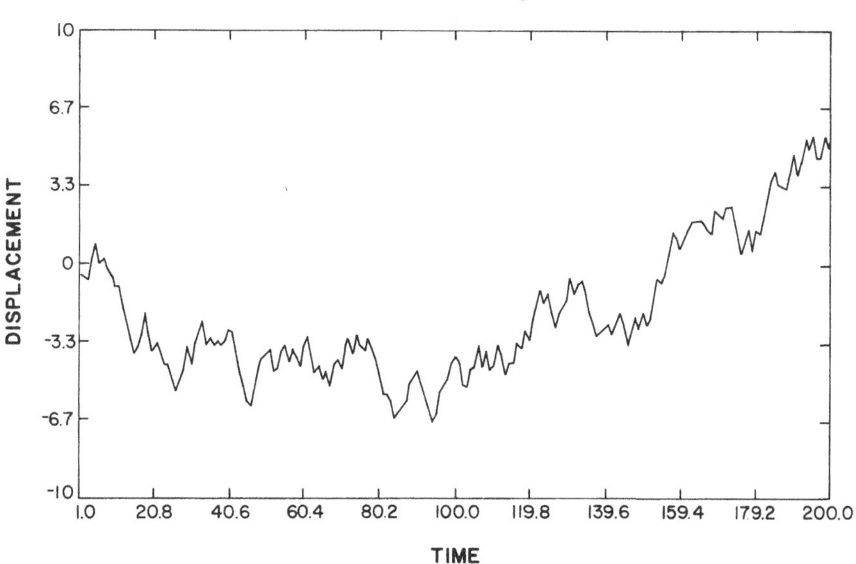

Table 8.10. *Case 4 of a nonstationary process*

Autocorrelations:			Lag							
Segment	Mean	Variance	1	2	3	4	5	6	7	8
1	.13	1.09	−.37	.06	−.08	−.02	.09	−.25	−.06	.21
2	.04	1.05	−.21	.01	−.05	.04	−.10	.00	−.05	−.13
3	.08	.64	−.05	−.04	.01	.09	.07	−.15	−.11	−.04
4	−.01	.92	.19	−.07	.02	.11	.13	.09	.04	−.02

Spectral density:	Selected frequencies									
Segment	.025	.075	.125	.175	.225	.275	.325	.375	.425	.475
1	.09	.16	.14	.29	.40	.16	.46	.54	.54	.69
2	.14	.27	.23	.29	.31	.37	.42	.34	.42	.54
3	.17	.23	.11	.18	.28	.24	.17	.21	.23	.20
4	.48	.27	.30	.38	.41	.30	.26	.18	.19	.14

variance would be near zero and 1, respectively. But note that the autocorrelations shift. For the first and second segments they are negative and oscillate in sign; for the third segment there is essentially no autocorrelation, and for the fourth segment the autocorrelation begins positive. The spectral density functions follow this pattern. For the first two segments they increase, suggesting high-frequency components, consistent with the negative lag 1 autocorrelation; for the third segment the spectral density is flat, suggesting white noise, consistent with zero autocorrelations; for the fourth segment the spectral density function is highest for the lowest frequencies, decreasing as frequency increases, consistent with the positive lag 1 autocorrelation.

Case 4 was generated by an autoregressive process,

$$z_t = a(t)z_{t-1} + e_t \tag{8.7}$$

where $a(t)$ changed across segments from negative to positive and e_t was normal white noise as before. This case represents a nonstationarity only on Condition 2; that is, means and variances are stable, but the correlogram is not stable.

The shape of the correlogram: a rule of thumb for stationarity

We discuss next a condition that will usually be satisfied by stationary time-series processes. This is not a *necessary* condition for stationarity, but if Condition 1 is satisfied, in most cases when the new condition is not satisfied, the process is nonstationary, and in all cases when the new condition is satisfied, the series will be stationary. The new condition will thus be a useful guide for most practical applications.

The new condition concerns the *speed* with which the sample autocorrelation function r_k goes to zero. For infinitely long series, where we can compute the autocorrelations for an infinite number of lags, the condition is stated as follows:[1]

$$\sum_{0}^{\infty} r^2_k < \infty \tag{8.8}$$

This means that the partial sums

$$\sum_{1}^{n} r^2_k \tag{8.9}$$

converge to a finite limit.

This convergence will be satisfied if r_k^2 goes to zero sufficiently fast. One fact about infinite series is that series of the form

$$\sum_{k=1}^{n} \frac{1}{k^a} \tag{8.10}$$

will converge as n increases if $a > 1$. Thus, one *sufficient* condition for convergence of the sums of squares of the autocorrelations is that the autocorrelations go to zero faster than the following:

$$r^2_k < \frac{1}{k} \qquad \text{or} \qquad kr^2_k < 1$$

at least for large k.

Note again that the series can be stationary and that this condition need not hold. But a general guide is that for most stationary series the square of the autocorrelations will go to zero very rapidly. If this is not the case, we might suspect the stationarity of the process.

In effect, the condition implies that, after some lag, the observations can be considered almost independent.

Social interaction: a practical example of examining nonstationarity

There are times when it can be extremely useful to *think* in terms of breaking a time series into separate chunks and analyzing patterns in each chunk separately. For example, suppose that we are studying mother–infant social interaction and we film or videotape-record the play of a mother with her infant. We can score each frame of the film using a coding system developed by Tronick et al. (1977) and add up behaviors that indicate positive involvement and excitement and similarly subtract those behaviors that indicate disengagement and withdrawal. We will obtain a graph similar to the one portrayed in Figure 8.4. These data appear to be cyclic and the interaction appears to be mutual and bidirectional; in fact, these aspects of mother–infant interaction have been noted by a number of developmental psychologists (see Stern, 1977).

Suppose that we wish to test the rival hypothesis that it is the mother who is adjusting to the baby's rhythms of attention and that the baby is, in fact, unresponsive. Call this the "robot-baby" hypothesis. Figure 8.5 illustrates fictional data to support this hypothesis. The baby's behavior looks very much as if it were created by an electronic sine-wave generator. In the first half of the interaction the mother appears to be out of synchrony with her infant, or searching for its rhythmicity. In the second half of the interaction the mother is in synchrony, slightly behind in phase (i.e., the infant is leading the interaction), and her amplitude, or energy, matches the infant's. How would this pattern be detected by a time-series analysis in the frequency domain?

Figure 8.4. Data obtained from coding the social play of three mother–infant dyads engaged in face-to-face interaction (from Tronick et al., 1977).

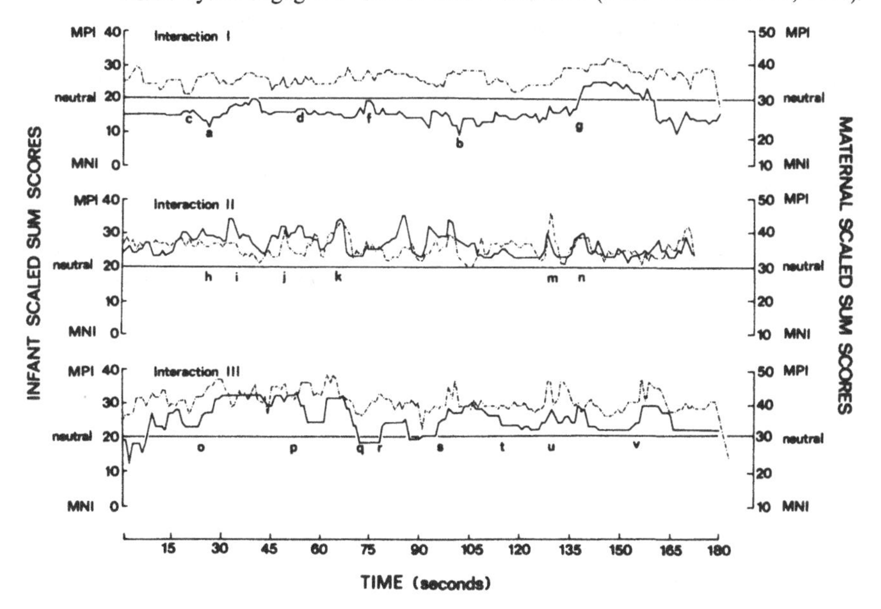

Figure 8.5. Fictional data illustrating a "robot" baby who is cyclic and a mother who eventually adjusts herself to her infant's rhythmicity.

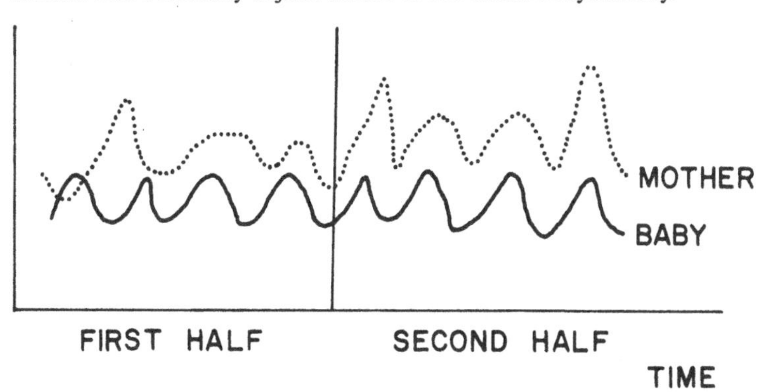

The robot-baby hypothesis would be detected by a specific form of nonstationarity as portrayed by Figure 8.6. The nonstationarity can be described as follows. If we examine the baby's spectral density function, it appears identical in the two halves and has one major peak. The mother's peak matches the baby's in the second half, but in the first half it is spread widely around that frequency. This reflects the mother's early search procedure. The mother's data have different spectral density functions in the two halves because of her search, and this is a form of nonstationarity. Also, the relationships between the mother's and the baby's data change from the first to the second half of the data. The amplitudes of mother and baby do not match well in the first half, illustrated by a low *coherence* in the first half and a higher coherence in the second half, when the amounts of mother's excitement or disengagement matches the baby's. Also, the phase spectrum is negative only in the second half, which indicates that the baby leads the interaction in this half.

In this example we can see how a specific hypothesis of interaction and causality can be tested by searching for nonstationarity of a particular type.

Figure 8.6. How spectral time-series analysis could be used to test the robot-baby hypothesis.

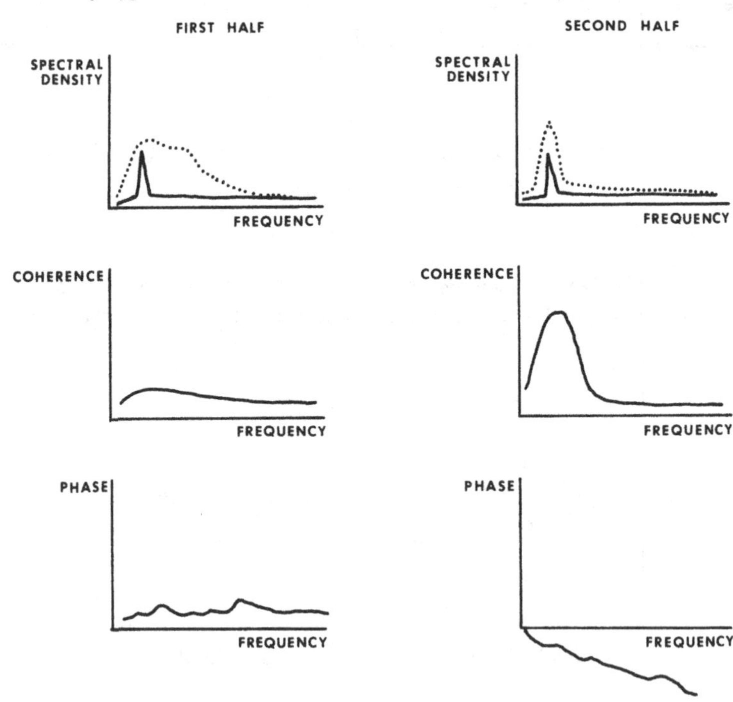

This example is also illustrative of establishing a steady state, which may be common after the initial phases of an experiment.

Summary

There were two definitions of a weakly stationary process introduced in this chapter. They differed in the definition of the covariance, the first definition involving an ensemble average over possible realizations, and the second definition involving an average over time within one realization (the Wiener definition). They are equivalent under certain reasonable conditions (the ergodic property).

In practice, we either assume that the process is stationary, or examine the mean, variance, and autocorrelation function (or spectral density function) for separate long chunks of the observed series and the shape of the autocorrelation function (looking for rapid decay to zero). If the process does not appear stationary, we can either transform the process to a new one that is stationary, or examine separate pieces of the series, each of which is more-or-less stationary.

This concept of transforming a series until the *transformed* series is stationary will be discussed extensively in the following chapters. A useful way of understanding this notion is to try to decompose the series into *independent components*, one of which is a stationary process. We then try to decompose the stationary process itself. The first basic decomposition of a stationary process is to express it as the sum of a *deterministic* and a stochastic, or probabilistic, or, more commonly, a nondeterministic process. This was shown always to be possible in a famous theorem proved by Wold (see Appendix 9A).

9 What if the data are not stationary?

This chapter considers what to do if the data are not stationary. There are two alternatives, modeling the nonstationarity and transforming the data so that they are no longer nonstationary. A comparison of these two approaches is instructive. This book recommends the former method, with the occasional cautious and wise use of the latter. The concept of a filter is introduced in this chapter.

If the data are not stationary, there are two alternatives. The first alternative is to *subtract* out components so that the residual is stationary. The second alternative is to employ a specific set of transformations to the data so that the transformed data are stationary.

These two alternatives are not necessarily equivalent. This chapter examines both approaches. In particular, this chapter examines the second alternative by studying how the most commonly used transformations affect the autocovariance and spectral density functions of the time series that is transformed.

The first alternative: multicomponent model

If our time series is x_t, we can attempt to represent x_t as the following sum:

$$x_t = \text{Trend} + \text{Deterministic cycles} + \text{Stationary stochastic process} \tag{9.1}$$

The first component of the model is *trend*. This usually means *linear* trend, but it need not be linear. It could be quadratic or a polynomial of higher order. However, one must be careful not to go to too high an order because each wiggle of a polynomial could be part of a cycle.

The second component consists of *deterministic cycles*. These are sinusoids that have extremely precise and regular cyclicities. Chapter 15 will be a discussion of the essential difference between deterministic cycles and probabilistic cycles. Appendix 9A is a general discussion of the difference between deterministic and nondeterministic processes. A nondeterministic process need not be cyclic, but if it is, a nondeterministic cycle meanders around a particular frequency, phase, and amplitude, whereas a deterministic cycle has a fixed frequency, phase, and amplitude. Also, as one ventures into the future on a nondeterministic cycle, the amount of predictability provided by the past decreases. This is not the case for a deterministic cycle. To use a frequency, or spectral analogy, a pure tone or a pure color is a deterministic cycle. Its spectral density would approximate a vertical *line* (i.e., all its power would be concentrated at one frequency). We will see later that the distinction between deterministic and nondeterministic is more of a dimension than a strict dichotomy.

The first step is to subtract trend by least-squares fitting a straight line. In some cases it will be necessary to subtract a polynomial trend that is of higher order than linear (e.g., a quadratic or a cubic). The second step is to identify the deterministic cycles, which are extremely regular periodicities, and would therefore show up as line spectra of the spectral density of the detrended data. It is possible to test for these using a procedure to be discussed shortly. These cycles can be estimated and the weights of the cycles can also be estimated by least-squares fitting.

After subtracting trend and deterministic cycles, we are left with what we hope is a stationary time series to model. Modeling this component is discussed in subsequent chapters of this book. In this alternative, at each step we have subtracted a component that we understand. Furthermore, we can reconstruct the original series by adding the components back together again.

How to detect trend

It is usually, but not always, possible to detect trend visually; however, the covariogram or the correlogram can be used as a guide. The covariogram of white noise, e_t, would be expected to be zero for every lag. However, suppose that a time series is noise plus a linear trend line; that is, suppose that the series is nonstationary of the following form:

$$x_t = a + bt + e_t \tag{9.2}$$

where a and b are constants and e_t is white noise with a constant variance and constant mean. Then the autocovariance function is approximately a linearly decreasing function of lag.[1] It is usually easier to plot the autocorrelation function or correlogram, but both functions have the same shape. If $b = 0$, however, the autocovariance function will decay to zero much more rapidly than will a straight line. Figure 9.1 illustrates this point. The solid line is the sample autocorrelation function for the case when $a = 2.0$ and $b = .5$, and the dashed line is the sample autocorrelation function for the

Figure 9.1. Autocorrelation as a function of lag for a series that is linear trend plus noise (solid line) and pure noise (dashed line).

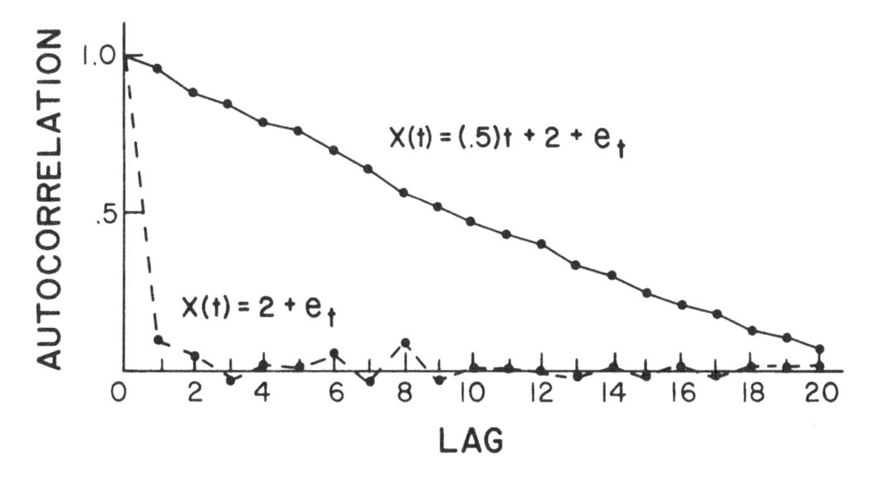

case when $a = 2.0$ and $b = .0$. Although the nonstationary correlogram is pure white noise, this rapid decay to zero with increasing lag is characteristic of all stationary stochastic series.

Unfortunately, a slow linear trend downward in the correlogram does not imply the existence of a *linear* trend in the data: the correlogram of nonlinear trends can also be linear. Another index of the existence of linear or nonlinear trend in the data is a nonzero value in the spectral density function at zero frequency.

The slope of the correlogram when the data contain trend will be affected by the number of observations in the series. When there are several hundred observations, for example, it will seem as if the autocorrelations stay relatively stable in the early lags. For example, for a relatively long series of 396 IBM stock prices, which have a linear trend, Box and Jenkins (1970) computed the correlogram, presented here as Table 9.1. Many books on time-series analysis simply assume that the series are long and claim that the correlogram does not decay for long series with trend; however, this is not the case. In fact, the reader can verify the fact that because the means are subtracted in computing the correlogram, for later lags the correlations will be negative if there is linear trend.

Table 9.1. *Correlogram for IBM stock prices*

k	1	2	3	4	5	6	7
r_k	.99	.99	.99	.98	.98	.97	.96

Table 9.2. *Pulse wave and setup for computing the autocovariances*

t	x_t	x_{t-1}	x_{t-2}	x_{t-3}	x_{t-4}
1	0	—	—	—	—
2	0	0	—	—	—
3	0	0	0	—	—
4	1	0	0	0	—
5	1	1	0	0	0
6	1	1	1	0	0
7	1	1	1	1	0
8	0	1	1	1	1
9	0	0	1	1	1
10	0	0	0	1	1

Another example of data for which the correlogram does not decay rapidly, as is the case for stationary stochastic data, are data that are deterministically cyclic. In this case the autocorrelation will die out for some lags but surface again later. The repeated increases will be related to the seasonality of the data. Note that for deterministically cyclic data with trend, r_k decreases linearly and then starts increasing again.

Unusual configurations of data can also have a trend-line covariogram; for example, Table 9.2 shows the "pulse wave," which is zero everywhere except for a small set of points. This example should be taken as an advertisement for the importance of plotting one's data. Visual inspection of the data is often useful, even if it is not sufficient. The computation of the covariogram follows.

$$\sum (x_t - \bar{x})^2 = 2.40$$

$$\sum (x_t - \bar{x})(x_{t-1} - \bar{x}) = 3(.6)^2 = 1.08; \qquad r_1 = \frac{1.08}{2.40} = .45$$

$$\sum (x_t - \bar{x})(x_{t-2} - \bar{x}) = 2(.6)^2 = .72; \qquad r_2 = \frac{.72}{2.40} = .30$$

$$\sum (x_t - \bar{x})(x_{t-3} - \bar{x}) = (.6)^2 = .36; \qquad r_3 = \frac{.36}{2.40} = .15$$

$$\sum (x_t - \bar{x})(x_{t-4} - \bar{x}) = 0; \qquad r_4 = .00$$

Note that r_k decreases linearly to zero for the pulse wave.

In general, it is sensible to always begin by computing and eliminating the mean and the trend. This is accomplished by using ordinary least-squares linear regressions or least-squares polynomial regression. If the trend is not significantly different from zero, this procedure will not alter the series very much. If the removal of trend yields a stationary residual series, then $\sum r_k^2$ should converge rapidly, as discussed in Chapter 8.

Detecting a deterministic cycle

If we were plotting average vegetable prices each month, we would find that prices were higher during the winter months. For some applications, if there is a steady 1-year cycle to these prices, we might be interested in computing a transformed series without this seasonal variation. Suppose that we have data for 5 years, 1965–9. We could compute the deviation of each month's prices from its monthly 5-year average. June prices for 1965 would then be replaced by (June 1965 minus the average for all Junes). If we observe a quarterly cycle, the transformation would give us (June 1965 minus the average for all summers), and so on. The transformed data are usually called "seasonally adjusted."

Figure 9.2. (a) Covariogram of a deterministic cyclic process; (b) periodogram (an estimate of the spectral density function) computed on half the data; (c) periodogram computed on all the data.

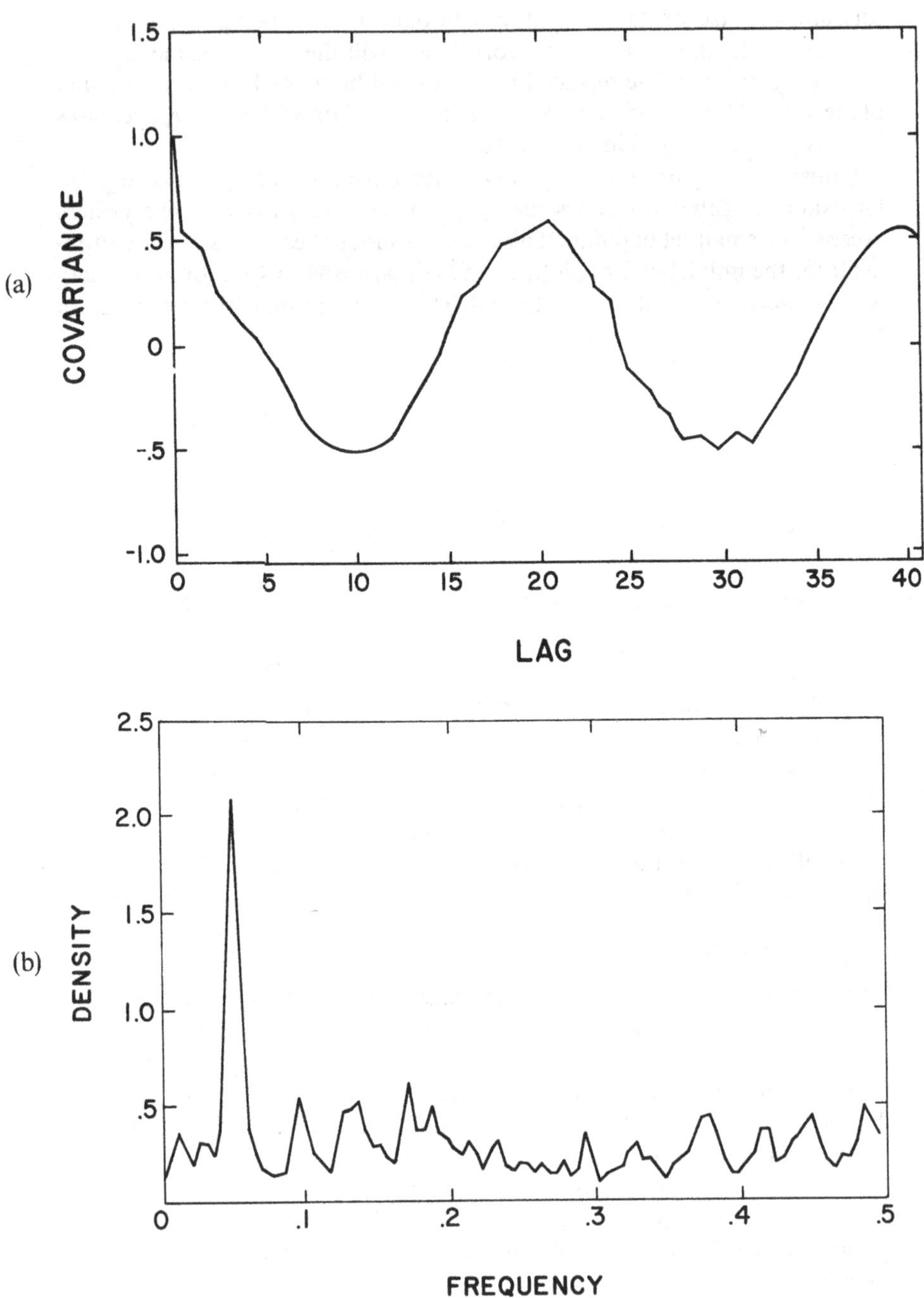

Once again, recall that if we were to compute the covariogram of white noise, e_t, we would expect it to be zero for every lag. Now consider the covariogram of a deterministic sine wave superimposed on white noise.

Figure 9.2a shows a covariogram that has a deterministic cyclic component. This provides an idea that the shape of the correlogram can be used as a guide for deciding whether components exist that are deterministically cyclic. Note that the covariogram is cyclic, and also that *its oscillations do not die off very rapidly*. If the correlogram has cyclic oscillations that damp down to zero rapidly, there is evidence of a cycle that is nondeterministic. An example of such a correlogram is given in Chapter 12 (see Figure 12.3c). Later it will be shown that another *test* of whether these oscillations are created by a *deterministic* sine wave or a probabilistic sine wave is the following: The periodogram, which is one estimate of the spectral density, *will have a peak whose height increases with the sample size if there is a frequency present that is deterministic.* Thus, if the periodogram is computed with half the data, it should have a peak about half as high as when it is computed with all the data (see Figure 9.2b and c).[2]

The last component

Most of this book is devoted to the study of this last component. Chapters 10 through 14 are devoted to the description of time-domain models of the

Figure 9.2. (*cont.*)

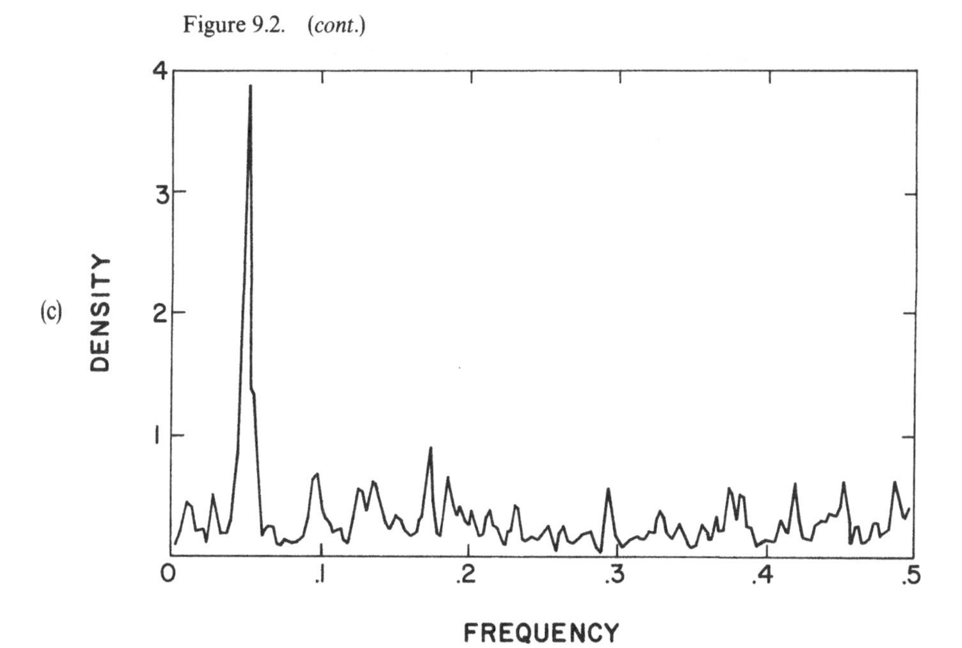

stationary stochastic component. Chapters 15 through 17 discuss frequency-domain models of this component. The next several chapters are devoted to estimating the parameters of these models.

The second alternative: transformations

An alternative to eliminating trend by subtracting a least-squares line was suggested by Box and Jenkins (1970). It is called *differencing*. We see in this chapter how differencing transformations are appropriate in different situations than the first alternative discussed, subtracting components. The discussion that follows has two objectives. First, it is a warning that transformations can radically alter the character of the data. For example, it can be shown (Yule, 1971b) that differencing a white-noise series can produce a cyclic series. In other words, transformations can *introduce* pattern where none existed before the transformation. They must therefore be used by an informed mind. Second, the way transformations alter the frequency content (spectral density) of the data is very instructive, and it is thus a good way to teach the reader about the frequency domain. Before discussing differencing, the opposite transformation will be introduced. It is called *smoothing*. The discussion thus begins with *smoothing*, which one would obviously consider for wildly erratic or highly variable data.

Smoothing

A common way of smoothing is to transform the series x_t using a moving-average transformation. For example,

$$x_t' = \tfrac{1}{2}(x_t + x_{t-1}) \tag{9.3}$$

Table 9.3. *Two-point moving-average transformation*

t	x_t	x_t'
1	−1	—
2	+1	0
3	−1	0
4	+1	0
5	−1	0
6	+1	0
Variance	1.2	.0

averages two points (x_t and x_{t-1}) that are next to one another. Actually, the term "moving average" is used for any weighted sum, whether or not the weights add to unity. For an illustration of the two-point moving average of equation (9.3), see Table 9.3. The original series, x_t, oscillated rapidly between $+1$ and -1. The two-point moving average was then used to produce the series x'_t. Note that the variance is obviously reduced by the two-point moving average, and this is generally the case for a moving-average transformation that is really an average (i.e., for which the weights sum to unity). The reduction of variability is accomplished by smoothing, and, in frequency-domain language, what smoothing accomplishes is the elimination, or at least the reduction in power, of the more erratic or faster frequencies. Smoothing is thus a "bandpass filter" because it reduces the intensity of the higher frequencies; that is, it does not "pass" the higher frequencies through the transformations. This point will now be discussed in detail and the notion of the transfer function of the filter will be introduced.

What smoothing does to the correlogram and the spectral density

This section uses simulated data to explain the concept of the transfer function of a filter. Whenever a time series is transformed, the frequency content of the series is usually altered. The term *transfer function* is used to describe this change. This use of the term "transfer function" is its common mathematical use in frequency-domain time-series analysis. It is used in a different way by McCleary and Hay (1980) and Hibbs (1977) in their treatment of the interrupted time-series experiment. This confusion is unfortunate. I have decided to stick to the wider use of the term in the mathematical literature. I apologize to the reader for any resulting problems. For example, averaging neighboring points smooths a series and reduces variability; the transfer function of this transformation thus reduces the power of high frequencies. Every transformation thus acts as a transfer function on the spectral density function. This discussion will visually display the transfer function of the transformation. You should try only to follow the idea of the transfer function conceptually. Subsequent chapters will teach you how to derive it.

This simulation began with the first-order autoregressive process (AR),

$$x_t = -.6x_{t-1} + e_t \tag{9.4}$$

The two-point smoothing transformation was then applied:

$$x'_t = \tfrac{1}{2}(x_t + x_{t-1}) \tag{9.5}$$

(see Table 9.4). Note in Table 9.4 (row 2) that this transformation made the first-order autocorrelation positive instead of negative, and it wiped out (or "filtered out") the high frequencies of the spectral density. This makes sense. Smoothing means reducing rapid oscillations, so the spectral density would have to change in this way. Note one extremely important contribution of smoothing: because it wipes out a wide *range* of high frequencies, smoothing can eliminate nonstationarity in the high-frequency range. For example, if the first half of a time series had component oscillations of $f = .38$ and in the second half this frequency shifted to $f = .42$, this series would be nonstationary. *To be stationary the spectral density cannot change with time.* A smoothing transformation has a transfer function that would filter out both frequencies and thus would eliminate the nonstationarity.

Another way of representing the effect of smoothing is to plot the transfer function of the filter that is a smoothing transformation. Recall that the transfer function is what the filter does to the spectral density function of the original data. If the transfer function is zero across a band of frequencies, for example, cycles in the original data in this range will be eliminated by a particular transformation. For example, for the two-point moving-average transformation of equation (9.3), the transfer function is graphed in Figure 9.3. Note that the effect of the filter is to diminish the contribution of the high-frequency components to the variance of transformed series. Subsequent chapters discuss the computation of the transfer function's shape from the equation of the transformation. At this point it is important to understand visually what the moving-average filter does and to note that one might inadvertantly eliminate high-frequency cycles of interest by smoothing.

Table 9.4. *Unsmoothed and smoothed autoregressive process*

Correlogram:	Lag							
	1	2	3	4	5	6	7	8
Unsmoothed	−.57	.41	−.31	.18	−.11	.10	−.03	.02
Smoothed	.15	−.22	.00	−.06	.07	−.01	−.22	−.12

Spectral density:	Frequencies									
	.025	.075	.125	.175	.225	.275	.325	.375	.425	.475
Unsmoothed	.12	.12	.15	.19	.16	.17	.27	.37	.90	1.51
Smoothed	.07	.12	.09	.14	.16	.09	.11	.09	.05	.02

To illustrate the usefulness of smoothing, the proportion of the time that each of two twin sisters looked at the other within a "moving window" of 7.5 sec was plotted. A moving window of 7.5 computes the probability of a gaze in the first 7.5 sec, then from seconds 2 to 9.5, then from seconds 3 to 10.5, and so on. The graph was so noisy it was impossible to read; it is plotted as Figure 9.4a. It was quite difficult to determine if there was any lead–lag relationship between the two sisters. The graph in Figure 9.4b represents the same proportions in a moving window of 60 sec. A 60-sec moving window first computes the proportion of time of a gaze in seconds 1 to 61, then in seconds 2 to 62, and so on. The data were smoother, and once these data were plotted it appeared that, for much of the interaction, the sister labeled L leads the interaction, in the sense that when she changed the probability of her gaze, her sister also eventually changed; thus, L's behavior was a lead indicator of R's behavior. In this case the moving-average filter eliminated the rapid changes in gaze and was a statement that the experimenter's interest was in the slow long cycles.

Removing trend by differencing

As you will recall, for a variety of applications it may be important to consider not the original time series but the deviations of the time series from some trend line. For example, an economist may wish to detect trends in an

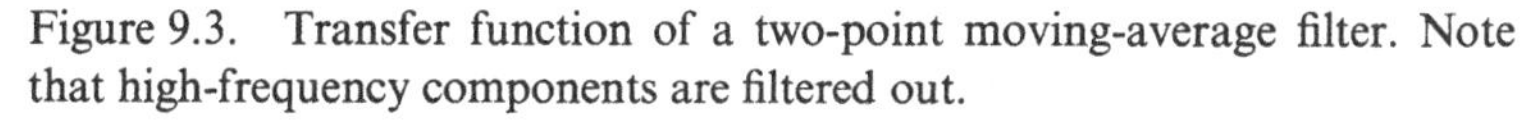
Figure 9.3. Transfer function of a two-point moving-average filter. Note that high-frequency components are filtered out.

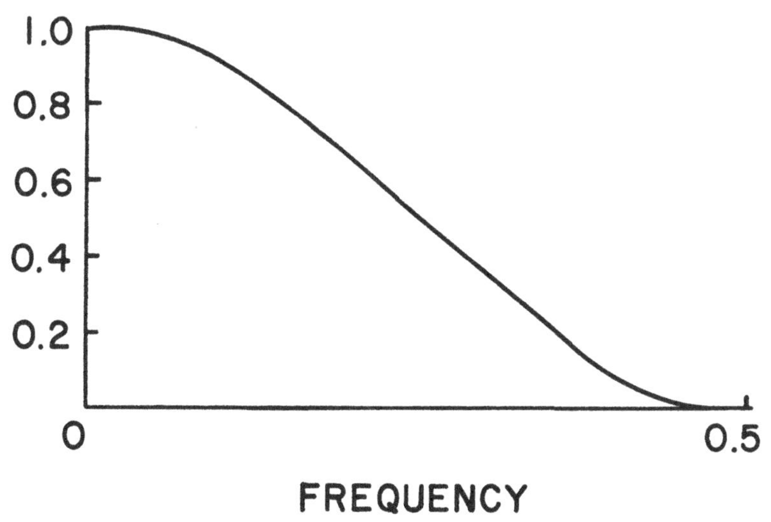

Figure 9.4. (a) Gaze probabilities of two sisters conversing, with a 7.5-sec moving-time window; (b) with a 60-sec moving-time window.

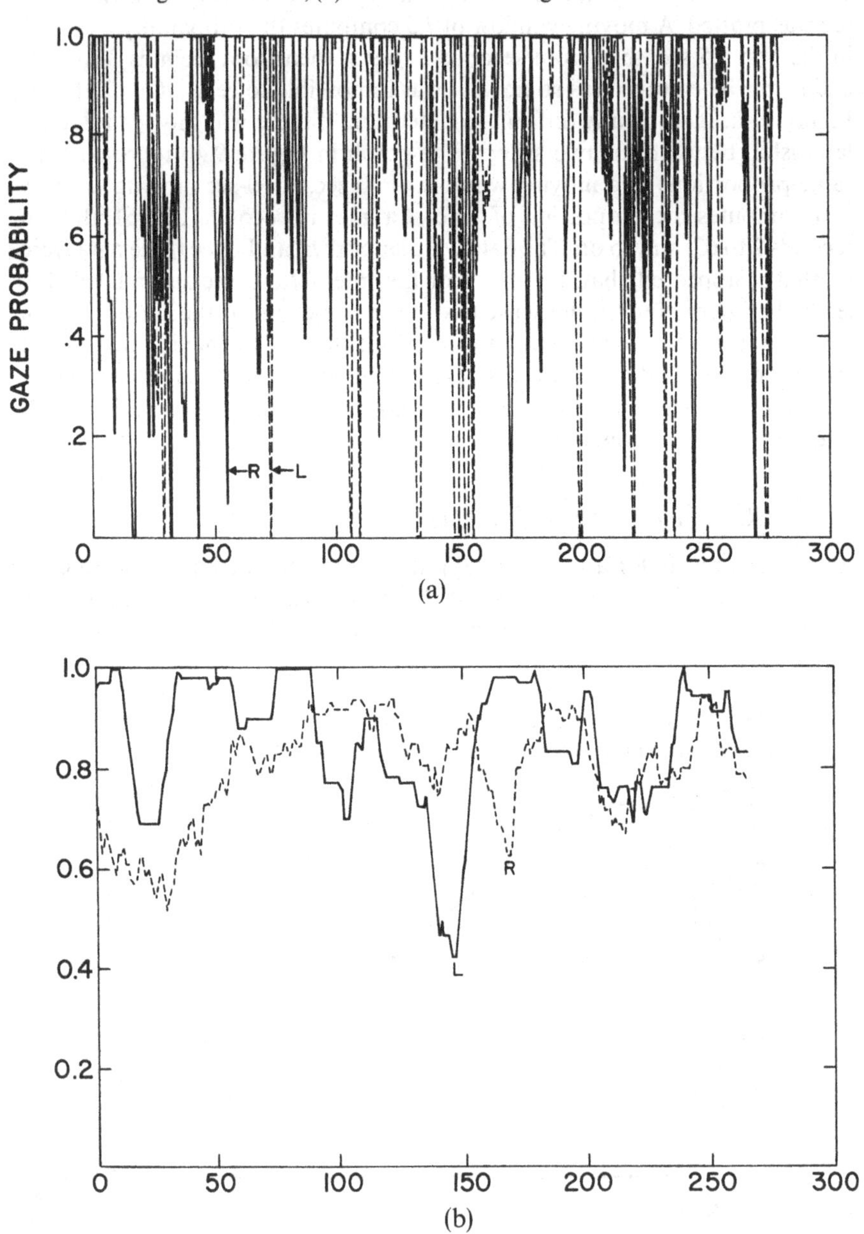

economic indicator over and above a linear estimate of inflation. In that case, the deviations from a linear trend line would be plotted. This is illustrated in Table 9.5. In Table 9.5, $x_t = \hat{x}_t + e_t$; $\hat{x}_t$ is the least-squares line $\hat{x}_t = (2.01)t + .05$. Thus, the data are $x_t = (2.01)t + .05 + e_t$.

Another procedure for reducing trend is called *differencing*. To eliminate linear trend the procedure is *first differencing*. It is defined as

$$x_t' = x_t - x_{t-1} = \nabla x_t \tag{9.6}$$

The *del* symbol (∇) is common notation for the difference. In the preceding example, first differencing would have resulted in the transformation depicted in Table 9.6. Note that there is far less trend in the transformed series resulting from either method (in either Table 9.5 or 9.6). Moreover, differencing can eliminate "local" temporal trends. The differencing transformation is essentially the slope of the original data. Hence, if the data tend upward

Table 9.5. *Eliminating trend by taking deviations from a least-squares line*

t	x_t	$\hat{x}_t$	$e_t = x_t - \hat{x}_t$
1	2.30	2.06	.24
2	4.20	4.07	.13
3	5.60	6.08	−.48
4	7.70	8.09	−.39
5	10.60	10.10	.50

Table 9.6. *First-differences transformation eliminates linear trend*

t	x_t	$x_t' = \nabla x_t = x_t - x_{t-1}$
1	2.30	—
2	4.20	1.90 = 4.20 − 2.30
3	5.60	1.40 = 5.60 − 4.20
4	7.70	2.10 = 7.70 − 5.60
5	10.60	2.90 = 10.60 − 7.70

for a while, then flatten out for a while, and then tend downward, the differenced series would be flat, except at the bends. This concept is illustrated in Figure 9.5. Although differencing eliminated trend in the top graph by turning the upward-sloping line into a flat line, the transformation is far more powerful than simply removing a trend line because it also eliminates the local trends of the bottom graph and amplifies the changes at the bends.

However, if the series exhibited a *quadratic* trend, first differencing would not eliminate trend (e.g., see Table 9.7). The solution in this case is to differ-

Table 9.7. *First-differences transformation will not eliminate quadratic trend*

t	x_t	∇x_t
1	1.20	—
2	4.30	3.10
3	8.70	4.40
4	15.80	7.10
5	25.00	9.20

Figure 9.5. How first differencing eliminates local as well as overall trends; this makes it a very powerful transformation, in the sense that it does many things at once.

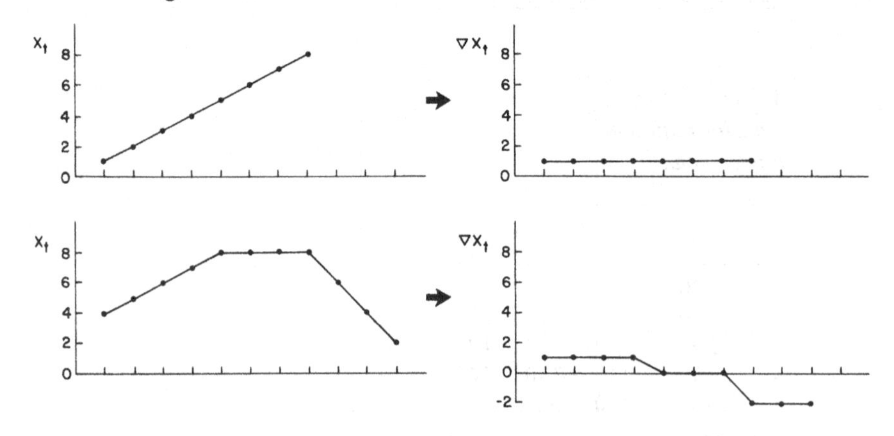

ence the series twice, or to difference the first difference (see Table 9.8). The second differenced series, denoted $\nabla^2 x_t$ (pronounced *del-squared* x_t), has no linear or quadratic trend; it should be reasonably constant. In general, to eliminate polynomial trend of the nth degree, we must difference n times, or apply ∇^n to x_t. The alternative is to fit an nth-degree polynomial to the data using least squares and to compute the residual from this nonlinear regression line. This is analogous to subtracting out a trend line in the linear case. But, once again, n-fold differences are more powerful; they can remove *different* polynomial trends in different stretches of the data.

Differencing a series changes the variable under consideration in an interpretable fashion. Since the data are plotted in time units, the first difference estimates the rate of change of the data. At times we might be considerably more interested in the way a variable changes than in its magnitude. In these cases a first-differences transformation of the data would be the transformation of choice. In a similar way, a second-differences estimates changes in the rate of change. In physics the analogs are position, velocity, and acceleration, respectively, for the data, the first differences, and the second differences.

It should be noted that differencing is likely to be appropriate when there is a high positive correlation between adjacent observations, whereas smoothing is likely to be appropriate if there is a high negative correlation. Each transformation moves the correlation of the transformed data closer to zero. Repeated differencing and many-point moving averages provide generalizations.

Effects of differencing on the correlogram

Eliminating linear trend by first differencing will dramatically affect the correlogram. For example, recall that it was claimed that the IBM stock price

Table 9.8. *Second differences eliminate a quadratic trend*

t	x_t	∇x_t	$\nabla(\nabla x_t) = \nabla^2 x_t$
1	1.20	—	—
2	4.30	3.10	—
3	8.70	4.40	1.30
4	15.80	7.10	2.70
5	25.00	9.20	2.10

data had a linear trend. If this is the case and if the data are purely linear trend plus a stationary stochastic component, the correlogram of the *differenced* data will rapidly decay to zero. Table 9.9 presents the correlogram of the first differences of the IBM series, and, this is indeed the case.

Table 9.9. *Correlogram of the first differences of the IBM data*

k	1	2	3	4	5	6	7
r_k	.08	.00	−.05	−.04	−.02	.13	.07

Figure 9.6. Transfer function of the first-differences filter (or transformation). Note that low-frequency components are filtered out.

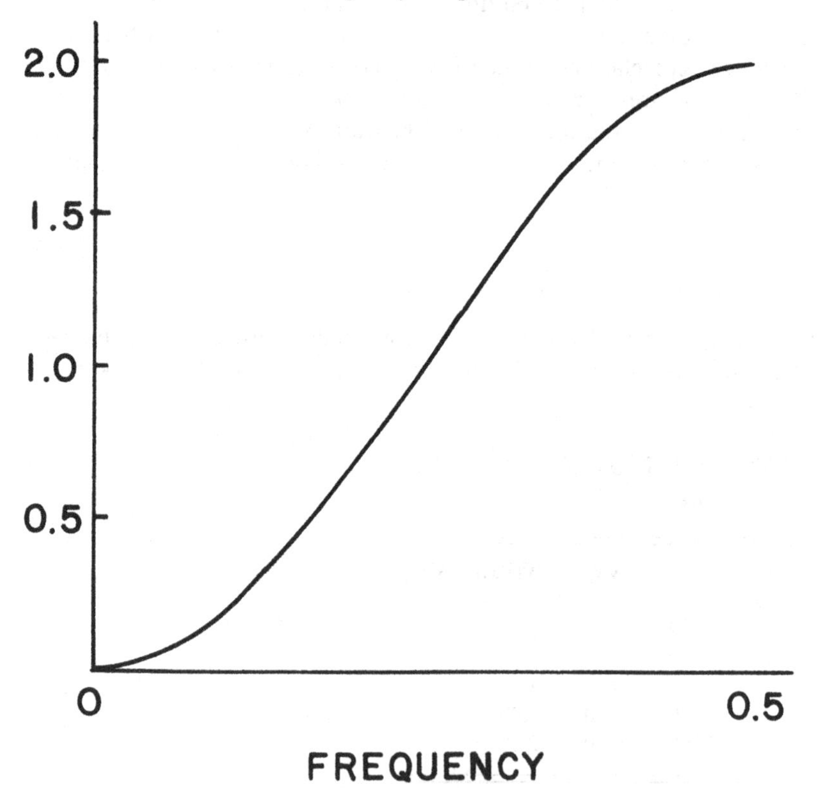

Effects of differencing on the spectral density

We noted that smoothing filters out higher frequencies from the spectral density. Differencing has quite the opposite effect – it filters out *low* frequencies from the spectral density and *amplifies* high frequencies (see Figure 9.6). Figure 9.6 illustrates the transfer function of the differencing transformation. Subsequent chapters will discuss how to compute the shape of the bandpass filter from its equation, but at this point it is important only to understand the effect of the filter. If we difference a series with linear trend and a cyclic component near $f = .125$, we will probably eliminate *both* the trend and the cyclic component. This may be an undesirable aspect of differencing, and one that we can avoid by subtracting a least-squares linear trend line from the series rather than differencing the series to eliminate trend.[3]

Step-by-step approach for nonstationary data

The chapter has addressed the question of what to do when the data are not stationary. The overarching suggestion is *not to* automatically transform the data by smoothing or differencing, because these transformations radically alter the spectral density function. Instead, it is wise to consider *modeling* the data as a sum of independent components, each of which can be understood and then subtracted. The chapter has also recommended modeling the nonstationarity before differencing or smoothing so that the work the transformation is doing can be known. One version of the model in equation (9.1) is reasonably general. It decomposes the series into linear trend, deterministic cycles, and a residual stationary stochastic process. The chapter has suggested the following:

1. *Linear trend* is detectable as a linear decrease in the covariogram (or correlogram). If trend is not present, the covariogram decreases faster than a straight line. Remove trend by computing residuals of linear least-squares fit (see Chapter 22).
2. *Deterministic cycles* are detectable by the fact that the periodogram (an estimate of the spectral density function discussed more fully in Chapter 15) increases with sample size if there is a deterministic cycle (proof is in Chapter 15). The periodogram is used to estimate the period, the amplitude, and the phase of the deterministic cycles, and then to remove them by least-squares fitting; the residuals are then computed. The third component is:
3. *Stationary stochastic process.* We are left with a stationary stochastic process. Its theoretical spectral density, which is the spectral density function of the process we are trying to model, no longer contains

the *line spectra*. We discuss how to model stationary stochastic processes in the following chapters.

Table 9.10 is a summary of the procedures recommended in this chapter for detecting each component. The table points out that many of the distinctions presented previously, such as deterministic/nondeterministic, are not

Table 9.10. *Summary of component analysis*

Trend	Deterministic cycles	Stationary stochastic process
Autocovariance Function (ACVF) decays linearly: implies trend in data	Periodogram test will detect deterministic cycles	Goal is usually to find models that leave: A small residual, compared to the original variance of the series and/ or A white-noise residual
Trend in data is also detectable by peak in the spectral density function at zero frequency	A "thin" spike in the spectral density function may suggest a deterministic cycle; "wide" spike suggests a nondeterministic cycle	White-noise residuals are stationary, have no autocorrelation, and have no peaks in spectral density
	If ACVF decays rapidly but cycles, this suggests a nondeterministic cycle; if it cycles but does not decay rapidly, this suggests a deterministic cycle	These goals are not always so clear cut. At times the best model is the one that is *theoretically* interesting; the rest is residual, whether it is white noise or not
	This illustrates that the deterministic–nondeterministic dichotomy is a continuum, not a sharp dichotomy	At times some very interesting process may be taking place in a small residual, as with the Porges fetus research

clear-cut dichotomies, but depend, for example, on the "thinness" of the peaks in the spectral density function. Table 9.10 also points out that the goals of model fitting may vary depending on the research application. Accounting for as much variance as possible is not always the goal of model fitting.

The concluding example of this chapter describes the work of Steve Porges on detecting hypoxia (not enough oxygen) in a fetus, a clinical problem often associated with infant mortality or brain damage. Porges's data are interesting for two methodological reasons. First, the data are interesting because Porges is concerned with detecting changes in cyclicity of a very small stochastic residual. Second, Porges solved a problem of eliminating a deterministic, nonlinear trend by a method this chapter did *not* discuss – the use of a moving-average transformation. This method is useful when the deterministic, nonlinear trend component cannot be adequately approximated by a polynomial (or by a weighted sum of deterministic cycles).

Removing a deterministic component by a moving-average transformation

In a series of investigations, Porges studied the problem of monitoring the adequacy of oxygen supply from the mother to the fetus. Even though the fetus does not breathe, the theory Porges proposed is that the fetus's respiratory system is sensitive to shifts in oxygen, so that, if the fetus is deprived of oxygen, there will be a signal in its brainstem that would represent the need to breathe more rapidly. Porges argued that even if the fetus's lungs are not moving, there would be detectable changes in physiological activity corresponding to the signal to breathe more rapidly. Porges studied periodicities in fetal heart rate because even in a newborn baby there is a functional brainstem mechanism that results in slight heart-rate increases during inhalation and slight heart-rate decreases during exhalation. Thus, there should be a periodicity in heart rate that is related to the periodicity of breathing. Conversely, in the case of the fetus, periodicities of increases or decreases in heart rate might correspond to a respiratory rhythm in the fetus. The fetal brainstem signal should thus modulate the rhythm of heart-rate activity, even if the fetus is not moving its lungs.

In the event of the sharply reduced oxygen supply to the fetus that follows uterine contractions, the distressed fetus that experiences transitory hypoxia will display what obstetricians call a type II dip. This drop in heart rate is similar to a "diving reflex" observed in many mammals. The diving reflex involves an adaptive physiological reaction to oxygen deprivation that decreases metabolic activity. In the fetus, this pattern is observed as a shift in the heart rate displayed in Figure 9.7. The type II dip is often used as a clinical index of fetal distress and contributes to the obstetric decision to surgically intervene during labor (i.e., Caesarian delivery).

Porges argued that the signal of interest was embedded as a small stochastic signal that has the characteristic type II dip superimposed. He thus argued that the brainstem signal to change the periodicity of heart rate would be detectable by the small oscillations that are masked by large deterministic shift represented by the type II dip. The statistical problem thus posed by Porges involves how to remove the smooth (deterministic) curve of the type II dip.

In solving this problem, Porges discovered that since the type II dip was not sinusoidal, neither a polynomial fit nor a sum of deterministic cycles could adequately represent the large deterministic component. However, a local cubic polynomial moving average applied to the data in Figure 9.7 produced the desired smooth curve. The moving average is a weighted sum with fixed weights applied locally and slowly along the curve in Figure 9.7 to obtain the smoothed curve plotted as Figure 9.8. The smoothed curve is then subtracted from the data in Figure 9.7, and the residual plotted as Figure 9.9 is obtained. In performing a spectral analysis of the residual, Porges discovered that the pattern of the hypoxic fetus's heart rate increased in periodicity and also increased in amplitude.

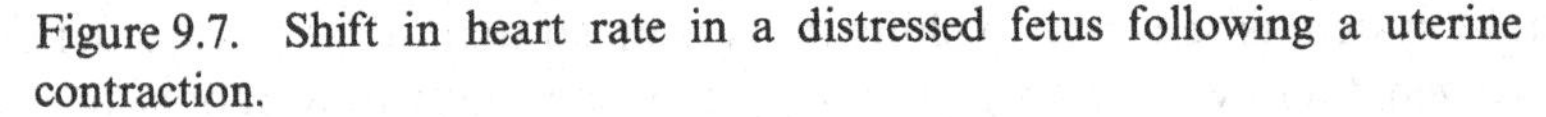

Figure 9.7. Shift in heart rate in a distressed fetus following a uterine contraction.

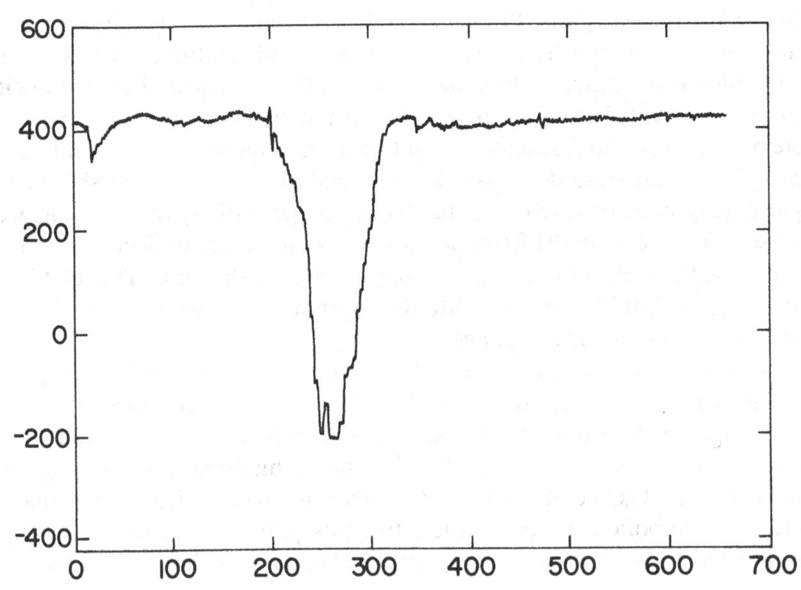

Figure 9.8. Moving-average transformation applied to the data in Figure 9.7 is used to estimate its deterministic component, which is displayed in this figure.

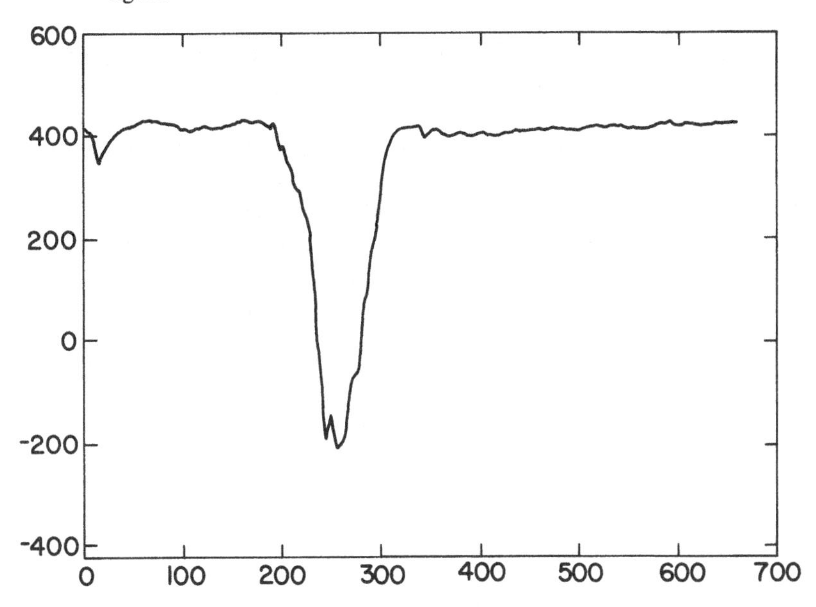

Figure 9.9. Residual obtained from subtracting the data in Figure 9.8 from the data in Figure 9.7.

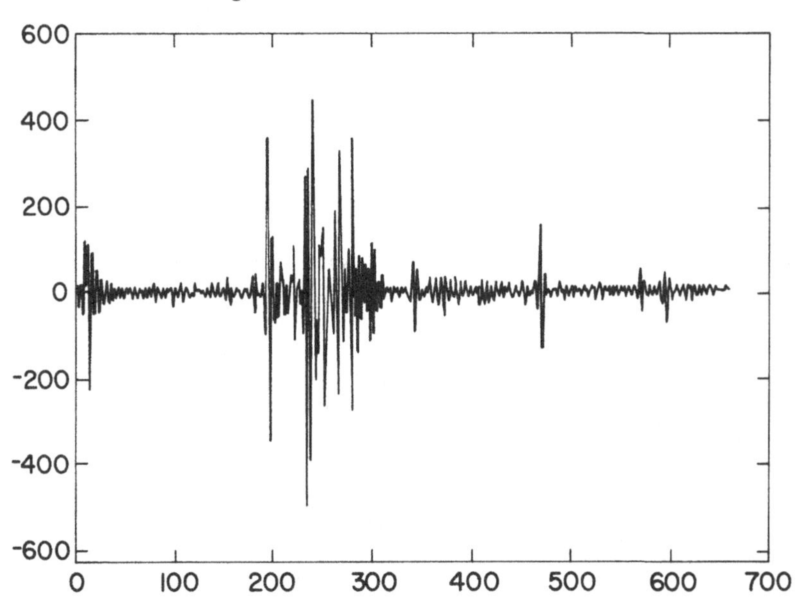

Porges was interested primarily in detecting changes in a small stochastic residual that resulted after a large deterministic component was removed from the data. The elimination of the large deterministic component was accomplished by using a moving-average transformation.

This example was reviewed because it illustrates a successful application of recommendations more general than those proposed by this chapter. The reader is referred to Kendall (1973) for a discussion of various smoothing procedures of the type that Porges employed in this example. It should be noted that each of these filters affects the spectral density function of the small residual, but because the filters are linear, they do so in predictable ways that we discuss in Chapter 17. In general, they tend primarily to cut off high-frequency oscillations, although they also distort low-frequency oscillations somewhat. The reader is also referred to Hamming's (1977) excellent and readable book on digital filtering.

APPENDIX 9A
Deterministic and nondeterministic components

What do we mean by a deterministic process? A useful definition is that the past of such a process completely determines its future forever. The implication of this is that in a linear sense the remote past is as good a predictor as the recent past (Cox and Miller, 1968). This is not true of nondeterministic processes, for which our ability to predict an event is best nearest the event and worsens as we recede from it.

Mathematically, how would this be expressed? If we regress x_t on its past, starting q observations back with x_{t-q}, x_{t-q-1}, and so on, we would have some residual error variance in the regression; call it ERVAR(q), since it may depend on q. If the prediction is abysmal, ERVAR(q) will be nearly equal to the variance of the series, var(x_t). For a nondeterministic process, as q increases, that is, as we go from the recent to the remote past, ERVAR(q) will approach a limit equal to var(x_t). However, if ERVAR(q) approaches a value less than var(x_t), then x_t has a deterministic component.

There is a theorem called the *Wold decomposition theorem* that shows that any discrete-time stationary process can be written as the sum of two independent processes, one of which is deterministic and one of which is nondeterministic. In our model we have already removed the deterministic parts and put them into the trend or cyclic components.

A deterministic seasonal or cyclic component will have a spectral density function that peaks sharply, nearly exactly at one frequency (as close as we can estimate). It is then said to have a line spectrum, similar to the spectrum of monochromatic light. It is derived from a series that cycles nearly like clockwork. It can be derived by measuring the swinging of a pendulum. On the other hand, a nondeterministically cycling series *more or less* repeats itself regularly. This means that there is some indeterminacy in its amplitude. It can be derived from a pendulum that is randomly struck by pellets shot at it from a peashooter run by a randomly motivated, mischievous small boy. The motion of the pendulum is still cyclic, but only in the statistical sense. The spectrum of the randomly perturbed pendulum will have a peak across a *band* of frequencies rather than just at one exact frequency (see Chapter 15 for more discussion).

Part III

Stationary time-domain models

10
Moving-average models

The next three chapters discuss time-domain models of the stationary stochastic component. This chapter discusses the moving-average models, beginning with a review of the remarkable Slutzky effect – that a moving average of white noise can produce almost any kind of stationary stochastic pattern (including pseudoperiodic oscillations). Moving-average models use noise as their input. They have finite memories, in the sense of autocorrelational structure. The correlograms of finite-order moving-average models are also discussed in this chapter. Equations are derived for estimating the parameters of a moving-average model from the autocovariances. These equations are nonlinear, making them somewhat difficult to solve. Chapter 11 introduces an alternative, the autoregressive models.

The Slutzky effect

The transformations we discussed in Chapter 9 all had an effect on the autocorrelation function and the spectral density. These effects were all intuitively clear and reasonable. However, recall that there are some unexpected results of transformations of time-series data, originally discovered by Slutzky in 1927. Recall that Slutzky was able to create a process that had a stochastic (not a deterministic) cycle simply by smoothing, or weighted averaging, of uncorrelated white-noise processes. In later chapters on moving-average models we see how this is possible. But for Slutzky's contemporaries, who were averaging time-series data with no expectations of *introducing* pattern into random data, the result came as a great surprise.

It turns out, in fact, that nearly all stationary processes that have a continuous spectrum (no sudden peaks or jumps) can be represented or approximated by finite moving averages of a noise process! This is part of the results of the Wold decomposition theorem.

It is important to point out that the "cycle" created from a moving average of noise process displays nonconstant amplitudes and a nonconstant peak-to-peak period. Recall that the process is only *probabilistically* cyclic. Also, the amount of predictability is zero after a certain number of points in the future, which is unlike the case of a deterministic sine wave.

Eugen Slutzky's 1927 paper titled "The summation of random causes as the source of cyclic processes" was translated into English and published in *Econometrica* in 1937; it was also updated by the author, who added some new results and some ideas that had been developed from some suggestions made by Khintchine. Slutzky's paper is remarkable because it appears to be derived from the same insight that G. Udny Yule arrived at independently at roughly the same time – the fundamentally different character of processes that were not merely a deterministic process plus white noise, but somehow *intrinsically* probabilistic. Yule's classic 1927 paper, which introduced the autoregressive models, is discussed in Chapter 15. Yule's concern with the intrinsically probabilistic nature of many processes was also addressed by Slutzky, when he wrote:

> Almost all of the phenomena of economic life, like many other processes, social, meteorological, and others, occur in sequences of rising and falling movements, like waves. Just as waves following each other on the sea do not repeat each other perfectly, so economic cycles never repeat earlier ones exactly in duration or in amplitude. [Slutzky, 1937, p. 105]

The problem Slutzky formulated was: To what extent was it possible that "a definite structure of a connection between random fluctuations could form them into a system of more or less regular waves"? (p. 106). Slutzky believed that it was likely that *order in nature could be produced by lawful operations on totally random fluctuations.* He wrote: "Wherein lies the source of the regularity? What is the mechanism of causality which, decade after decade, reproduces the same sinusoidal wave which rises and falls on the surface of the social ocean with the regularity of day and night" (pp. 106–7).

Slutzky was rhetorically puzzled. He asked: "What means of explanation, however, would be left to us if we decided to give up the hypothesis of the superposition of regular waves complicated only by purely random components?" (p. 107).

He began by noting that Schuster's (1898) periodogram assumed that successive observations in a time series are independent, but that "as a general

Figure 10.1. Slutzky's simulation (dashed line), using a moving-average transformation of noise, of business-cycle data (solid line).

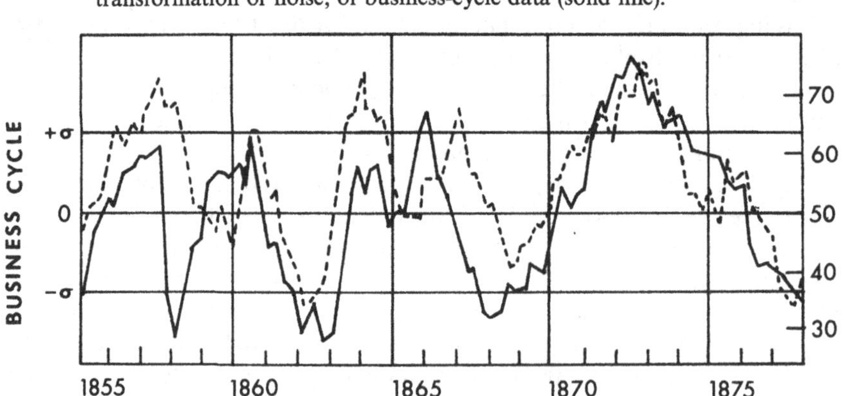

rule we find that the terms of an empirical series are not independent but correlated and at times correlated very closely" (p. 106). He thus suggested that there were two kinds of "chance series," those in which an observation depends on previous ones, and those in which it does not. To generate order from pure randomness required constructing equations by which the correlated series were constructed from the uncorrelated. To accomplish this, Slutzky used the *moving summation* of the purely random process. For example, initially the first 10 points of the random series are summed to give the first point of the correlated series; then points 2 to 11 are summed to give the second point of the correlated series; then points 3 to 12; and so on. The derived series can then become the "cause" of another process by a similar moving summation. In this way Slutzky derived various models from a purely random series.

One of these models is presented as Figure 10.1. The solid line is the actual business cycle data and the dashed line is derived from a 10-point moving summation of a random series; the fact that the two series were so similar to one another provided a provocative display of the usefulness of what later became known as the class of moving-average models.

First-order moving-average models

A first-order moving-average process, written as MA(1), has the general equation

$$x_t = e_t + be_{t-1} \tag{10.1}$$

where e_t is a white-noise series distributed with constant variance σ_e^2. An example of this process would be

$$x_t = e_t + .5e_{t-1} \tag{10.2}$$

We will see how a time series that has a nonzero autocorrelation (i.e., is *not* noise) can be obtained as a two-point weighted sum of a white-noise process. In the case of the model represented in equation (10.2), suppose that the first four terms of the noise series e_t were $e_0 = .5$, $e_1 = .0$, $e_2 = .2$, $e_3 = .6$. We can then compute x_t as

$$x_1 = e_1 + (.5)e_0 = .0 + (.5)(.5) = .25$$
$$x_2 = e_2 + (.5)e_1 = .2 + (.5)(.0) = .20$$
$$x_3 = e_3 + (.5)e_2 = .6 + (.5)(.2) = .70$$

In general, we wish to go the other way around. Note that the word "average" in the term *moving average* is only historical. A moving average is a weighted sum, with any set of weights we choose; they need not add to unity.

Later in the chapter we will see that the MA(1) process has only one nonzero autocovariance, the one at lag 1, which can be shown to be

$$\gamma_1 = b\sigma_e^2 \tag{10.3}$$

The lag k autocovariance is defined as

$$\gamma_k = E[(x_t - \mu)(x_{t+k} - \mu)]$$

which is usually estimated from the sample of N data points as

$$c_k = \frac{1}{N} \sum_{t=1}^{N-k} (x_t - \bar{x})(x_{t+k} - \bar{x})$$

The symbol γ_k will always stand for the lag k autocovariance, and its estimate from the data will always be denoted c_k. Note that γ_0, the zero-lag autocovariance, is just the autocovariance of the series with itself, which is an estimate of the variance of the series. Thus, in general,

$$\gamma_0 = \text{var}(x_t) = \sigma_x^2$$

The variance of the MA(1) series (var $x_t = \gamma_0$) can also be shown to be

$$\gamma_0 = (1 + b^2)\sigma_e^2 \tag{10.4}$$

These facts can be very useful in fitting models from the data. If we have a time series with only the lag 1 autocorrelation nonzero, we can fit the data using these two equations. That is, given the data x_t and a knowledge that the process is an MA(1), we will be able to determine the value of b. However, unfortunately, most of the time we do not have knowledge of the fact that the process is MA(1). We thus need to estimate the *order* of the process as well the parameters, such as b. How are these feats to be accomplished?

Moving-average models of any order

The Slutzky effect is disconcerting if one is interested in being able to transform data freely. However, the positive side of the effect is the creation of a class of time-series models, called *moving-average models*, or MA models. The MA models are simply moving averages of a white-noise process e_t, such as

$$x_t = b_0 e_t + b_1 e_{t-1} + b_2 e_{t-2} + \cdots + b_q e_{t-q}$$

The first point, x_1, is a weighted sum of the first $q + 1$ points of the noise series, points $e_1, e_0, e_{-1}, \ldots, e_{-q+1}$, the second point, x_2, is a weighted sum (using the same weights) of points e_2 to e_{-q+2}, and so on. The moving average is thus obtained from the noise series by a weighted sum of the noise points within a window that is $q + 1$ time units wide. The model can be written succinctly in summation notation as

$$x_t = \sum_{s=0}^{q} b_s e_{t-s} \tag{10.5}$$

where the b_s are constants. For simplicity, we take $b_0 = 1$, and modify the variance of the e's accordingly, if necessary. In equation (10.5) we assume that we have already subtracted the mean from x_t.

It can be shown that for any constants b_s, the resulting time series is stationary, so there are no constraints on the constants. The moving-average models, then, are used to simulate or "model" another time-series process.

How will the estimation of the b's be accomplished? In one of two ways. First, we can estimate the *autocovariance structure* of the process that generated the observed series. Second, we can examine the *spectral density* of the series. A moving average is a linear filter of white noise, and its spectral density is a key to identifying the filter. This chapter will examine only the first method. In Chapter 18 the second method will be examined.

Our goal is to be able to derive the weighs b_s and the variance of the white-noise process, σ_e^2. These are the parameters of our model. To proceed, let us first assume that the model is MA(1) and derive equations (10.3) and (10.4). We must compute γ_k, which is defined as the autocovariance of the process at lag k. For simplicity, assume that the mean has been subtracted from our data, so that x_t has zero mean. Then

$$\boxed{\gamma_k = E(x_t x_{t-k})} \tag{10.6}$$

This is a basic equation. Now we substitute values for x_t and x_{t-k} in terms of the e_t process, and perform some algebra, obtaining

$$\begin{aligned}\gamma_k &= E[(e_t + be_{t-1})(e_{t-k} + be_{t-k-1})]\\ &= E(e_t e_{t-k} + be_t e_{t-k-1} + be_{t-1}e_{t-k} + b^2 e_{t-1}e_{t-k-1})\\ &= E(e_t e_{t-k}) + bE(e_t e_{t-k-1}) + bE(e_{t-1}e_{t-k}) + b^2E(e_{t-1}e_{t-k-1})\end{aligned} \tag{10.7}$$

The last result follows from the properties of the expectation operation and from the fact that b is a constant. Now set $k = 0$ and recall that $\gamma_0 = \sigma_x^2$, the variance of the series. This gives us the expression

$$\gamma_0 = \sigma_x^2 = E(e_t^2) + bE(e_t e_{t-1}) + bE(e_{t-1}e_t) + b^2E(e_{t-1}^2)$$

Since the process e_t has constant variance, and since it is a noise process, we obtain

$$\begin{aligned}\sigma_x^2 &= \sigma_e^2 + 0 + 0 + b^2\sigma_e^2\\ &= (1 + b^2)\sigma_e^2\end{aligned}$$

In general, we will use this equation to estimate σ_e^2 once we have an estimate of b, using the sample variance to estimate σ_x^2. Note that, in general, $E(e_t e_{t-k}) = 0$ for any nonzero k, and that $E(e_{t-k} e_{t-k}) = \sigma_e^2$ for any k. These assumptions were important in the derivation.

Now set $k = 1$ in equation (10.7). This yields

$$\begin{aligned}\gamma_1 &= E(e_t e_{t-1}) + bE(e_t e_{t-2}) + bE(e_{t-1}^2) + b^2 E(e_{t-1} e_{t-2}) \\ &= b\sigma_e^2\end{aligned}$$

which is equation (10.3).

For $k > 1$, we will obtain $\gamma_k = 0$, since equation (10.7) will contain only terms whose expected value is zero. We have thus derived the important result that *for an $MA(1)$, the autocovariance function truncates (i.e., it is zero) after lag 1.*

Note that if c_0 and c_1 are the sample estimates of γ_0 and γ_1, we can find estimates of b and σ_e^2 by solving

$$\begin{aligned}c_0 &= (1 + b^2)\sigma_e^2 \\ c_1 &= b\sigma_e^2\end{aligned}$$

As an example, consider the baseline period for the "talking-out" data we discussed in Chapter 7. The computations were performed using the fact that from the data we can compute γ_0 and γ_1 as

$$\begin{aligned}c_0 &= \text{var}(x_t) = 4.45 \\ c_1 &= 1.25\end{aligned}$$

Using these values it is possible to solve the following nonlinear equations for b and σ_e^2:

$$\begin{aligned}4.45 = c_0 &= \sigma_e^2(1 + b^2) \\ 1.25 = c_1 &= \sigma_e^2 b\end{aligned}$$

Solving these equations gives two possible values for b, .31 and 3.26. The former value is essentially (except for rounding) the Glass et al. (1975) estimate of b.

For the qth-order MA process, we can use a similar derivation to show that *the autocovariance function, γ_k, truncates after lag q.* Once again,

$$\gamma_k = E(x_t x_{t-k})$$

and we substitute for x_t, and x_{t-k} to obtain

$$\begin{aligned}\gamma_k = E[&(b_0 e_t + b_1 e_{t-1} + \cdots + b_q e_{t-q}) \\ &\times (b_0 e_{t-k} + b_1 e_{t-k-1} + \cdots + b_q e_{t-k-q})]\end{aligned} \tag{10.8}$$

For $k = 0$, in equation (10.8), we obtain

$$\gamma_0 = E[(b_0 e_t + b_1 e_{t-1} + \cdots + b_q e_{t-q})(b_0 e_t + b_1 e_{t-1} + \cdots + b_q e_{t-q})]$$
$$\sigma_x^2 = b_0^2 \sigma_e^2 + b_1^2 \sigma_e^2 + \cdots + b_q^2 \sigma_e^2$$

or

$$\boxed{\sigma_x^2 = (1 + b_1^2 + b_2^2 + \cdots + b_q^2)\sigma_e^2}$$

For $k = 1$, we obtain

$$\begin{aligned}\gamma_1 &= E[(b_0 e_t + b_1 e_{t-1} + \cdots + b_q e_{t-q}) \\ &\quad \times (b_0 e_{t-1} + b_1 e_{t-2} + \cdots + b_q e_{t-q-1})] \\ &= b_1 b_0 \sigma_e^2 + b_2 b_1 \sigma_e^2 + \cdots + b_q b_{q-1} \sigma_e^2\end{aligned}$$

For $k = 2$, we obtain

$$\begin{aligned}\gamma_2 &= E[(b_0 e_t + b_1 e_{t-1} + \cdots + b_q e_{t-q}) \\ &\quad \times (b_0 e_{t-2} + b_1 e_{t-3} + \cdots + b_q e_{t-q-2})] \\ &= (b_2 b_0 + b_3 b_1 + b_4 b_2 + \cdots + b_q b_{q-2})\sigma_e^2\end{aligned}$$

In general, we obtain the basic equation

$$\gamma_k = \sigma_e^2 (b_0 b_k + b_1 b_{k-1} + \cdots + b_q b_{q-k}) \tag{10.9}$$

or

$$\boxed{\gamma_k = \sigma_e^2 \sum_{s=0}^{q} b_s b_{s-k}} \tag{10.10}$$

Replacing γ_k by c_k in equations (10.9) and (10.10) gives us $(q + 1)$ *nonlinear equations* for the $(q + 1)$ unknowns $b_1, b_2, \ldots, b_q$ and σ_e^2. In general, they are quite difficult to solve; furthermore they have unique solutions only under a certain set of conditions, called the *invertibility conditions*, which we discuss later when we study the duality between MA and autoregressive (AR) models. In Chapter 11 we introduce autoregressive models that have a set of linear equations, called the Yule–Walker equations, that greatly simplify the process of estimating the model parameters.

Second-order moving-average models

To illustrate the equations of the preceding section, consider the MA(2) process, which is given by $x_t = e_t + b_1 e_{t-1} + b_2 e_{t-2}$, where e_t is again a white-noise process. Using equation (10.10) it is easy to derive the facts that

$$\begin{aligned}\sigma_x^2 &= (1 + b_1^2 + b_2^2)\sigma_e^2 \\ \gamma_1 &= (b_1 + b_1 b_2)\sigma_e^2 \\ \gamma_2 &= b_2\sigma_e^2 \\ \gamma_k &= 0 \qquad \text{for } k > 2\end{aligned}$$

Using the fact that the lag k autocorrelation $\rho_k = \gamma_k/\gamma_0$, it is also easy to show that the autocorrelation function (ACF) is

$$\rho_1 = \frac{b_1 + b_1 b_2}{1 + b_1^2 + b_2^2}$$

$$\rho_2 = \frac{b_2}{1 + b_1^2 + b_2^2}$$

$$\rho_k = 0 \qquad \text{for } k > 2$$

Thus, we see that the ACF for an MA(2) process truncates after *two* lags.

Intuitive properties of the MA(q) process

Recall that for the MA(1) process the lag 1 autocovariance was nonzero but that all other autocovariances were zero. This fact demonstrates that the MA(1) had dependency only for successive observations. For an MA(q) process, the autocovariances truncate to zero after q lags. Thus, the general MA(q) process is called a *q-dependent process*, because the dependency only lasts for q successive time units, and is zero thereafter. This is different from the autoregressive process, where the autocovariances decreased exponentially (see Chapter 5).

11 Autoregressive models

Autoregressive models are clearer and easier to handle than moving-average models, but they are not always stationary. Thus, we have to discuss "stationarity conditions" because we are interested in modeling the stationary stochastic component of equation (9.1). The Yule–Walker equations are derived, making it possible to estimate the model parameters from the autocovariance function using a set of linear equations. These equations make the autoregressive models easier to use than the moving-average models.

Any stationary time-series process can be modeled with an infinite moving average and approximated by a finite MA(q) model. However, this representation is not unique, and it is not always the most efficient model. An extremely useful and intuitively appealing extension of regression involves a group of models called the *autoregressive*, or AR, models. We can approximate most stationary time series as either a moving average or an autoregressive model. We will build the concepts necessary to understand this duality between AR and MA models. The conditions needed to guarantee that we can transform from an AR representation to an MA representation are called *stationarity* conditions, and from MA to AR representation are called *invertibility* conditions.

Box and Jenkins (1970) suggested that there are three stages in building a model for a time-series process from a realization of that process: (1) identification, (2) estimation, and (3) diagnostic checking. *Identification* finds the most efficient (in terms of computational ease or conceptual parsimony) form of the model. In this chapter we see how the autocorrelations assist in this process. Later we see that the spectral density function is also helpful in this stage of model building. *Estimation* is the process of calculating best estimates (unbiased, consistent, and efficient) of model parameters (denoted by Greek letters such as γ_k for the autocovariances) from sample estimates (denoted by English letters such as c_k for the sample autocovariances). Estimation includes significance testing under varying sets of assumptions of the distributional properties of the data. We see that for autoregressive models this involves a simple system of linear equations. *Diagnostic checking* is an examination of the residuals from a model fit, which ought to be white noise, or comparing a model that is a bigger, more general model than the one we are considering. We pursue this last step later in the book. These three stages of model building – identification, estimation, and diagnostic checking – are obviously related.

Autoregressive models

In a familiar multiple regression equation we have a dependent variable, y, and we might regress y on two independent variables x_1 and x_2, finding the beta weights, β_1 and β_2, that minimize the square error, $\sum e^2$, in the familiar equation

$$y = \beta_1 x_1 + \beta_2 x_2 + e$$

In a time series, we may similarly pick a point, x_3, and plan to use the two previous time points, x_2 and x_1, as dependent variables, but we will soon realize that we cannot accomplish the familiar minimization unless we slide the point x_3 along the time series so that we obtain sets of three neighboring points, x_t, and its two previous ancestors, x_{t-1} and x_{t-2}, where t varies from 3 to N, the number of observations in the series. Then we can proceed as usual. Analogous to a multiple regression is the *autoregression:*

$$x_t = a_1 x_{t-1} + a_2 x_{t-2} + e_t$$

and the model written above is called an autoregressive model of the second order (because we are going back into the past two time units). The autoregressive model is fairly easy to translate into English. It says that an observation at time t is predictable (to within an un-autocorrelated residual with zero mean and fixed variance) from a weighted sum of the p previous observations [called an AR(p) process]. In other words, the series is predictable from its immediate past:

$$(x_t - \bar{x}) = a_1(x_{t-1} - \bar{x}) + a_2(x_{t-2} - \bar{x}) + \cdots + a_p(x_{t-p} - \bar{x}) + e_t \tag{11.1}$$

where e_t has variance σ_e^2, has zero mean, is uncorrelated with $e_{t'}$ for $t' \neq t$, and as a consequence, is more generally uncorrelated with the past; that is, $\mathrm{cov}[(x_{t-k}, e_t)]$ is zero for all $k > 0$. The e_t process is called an "innovation" process for this reason. The autoregressive model above is usually written with the deviations from the mean, $\bar{x}$, assumed as

$$x_t = \sum_{i=1}^{p} a_i x_{t-i} + e_t \tag{11.2}$$

How can we *identify* the AR(p) process? Once again we use the autocovariance structure of the time series. In the next section we derive the autocovariances γ_k, defined as usual by

$$\gamma_k = E(x_t x_{t-k}) \tag{11.3}$$

We begin with the first-order autoregressive, AR(1), process.

First-order autoregressive process

The first-order autoregressive process is written

$$x_t = a_1 x_{t-1} + e_t \tag{11.4}$$

We can calculate the autocorrelation coefficient by multiplying this equation by x_{t-k} and then taking expected values of both sides of the resulting equation:

$$x_{t-k} x_t = a_1 x_{t-k} x_{t-1} + e_t x_{t-k} \tag{11.5}$$

$$\mathrm{cov}(x_t, x_{t-k}) = a_1 \,\mathrm{cov}(x_{t-k}, x_{t-1}) + \mathrm{cov}(x_{t-k}, e_t) \tag{11.6}$$

The covariance between e_t and x_{t-k} is zero because x_{t-k} depends only on $e_{t-k}, e_{t-k-1}, \ldots$, which are not correlated with e_t as long as $k > 0$. Hence, we get

$$\gamma_k = a_1 \gamma_{k-1} \tag{11.7}$$

Dividing through by γ_0 and using the definition of autocorrelation,

$$\rho_k = \frac{\gamma_k}{\gamma_o} \tag{11.8}$$

gives the result

$$\rho_k = a_1 \rho_{k-1} \tag{11.9}$$

For $k = 1$ this reduces to

$$\rho_1 = a_1 \rho_0 = a_1 \qquad \text{(since } \rho_0 = 1\text{)} \tag{11.10}$$

For $k = 2$,

$$\rho_2 = a_1 \rho_1 = a_1(a_1) = a_1^2 \tag{11.11}$$

For $k = 3$,

$$\rho_3 = a_1 \rho_2 = a_1(a_1^2) = a_1^3 \tag{11.12}$$

And in general,

$$\rho_k = a_1^k \tag{11.13}$$

This means that the autocorrelation function of an AR(1) process does not truncate as was the case for an MA(1) process. We will soon explore the implications of this fact. Figure 11.1 compares the autocorrelations of the MA(1) and the AR(1) processes for the example when $b = .5$ and $a_1 = .5$.

Figure 11.1. Autocorrelations for the MA(1) and AR(1) models.

MA(1)	AR(1)
General: $X_t = E_t + b_1 E_{t-1}$	$X_t = a_1 X_{t-1} + E_t$
Example: $X_t = E_t + .5E_{t-1}$	$X_t = .5X_{t-1} + E_t$
AUTOCORRELATIONS	AUTOCORRELATIONS
Lag 1: $r_1 = \frac{b_1}{1+b_1^2}$	Lag 1: $r_1 = .5$
$r_1 = \frac{.5}{1+(.5)^2} = \frac{.5}{1.25} = .4$	
Lag 2 and above: $r_K = 0$	Lag 2: $r_1 = (.5)^2 = .25$
	Lag K: $r_K = (.5)^K$
MA(1) TRUNCATES	AR(1) DIES OUT RAPIDLY

The AR(1) process is not always stationary

Note that all MA(1) processes are stationary. This is not the case for AR(1) processes. To see that this is so, consider the variance of the AR(1) process. Later we will show that the AR(1) process can be written as an infinite moving-average process. We will use this fact to compute the variance of the AR(1) process.

The variance of the AR(1) process can be shown to be

$$\sigma_x^2 = \sigma_e^2[1 + a_1^2 + (a_1^2)^2 + (a_1^2)^3 + (a_1^2)^4 + \cdots]$$

If $a_1 = 1$ or if a_1 is larger, this variance will increase without bound. As an example, let us consider the case when $a_1 = 1$, which is usually called the drunkard's walk.

Drunkard's walk

Imagine a drunkard at a lamp post who will move a random distance e_t at time t, and who continues this walk indefinitely. It can be shown that the drunkard will eventually drift farther and farther away from the lamp post. This process can be represented as an AR(1) process with $a_1 = 1$.

$$x_t = x_{t-1} + e_t \tag{11.14}$$

Table 11.1. *Drunkard's walk*[a]

t	$x_t = x_{t-1} + e_t$
1	0
2	.63
3	.09
4	.88
5	.94
6	1.43
7	1.00
8	2.71
9	3.87
10	4.72
11	4.42
12	4.79

[a] where $e_t \sim N(0, 1)$. This notation is read "e_t is normally distributed with mean zero and variance 1.0."

Consider one realization with $N = 12$ observations (Table 11.1). We can see that this realization x_t appears to have a trend. The theoretical means, however, are zero. Furthermore, notice that $\rho_k = a_1^k = (1)^k = 1$, for all k, so the theoretical correlogram does not die out. It does not even decrease in linear fashion to zero.

The variance of x_t equals

$$\text{var}(x_t) = \sigma_e^2(1 + 1 + 1 + \cdots \text{out to infinity})$$

and therefore is infinite. Almost all realizations will increase without bound as N increases. Refer back to stationarity conditions 1 and 2 (Chapter 8).

To deal with this nonstationarity problem about AR(1) models, it is necessary to impose the conditions that $|a_1| < 1.0$ for the AR(1) model to be stationary. It will also be necessary to impose similar conditions for general models; however, discussion of these conditions will be deferred.

Correlograms for first-order autoregressive models

For the process $x_t = .5x_{t-1} + e_t$, the correlogram will be as shown in Table 11.2. Suppose that we had a series with 100 observations. The Bartlett estimate for the autocorrelation estimates gives a confidence interval of $\pm 2/\sqrt{N} = \pm 2/10 = \pm .2$. Thus, the noise introduced in estimating autocorrelations would mask many of these smaller autocorrelations. Even with 100 observations, this exponential decay might not be clear; beyond lag 2 we might be unable to distinguish between exponential decay and zero autocorrelations. This illustrates the need for large samples when we need to examine the internal structure of the data as opposed to the analysis-of-variance situation, where we can often impose some external structural models.

Table 11.2. *Correlograms for first-order autoregressive models*

	r_k	
Lag, k	$x_t = 5x_{t-1} + e_t$	$x_t = -.5x_{t-1} + e_t$
0	1.00	1.00
1	.50	−.50
2	.25	.25
3	.13	−.13
4	.06	.06
5	.03	−.03
6	.02	.02

If r_1 is negative, the correlogram will still decay, but it will oscillate around zero. For example, if $r_1 = -.5$, $x_t = -.5x_{t-1} + e_t$, the terms are alternately negative and positive (see Table 11.2).

AR(1) example

An example of an AR(1) process was obtained during the preliminary analysis of the visual gaze of two identical twin sisters during a conversation. The twins were videotaped and the tape coded in slow motion; there were 60 frames per second (a frame is a complete set of video scan lines). On every frame each person was coded as either looking toward or away from her sister. This dichotomous variable was used to compute an observed probability of a gaze toward the other during an arbitrary time window. The window initially selected was 150 frames, about 2.50 sec, because it seemed that the dichotomous data did not change more rapidly than this. The first probability is thus the probability of a gaze in frames 1 to 150, the second probability is the probability of a gaze in frames 151 to 300, and so on. Figure 11.2 presents a plot of part of these data. The autocorrelations of the data are presented in Table 11.3. These correlations are based on 282 windows, or 42,300 frames. Note how the lag 2 autocorrelation is nearly a third of the lag 1 autocorrelation, that the lag 3 autocorrelation is nearly a third of the lag 2 autocorrelation, and so on up to lag 5.

Figure 11.2. Moving probability of a gaze of one sister at the other sister, obtained from coding videotapes of the conversation of identical twins.

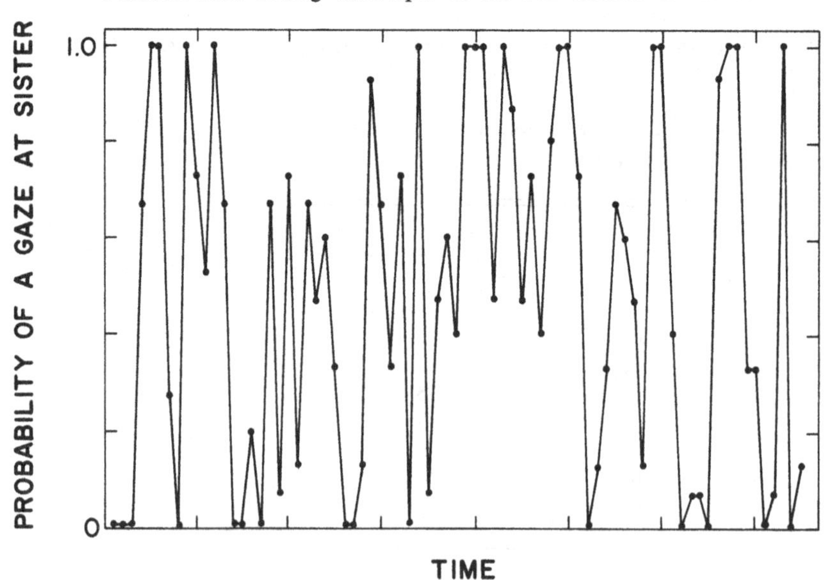

The Yule–Walker equations

The important fact about AR models is that it is possible to obtain a simple set of linear equations that expresses the parameters of the model in terms of the autocorrelations and variance of the data. These linear equations are called the Yule–Walker equations.

The development of these useful equations is not difficult. Write the general AR(p) model as

$$x_t = a_1 x_{t-1} + a_2 x_{t-2} + \cdots + a_p x_{t-p} + e_t \quad (11.15)$$

where once again we assume that x_t is a zero-mean process (or that the mean has been subtracted) and that e_t is a white-noise process and that $E(e_t x_{t-k}) = 0$ for $k > 0$. Once again, compute γ_k:

$$\gamma_k = E(x_t x_{t-k}) \quad (11.16)$$

This time a different trick is used; we substitute only the expression in equation (11.15) for x_t and leave x_{t-k} alone:

$$\gamma_k = E(x_t x_{t-k}) = E[(a_1 x_{t-1} + a_2 x_{t-2} + \cdots + a_p x_{t-p} + e_t)x_{t-k}] \quad (11.17)$$

$$\gamma_k = a_1 E(x_{t-1} x_{t-k}) + a_2 E(x_{t-2} x_{t-k}) + \cdots + a_p E(x_{t-p} x_{t-k}) + E(e_t x_{t-k}) \quad (11.18)$$

Let us consider equation (11.18) carefully. From the definition of the autocovariance γ_k of a stationary process, it is a function only of the lag between observations, not their starting point. Thus,

$$\gamma_k = E(x_t x_{t-k}) = E(x_{t+s} x_{t+s-k})$$

for any value s, since $t + s - (t + s - k) = k$. We can use this fact to simplify equation (11.18) to obtain

$$\gamma_k = a_1 \gamma_{k-1} + a_2 \gamma_{k-2} + \cdots + a_p \gamma_{k-p} + E(e_t x_{t-k}) \quad (11.19)$$

Table 11.3. *Autocovariances and autocorrelations for the gaze data (for seven lags)*

Lag	Autocovariance	Autocorrelation
0	.135	1.000
1	.045	.337
2	.016	.119
3	.006	.046
4	.002	.014
5	.003	.020
6	.001	.007
7	.008	.064

For $k > 0$ we know that the last term is zero. Hence, for $k > 0$ we have

$$\gamma_k = a_1\gamma_{k-1} + a_2\gamma_{k-2} + \cdots + a_p\gamma_{k-p} \tag{11.20}$$

If we divide by the variance of the series $\gamma_0 = \sigma_x^2$ and recall the definition of the autocorrelation $\rho_k = \gamma_k/\gamma_0$, we obtain the Yule–Walker equations

$$\boxed{\rho_k = a_1\rho_{k-1} + a_2\rho_{k-2} + \cdots + a_p\rho_{k-p}, \qquad k > 0} \tag{11.21}$$

These are extremely important equations.

For $k = 0$, we obtain

$$\gamma_0 = \sigma_x^2 = a_1\gamma_{-1} + a_2\gamma_{-2} + \cdots + a_p\gamma_{-p} + E(e_tx_t) \tag{11.22}$$

But for the autocovariance, $\gamma_{-s} = \gamma_s$. Hence, we can drop the minus signs from equations (11.22). To compute $E(e_tx_t)$, use equation (11.5) as follows:

$$\begin{aligned} E(e_tx_t) &= E(a_1x_{t-1}e_t + a_2x_{t-2}e_t + \cdots + a_px_{t-p}e_t + e_t^2) \\ &= E(e_t^2) = \sigma_e^2 \end{aligned}$$

because $E(x_{t-k}e_t) = 0$ for $k > 0$. Thus, equation (11.22) reduces to

$$\gamma_0 = \sigma_x^2 = a_1\gamma_1 + a_2\gamma_2 + \cdots + a_p\gamma_p + \sigma_e^2$$

and substituting $\rho_k = \gamma_k/\gamma_0$, we obtain

$$1 = a_1\rho_1 + a_2\rho_2 + \cdots + a_p\rho_p + \frac{\sigma_e^2}{\sigma_x^2}$$

or

$$\boxed{\sigma_e^2 = \sigma_x^2(1 - a_1\rho_1 - a_2\rho_2 - \cdots - a_p\rho_p)} \tag{11.23}$$

We can use equations (11.21) and (11.23) to estimate, say, a_1, a_2, σ_e^2 from the autocorrelations if we knew that the order of the model is $p = 2$. Suppose further that we computed $\hat{\sigma}_x^2 = 1.0$ and $\hat{\rho}_1 = .5$, $\hat{\rho}_2 = .1$; then to find a_1 and a_2, we write the Yule–Walker equations as

$$k = 1: \quad \hat{\rho}_1 = a_1\hat{\rho}_0 + a_2\hat{\rho}_{-1} = a_1\hat{\rho}_0 + a_2\hat{\rho}_1$$

But since $\rho_0 \equiv 1$, we obtain

$$\hat{a}_1 + .5\hat{a}_2 = .5 \tag{11.24}$$

For $k = 2$, we obtain

$$\hat{\rho}_2 = \hat{a}_1\hat{\rho}_1 + \hat{a}_2\rho_0 \qquad \text{or} \qquad .5\hat{a}_1 + \hat{a}_2 = .1 \tag{11.25}$$

Equations (11.24) and (11.25) can be solved easily for $\hat{a}_1$ and $\hat{a}_2$: $\hat{a}_1 = .6$, $\hat{a}_2 = -.2$, and the model is

$$x_t = .6x_{t-1} - .2x_{t-2} + e_t$$

and the remaining parameter is

$$\begin{aligned}\hat{\sigma}_e^2 &= \hat{\sigma}_x^2(1 - \hat{a}_1\hat{\rho}_1 - \hat{a}_2\hat{\rho}_2)\\ &= 1.0[1 - (.6)(.5) - (-.2)(.1)] = .72\end{aligned}$$

The original variance was 1.00, and the residual variance is .72, so there is quite a bit of noise in these data. Later we learn how we can do significance tests on the model parameters, which will make it possible to estimate the order of the AR(p) process as well as the model parameters. Furthermore, we can keep fitting larger models until we do not find a significant model parameter, which also provides a test that the residual is pure noise. In fact, as the number of observations, N, increases, if we know the process is AR(2), for all reasonable distributions of the noise process, under the null hypothesis that all the a_i are zero, the $\hat{a}_i$ are normally distributed with mean zero and variance $\sigma_e^2/N(1 - \hat{\rho}_1^2)$. Chapter 18 discusses the problem of estimation of the parameters of a general AR(p) process.

The reader should note at this juncture that AR models are considerably easier to work with than MA models because of the linearity of the Yule–Walker equations. This fact is the basis of some of the recommendations of this book.

The stationary AR(1) is an MA model of infinite order

This section introduces the fundamental duality between AR and MA models. We can keep going backward in time using the first-order autoregressive model:

$$x_t = a_1x_{t-1} + e_t \tag{11.26}$$

$$x_t = a_1(a_1x_{t-2} + e_{t-1}) + e_t \tag{11.27}$$

or

$$\begin{aligned}x_t &= a_1^2x_{t-2} + (a_1e_{t-1} + e_t)\\ &= a_1^2(a_1x_{t-3} + e_{t-2}) + (a_1e_{t-1} + e_t)\\ &= a_1^3x_{t-3} + a_1^2e_{t-2} + a_1e_{t-1} + e_t\end{aligned} \tag{11.28}$$

Continuing back to minus infinity we would get

$$x_t = e_t + a_1e_{t-1} + a_1^2e_{t-2} + a_1^3e_{t-3} + \cdots \tag{11.29}$$

which makes sense if it is the case that $a_1^k \to 0$ as $k \to \infty$ rapidly enough for the series to converge to a finite limit. This is our stationarity condition.

The expression for x_t is then

$$x_t = \sum_{i=0}^{\infty} a_1^i e_{t-i} \tag{11.30}$$

The AR(1) model can thus be written as an MA(∞) model.

The MA(1) is sometimes an AR(∞)

Suppose that we have an MA(1) model

$$x_t = e_t + be_{t-1} \tag{11.31}$$

It follows that

$$x_{t-1} = e_{t-1} + be_{t-2} \tag{11.32}$$

Solve this equation for e_{t-1} and substitute the result back into equation (11.31). This gives

$$\begin{aligned} x_t &= e_t + b(x_{t-1} - be_{t-2}) \\ &= bx_{t-1} + e_t + (-b^2)e_{t-2} \end{aligned} \tag{11.33}$$

This looks almost like an AR(1), except for the e_{t-2} term. But we can write equation (11.31) as

$$x_{t-2} = e_{t-2} + be_{t-3} \tag{11.34}$$

solve for e_{t-2}, and substitute this back into equation (11.33):

$$\begin{aligned} x_t &= bx_{t-1} + e_t + (-b^2)(x_{t-2} - be_{t-3}) \\ &= bx_{t-1} - b^2x_{t-2} + e_t + b^3e_{t-3} \end{aligned} \tag{11.35}$$

But this resembles an AR(2) process. We can continue indefinitely as long as b^s goes to zero (i.e., $|b| < 1$) to obtain

$$x_t = bx_{t-1} - b^2x_{t-2} + b^3x_{t-3} - + \cdots \tag{11.36}$$

This is an AR(∞) process, but it only holds under the *invertibility condition* that $|b| < 1$. But note now that the invertibility condition for the MA(1) to be written as an AR(∞) is similar to the stationarity condition on an AR(1) process. This is true in the general case; that is, the invertibility condition of the MA(s) process is similar in form to the stationary condition of the AR(s) process.

What are the practical implications of these last two sections? Suppose that we have data generated by an MA(1) process. This section shows that it can be estimated using an autoregressive process of high order. In practice, b to some power will eventually be effectively zero, so we might obtain a good AR approximation of an MA(1) with five or six AR terms, depending on how large b is. Nonetheless, the most compact model we can get is still the MA(1), with only one parameter. Although the AR(∞) form of the MA(1) model still has one parameter in theory, in practice we will not know this, and it is in this sense that an AR representation may not always be the most parsimonious. However, for stationary data with an infinite number of data points, we could obtain, at least in theory, any degree of precision we wish in approximating the data with either an MA or an AR model.

In addition, higher-order MA models can sometimes be inverted to infinite AR and approximated by finite ARs. Thus, at the price of using models with, say, 5 to 10 parameters instead of 3 or 4, we can generally use the more tractable AR models.

12 The complex behavior of the second-order autoregressive process

The autoregressive models have great flexibility. This chapter introduces the reader to the capability of autoregressive models to simulate pseudoperiodic oscillations. Because the details of pseudoperiodic oscillations are reasonably technical, they have been examined in this short, separate chapter of their own. The precise meaning of the term "pseudoperiodic" will become clear in Chapter 15.

This chapter discusses a simple extension of autoregressive models to the case where the form of the autoregression extends back two points in time rather than just one. These are called *second-order*, or AR(2), *processes*. It is remarkable how much descriptive power this simple extension provides. We will divide AR(2) processes into four types. Note that the division at this point is somewhat arbitrary. Two of them are called "pseudoperiodic" because they resemble periodic series, but they are not deterministically periodic.

How are the four types of AR(2) series distinguished? They are distinguished by four regions of stationarity, that is, four patterns of constraints on the model parameters. Only in these four regions is an AR(2) stationary. A visual scan of a plot of four realizations of these four types of processes would be of little help in discriminating the four types of series; although of different autocovariance structures, the visual distinctions are sometimes subtle.

These four types are distinguished by different forms of the autocorrelation function. Thus, an examination of the autocorrelation function can identify the type and give, as a result, the approximate position of the model parameters.

The AR(2) process is capable of representing processes whose spectral density function has one peak. For processes with two cycles (i.e., two peaks in the spectral density function) we need an AR(4) process; for three peaks, an AR(6) process, and so on. The AR(2) process thus brings back the discussion of cycles.

The autocovariances of an AR(2) process

This process is defined by

$$x_t = a_1 x_{t-1} + a_2 x_{t-2} + e_t \tag{12.1}$$

where, once again, the x_t are deviations from the mean or are a zero-mean process, the e_t involve an innovation process uncorrelated with observations previous to x_t, and the e_t are independently distributed with constant variance σ_e^2.

We will see once again that this process is not stationary for all values of a_1 and a_2, and this fact defines a *region of stationarity.*

The Yule–Walker equations for the AR(2) were

$$\begin{aligned} \rho_1 &= a_1 + a_2 \rho_1 \\ \rho_2 &= a_1 \rho_1 + a_2 \end{aligned} \tag{12.2}$$

The first equation can be solved for ρ_1:

$$\rho_1 = \frac{a_1}{1 - a_2} \tag{12.3}$$

The second Yule–Walker equation can now be solved for ρ_2:

$$\begin{aligned} \rho_2 &= a_1 \left(\frac{a_1}{1 - a_2} \right) + a_2 \\ &= \frac{a_1^2}{1 - a_2} + a_2 \end{aligned} \tag{12.4}$$

A more useful set of expressions is the solutions for a_1 and a_2 in terms of ρ_1 and ρ_2, because we can use these expressions to estimate a_1 and a_2:

$$a_1 = \frac{\rho_1(1-\rho_2)}{1-\rho_1^2}$$

$$a_2 = \frac{\rho_2-\rho_1^2}{1-\rho_1^2} \tag{12.5}$$

Notice that this means that given the sample time series, which we know to be second-order autoregressive, we can estimate a_1 and a_2 as

$$\hat{a}_1 = \frac{r_1(1-r_2)}{1-r_1^2}$$

$$\hat{a}_2 = \frac{r_2-r_1^2}{1-r_1^2} \tag{12.6}$$

This makes clear the advantage of being able to determine the order of the autoregressive process before estimating the parameters, a_1 and a_2.

As we learned in Chapter 11, if we knew that the process was a first-order autoregressive process, the best estimate of a_1 would be simply r_1; for the first-order autoregressive process, $r_2 = r_1^2$; substituting these values into equation (12.6) would give $\hat{a}_1 = r_1$ and $\hat{a}_2 = 0$.

Stationarity conditions for an AR(2) process

Because we know that ρ_1 and ρ_2 must be less than 1 in absolute value, we can determine the stationarity conditions, or constraints, on a_1 and a_2 as the region inside the striped triangle of Figure 12.1. For values of a_1 and a_2 outside this triangle, the series is not stationary.

Recall that for the stationary first-order autoregressive process, there were only two possible patterns of the correlogram (see Table 11.2). However, the

Figure 12.1. Stationarity region for the second-order autoregressive AR(2) process.

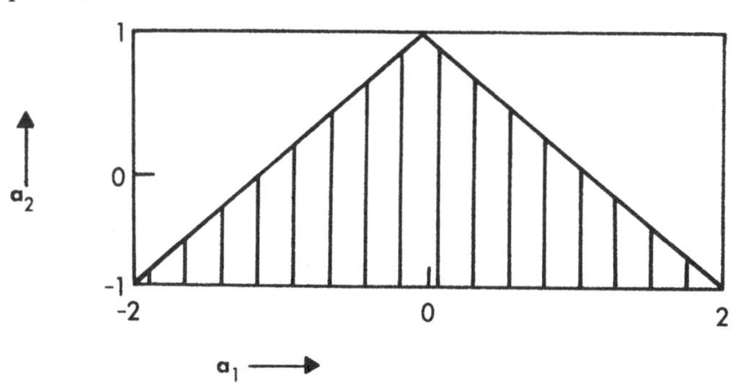

potential behavior of the correlogram of the stationary *second-order* autoregressive process is much more varied.

It is difficult to imagine the complexity introduced simply by proposing that the series is a second-order autoregressive, AR(2), series instead of a first-order autoregressive, AR(1), series. The best way for us to begin this introduction is to sharpen the reader's ability to visually examine the correlogram of a series. Realizations of the four types of stationary AR(2) series will be plotted. Then the mathematics will be derived to explain this behavior as a function of the possible values of the two model parameters of the AR(2) process.

Stationarity region for AR(2)

In Chapter 11 we discovered that the equation for the variance of an AR(p) process was

$$\sigma_x^2 = \frac{\sigma_e^2}{1 - \rho_1 a_1 - \rho_2 a_2 - \cdots - \rho_p a_p} \tag{12.7}$$

For an AR(2) process this reduces to

$$\sigma_x^2 = \frac{\sigma_e^2}{1 - \rho_1 a_1 - \rho_2 a_2} \tag{12.8}$$

Substituting our expressions derived above for a_1 and a_2 can be shown to yield (after some algebra)

$$\sigma_x^2 = \frac{(1 - a_2)\sigma_e^2}{(1 + a_2)(1 - a_1 - a_2)(1 + a_1 - a_2)} \tag{12.9}$$

Using the facts that autocorrelations must be less than unity, and each factor in the denominator and numerator must be positive, we can derive the conditions

$$\begin{aligned} -1 < a_2 < 1 \\ a_2 + a_1 < 1 \\ a_2 - a_1 < 1 \end{aligned} \tag{12.10}$$

The inequalities define the region inside the striped triangle of Figure 12.1.

The four series

Figure 12.2 illustrates realizations of the four types of stationary AR(2) series. For the time being, ignore the notations on the figure that indicate regions. Offhand, it would appear that the major visual distinction is that Figure 12.2b and c are more jagged than Figure 12.2a and d. However, that is not the important distinction. That distinction exists only because the coefficients of x_{t-1} are negative for Figure 12.2b and c. This implies that, for

Figure 12.2. Realizations of AR(2) time-series processes from the four regions of stationarity: (a) region 1; (b) region 2; (c) region 3; (d) region 4.

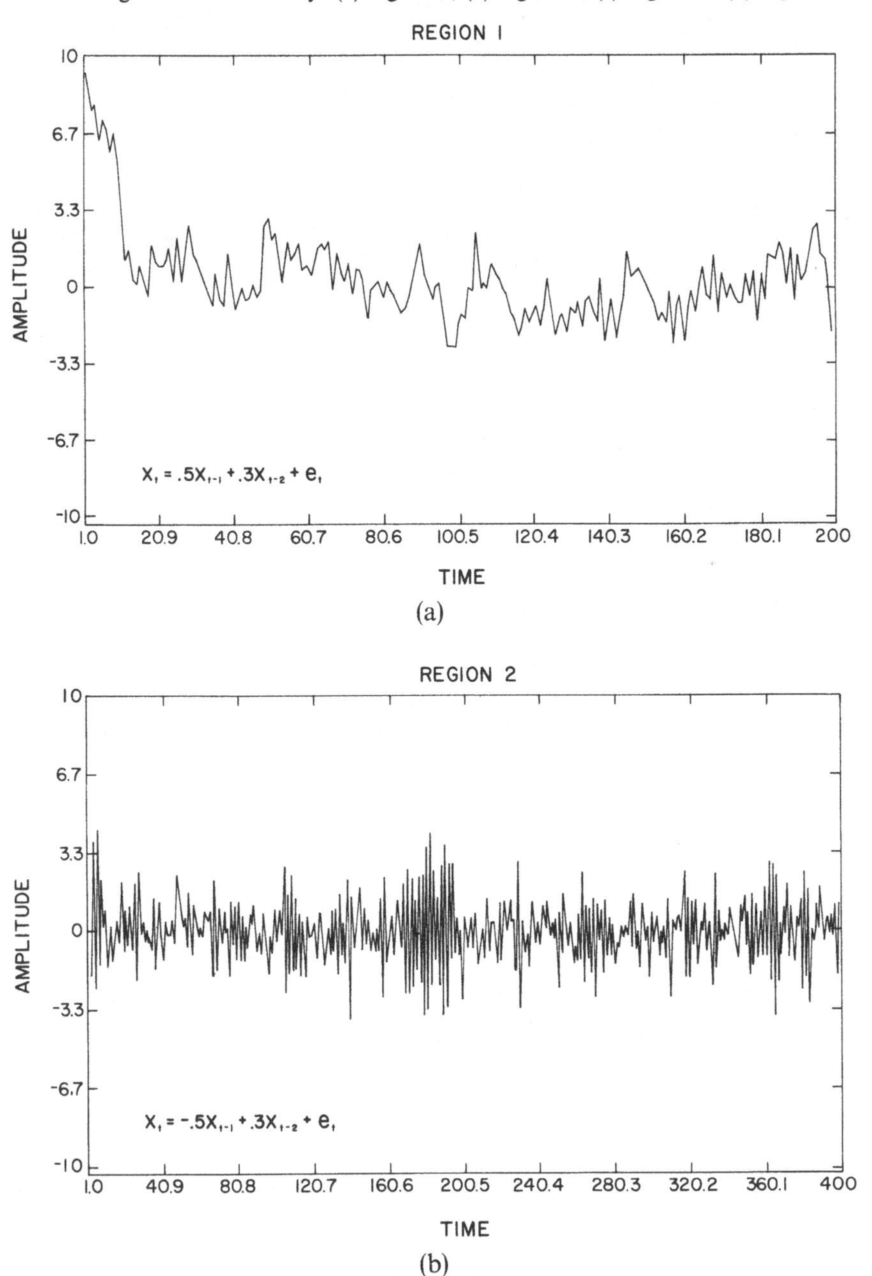

(a)

(b)

Figure 12.2. (*cont.*)

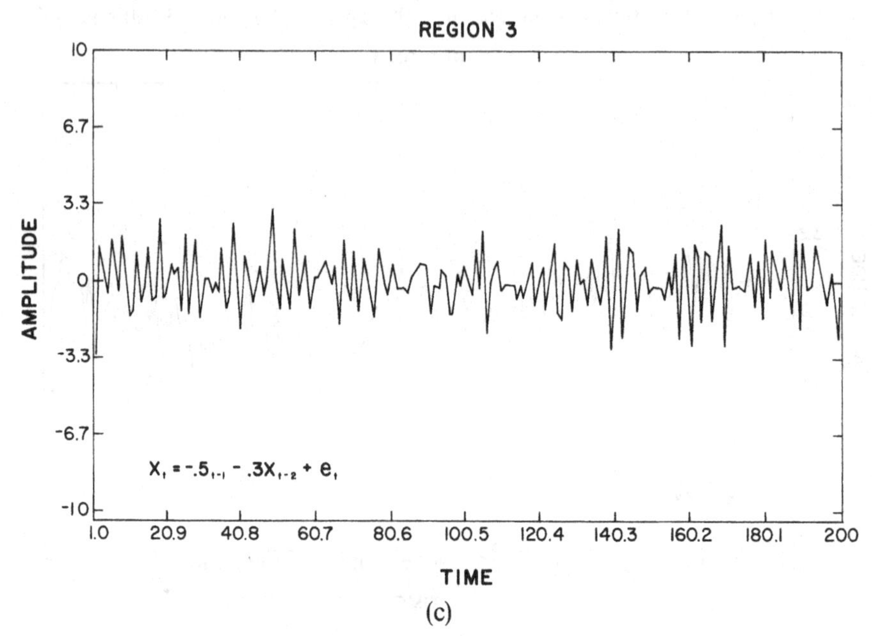

(c)

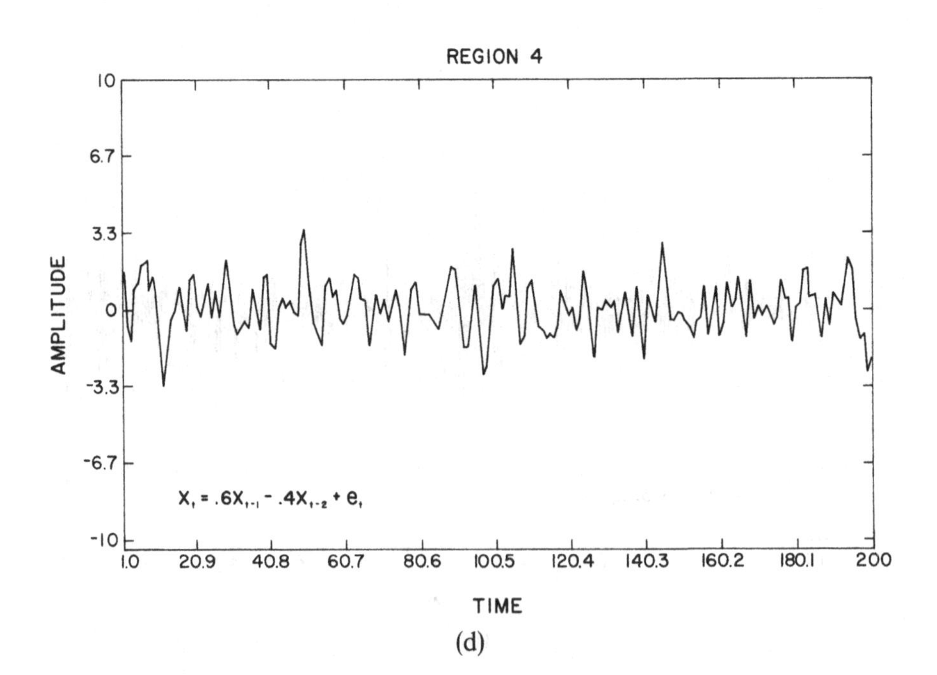

(d)

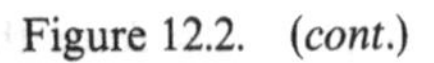

Figure 12.3. Autocovariance functions (ACVF) of realizations from each of the regions of stationarity and the theoretical ACVFs: (a) region 1; (b) region 2; (c) region 3; (d) region 4.

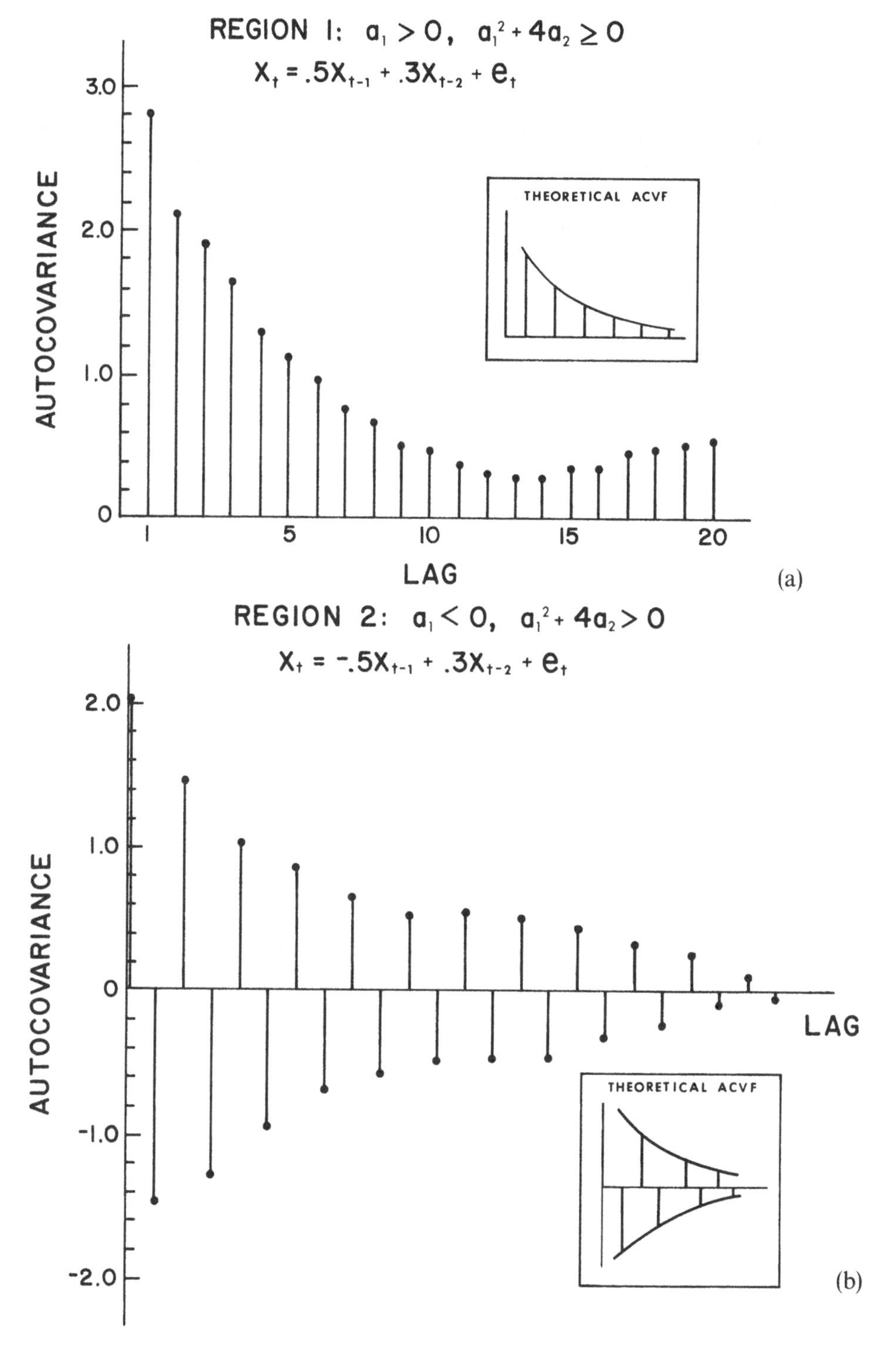

these two series, increases in x_t must immediately be followed by decreases, which will make the series jagged; if a_1 is positive, the series will be smoother.

To understand the variety in these four series we need to look at their autocovariances. The sample autocovariances of these four realizations are plotted in Figure 12.3 to dramatize the patterns in the data. The theoretical autocovariance functions (ACVFs) are in the boxes. Note that the ACVF in Figure 12.3d is alternately positive and negative in bundles that decrease in magnitude. This is like a seasonal process, but it is somewhat more complicated because it dies out. In fact, *it is the covariogram of a nondeterministic seasonal process.* The same kind of pattern is evident in Figure 12.3c. The

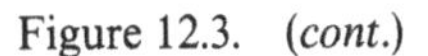

Figure 12.3. *(cont.)*

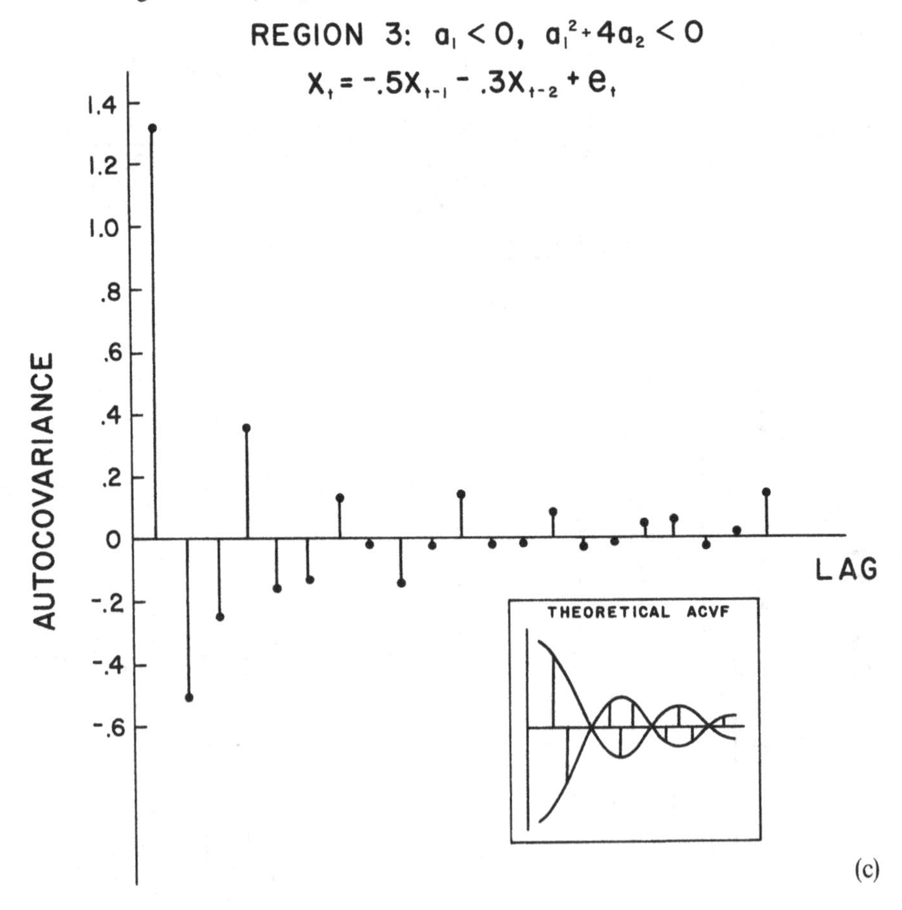

Figure 12.3. (*cont.*)

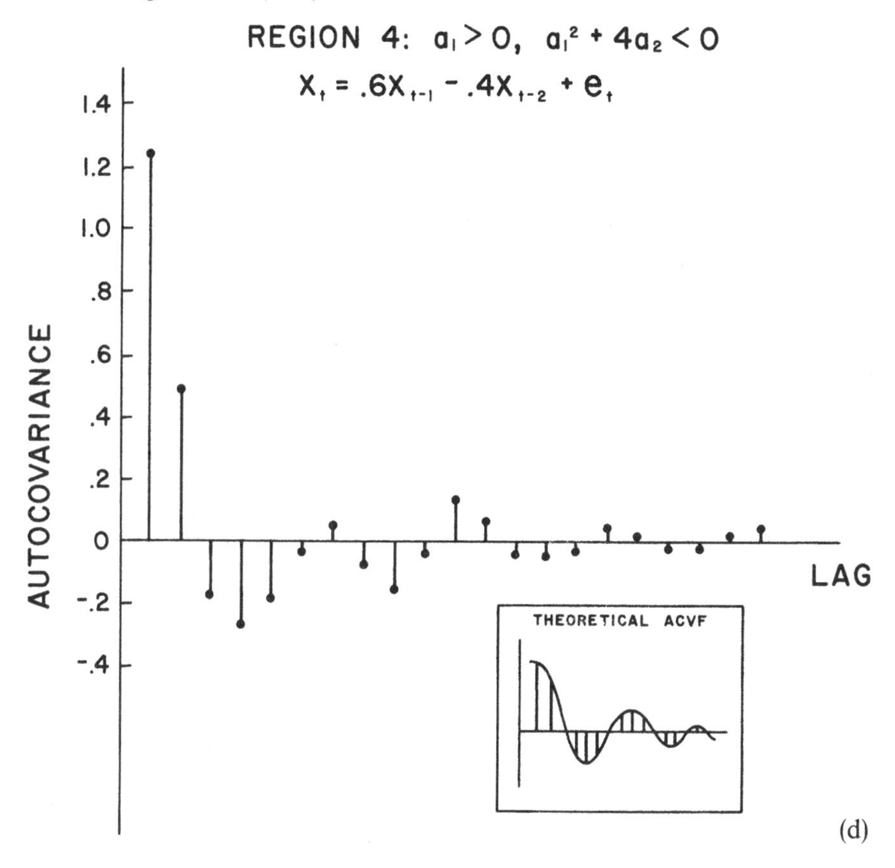

(d)

alternating signs of the ACVF are a function of a_1 being negative. However, notice that Figure 12.3a and b have correlograms that are not very different from the two types possible for the AR(1) process. The mathematics of the AR(2) process will explain how these four types of time series arise.

The triangular region revisited

The second of equations (12.2) for the autocorrelations can be rewritten

$$\rho_2 - a_1\rho_1 - a_2 = 0 \tag{12.11}$$

This equation can be solved to find all possible values of ρ_1 and ρ_2 in terms of the a_1 and a_2. At this point it is stated without proof that the triangular

Figure 12.4. Four regions of stationarity for AR(2) processes.

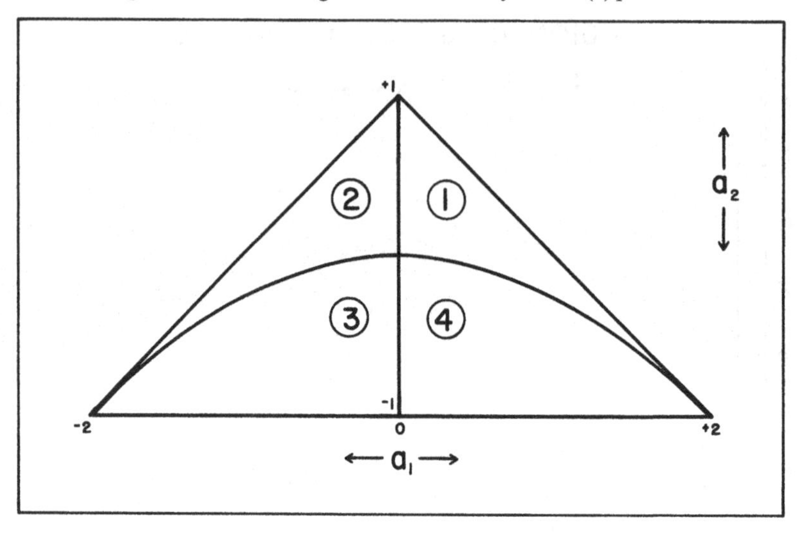

Figure 12.5. First 200 observations of a realization of a nonstationary AR(2) process.

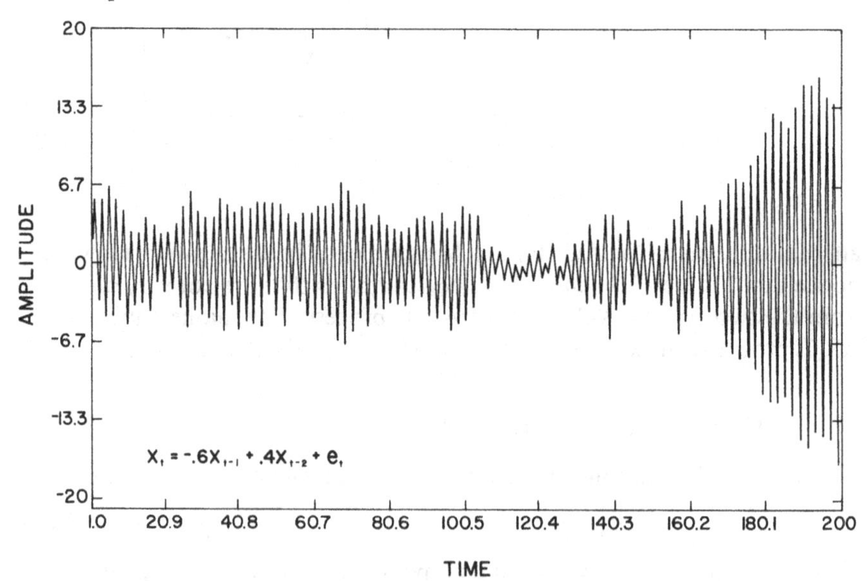

stationarity region can be partitioned as shown in Figure 12.4. The curve is the arc $a_1^2 + 4a_2 = 0$. Parameters inside the arc have $a_1^2 + 4a_2 < 0$. It can be shown that inside this circle the autocorrelation function will look much like a damped sine wave and the time series will be nondeterministically periodic. This fact, its ability to represent one nondeterministic cycle, is the strength of the AR(2) model. It turns out to be the case that an AR(4) can represent two cycles, an AR(6) can represent three cycles, and so on.

In regions 1 and 2 the correlograms are quite similar to the two possible correlograms for a first-order autoregressive process, AR(1). In fact, they are hard to distinguish from the AR(1) correlograms without the aid of another function, called the *partial autocorrelation function.* The introduction of this useful function will be deferred until the discussion of the duality between autoregressive and moving-average processes.

An example of a nonstationary AR(2) process

What would a time series look like if it were the realization of a *nonstationary* AR(2) process? I selected the process

$$x_t = -.6x_{t-1} + .4x_{t-2} + e_t$$

Note that this violates the third condition of equation (12.10) because $a_2 - a_1 = 1$. Figure 12.5 is a realization of the first 200 points of this process. It looks as if the nonstationarity is in the variance and that the series will oscillate violently as time increases. Figure 12.6 is a plot of the first 400 points and we see that the oscillations eventually come back into line. The series appears stationary in the mean but not in its variance. Also, Figure 12.7 illustrates the autocovariance of this series. Note that they do not die out (go to zero) rapidly as do the autocovariances of the stationary processes in each of the other four regions.

Spectral density of the AR(2) process

Since I remarked that region 3 and region 4 AR(2) processes (see Figure 12.3) are "pseudoperiodic," what would we expect the spectral density function of these processes to indicate? The spectral density function of AR(2) processes in regions 3 and 4 will show a major peak across a band of frequencies. Figure 12.8 presents the sample spectral densities for realizations of the four.

Figure 12.6. First 400 observations of a realization of the same process plotted in Figure 12.5.

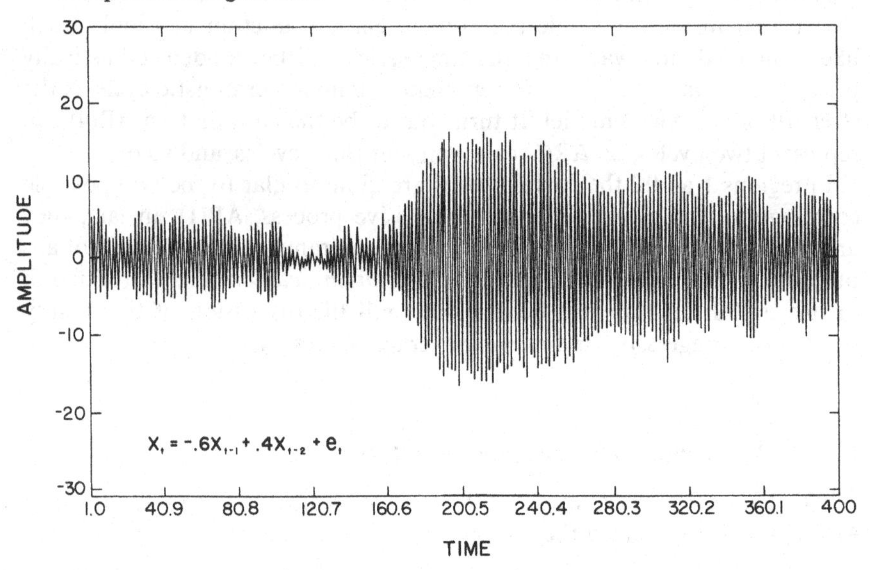

Figure 12.7 ACVF of the series plotted in Figure 12.6.

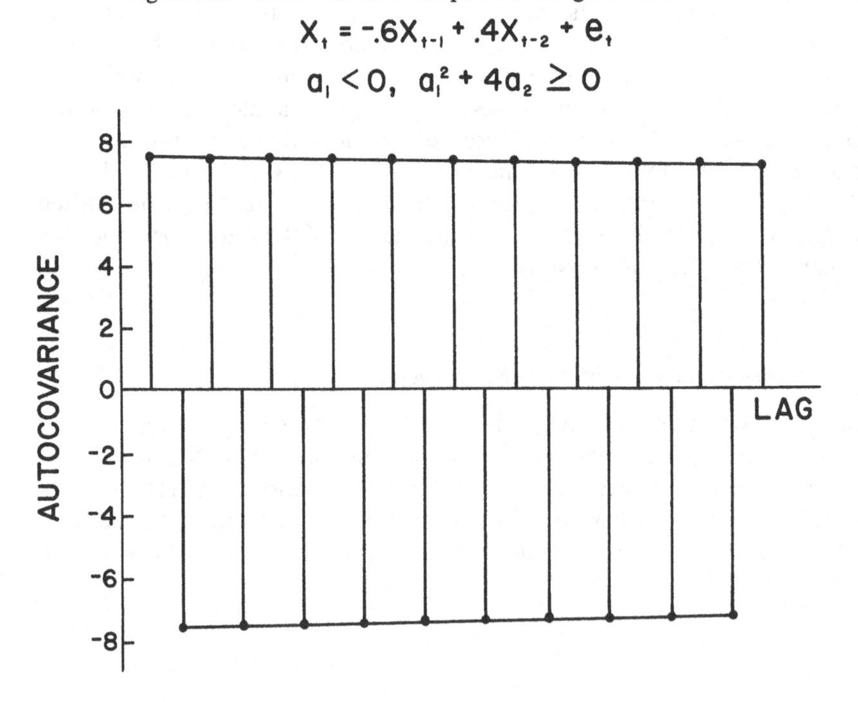

Figure 12.8. Spectral density functions of realizations of processes from each of the four regions of stationarity: (a) region 1; (b) region 2; (c) region 3; (d) region 4.

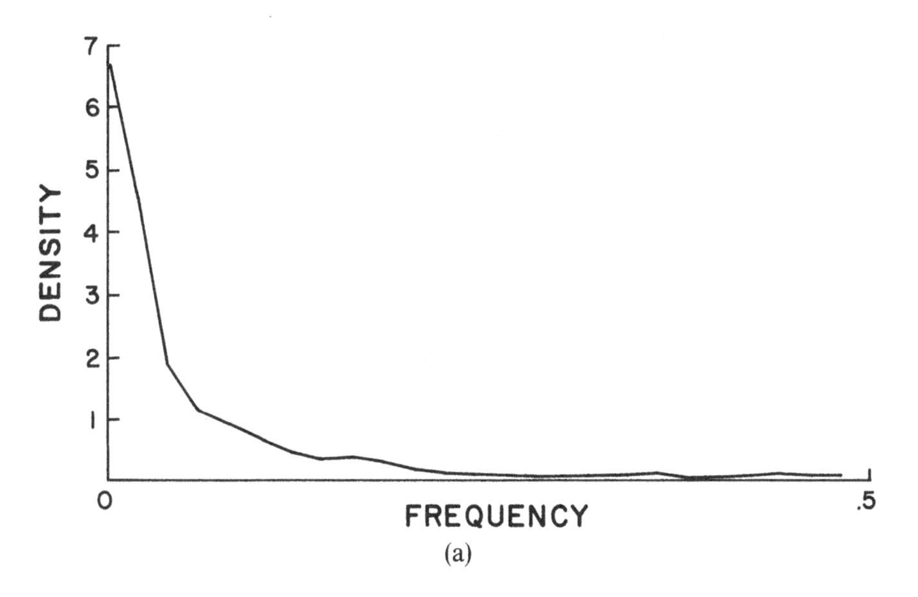

(a)

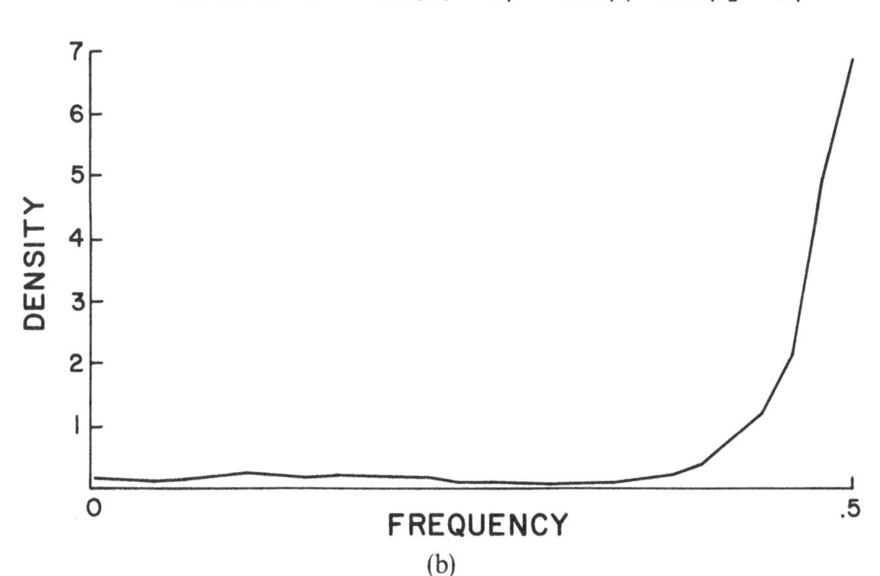

(b)

Figure 12.8. (*cont.*)

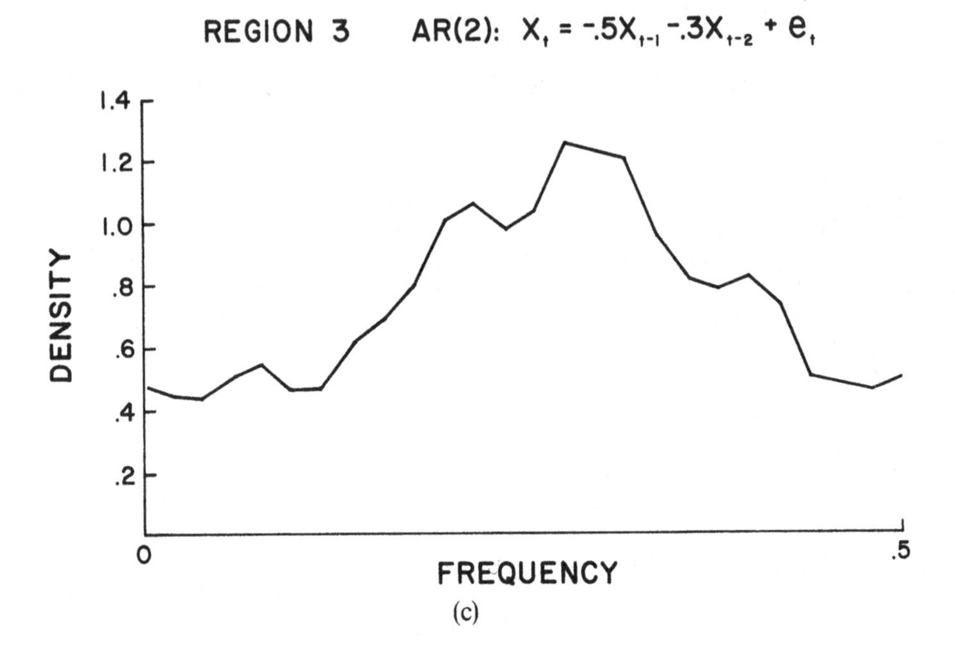

(c)

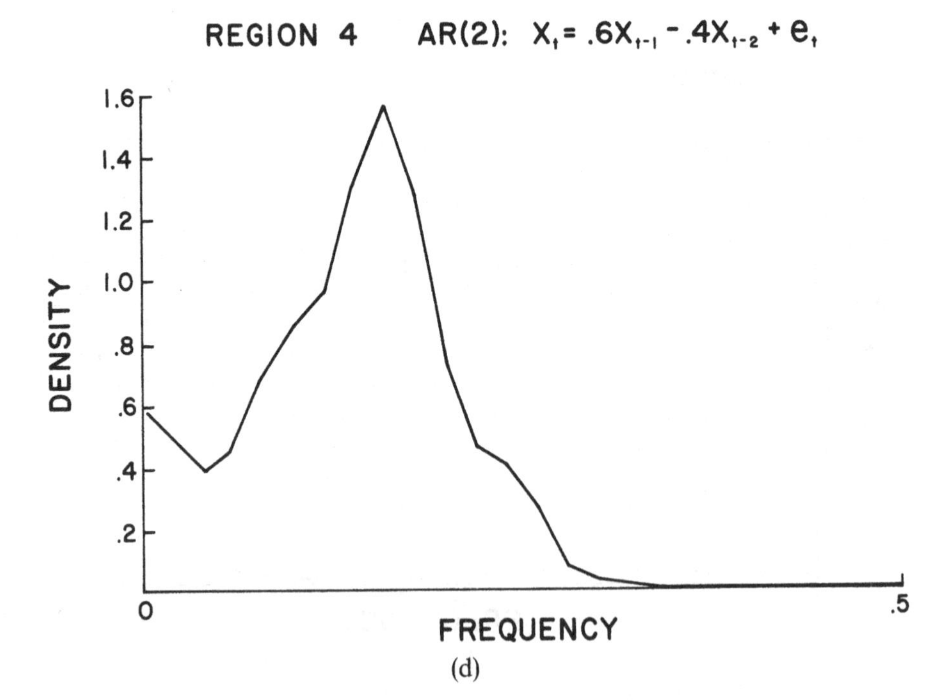

(d)

Example of an AR(2) process: migraine-headache data

The following data are from an experiment conducted by Budzynski and Stoyva with a subject who complained of migraine headaches. The dependent variable is the subject's average daily rating on a scale 0 = no pain to 5 = severe pain. These records were very well kept; most daily readings were based on averages of 8 or more hours (see Figure 12.9). These data strongly resemble an AR(2) process. In Figure 12.9 we can see the characteristic damped sine wave we described for the pseudoperiodic AR(2) process, and the spectrum, which shows one major peak.

Note that we can use the sample autocorrelations, the variance of the data, and the Yule–Walker equations to estimate the parameters of an AR(2) model for the headache data. We can already suggest that the series is of type 4 and thus have some approximate idea of the autoregressive coefficients. The exact computations however, will be deferred to Chapter 18, where AR parameter estimation and significance testing are discussed.

Figure 12.9. (a) Migraine-headache patient's ratings of the severity of his headache pain; (b) estimate of the correlogram of the headache data; (c) estimate of the spectral density function of the headache data.

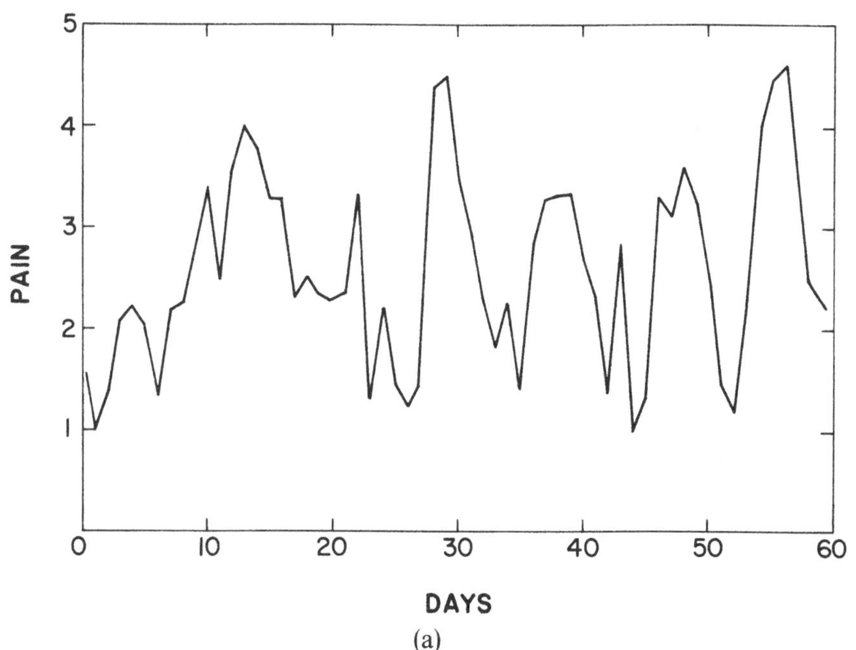

(a)

Figure 12.9. (*cont.*)

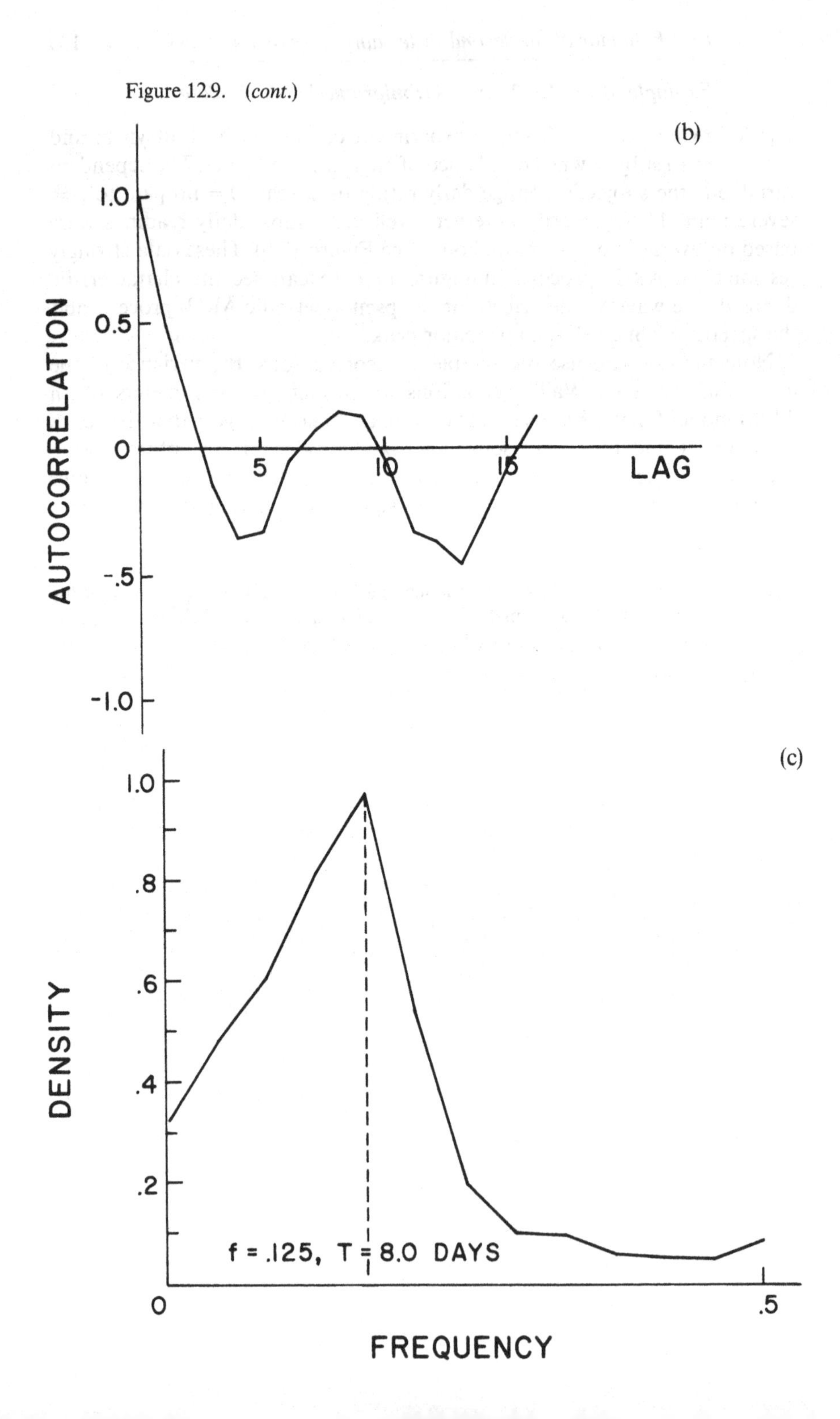

Summary

The major point of this chapter is simply that by going back into the past two time units, the autoregressive process of second order is capable of simulating a nondeterministic periodicity. There are four types of autoregressive processes, two of which are similar to AR(1) processes, but two of which are not. These four processes can be distinguished primarily by their ACVFs (or ACFs). The two nondeterministically cyclic AR(2) processes have ACVFs that appear sinusoidal but die out rather quickly.

It is now obvious that only two lags of the autocorrelation function are needed to describe a nondeterministic cycle *of any frequency.* Many students have thought that if the data are gathered daily, a 28-day cycle requires computation of 28 lags, but this is not the case; only two lags are necessary.

13
The partial autocorrelation function: completing the duality

This chapter addresses the idea of moving-average and autoregressive models as "dual" processes, linked to one another by the autocorrelation and "partial autocorrelation" functions. The partial autocorrelation function is defined; the mysteries of the duality are explored; and the application of these facts is discussed for the model identification stage of model building.

In this chapter we introduce a function much like the autocorrelation function (ACF), called the *partial autocorrelation function* (PACF). It is, in a sense, the "dual function," analogous to the ACF. More important, these two functions, the ACF and the PACF, are a useful team in identifying time-series models in practice.

In general, MA processes and AR processes are considered *dual* processes. We shall see that by defining the partial autocorrelation function, the duality will be complete. Just as the autocorrelation function dies out for a finite-order AR process and truncates for a finite-order MA process, the reverse is true for the PACF (see Table 13.1). The PACF is a complicated function to define, but its properties are very useful.

What is the PACF?

Suppose that we consider the "memory" in an AR(1) process. We know that its autocorrelation function is given by $\rho_k = (a_1)^k$. Consider the dependency of observations one lag apart; they are correlated $\rho_1 = a_1$ for the AR(1) model. Now consider observations two lags apart. You will be prone to answer that they are correlated $\rho_2 = (a_1)^2$. Hence, observations x_t are correlated with observations x_{t+2} to the extent a_1^2. But is x_{t+2} dependent on x_t after considering the intermediate link with x_{t+1}? The question can be answered by partial correlation. If the terms are denoted 1, 2, and 3 (for x_t, x_{t+1}, and x_{t+2}, respectively), we want to know if $\rho_{13.2}$ is zero, where $\rho_{13.2}$ is the correlation of x_t and x_{t+2} given (conditional on) x_{t+1}. The standard equation for partial correlation is

$$\rho_{13.2} = \frac{\rho_{13} - \rho_{12}\rho_{32}}{[(1 - \rho_{12}^2)]^{1/2}[(1 - \rho_{32}^2)]^{1/2}} \tag{13.1}$$

We know that for the AR(1), $\rho_{13} = \rho_2 = a_1^2$ and $\rho_{12} = \rho_{32} = a_1$. Hence, the numerator is $a_1^2 - (a_1)(a_1)$, which is zero. So the answer is: no, there is no relationship between x_{t+2} and x_t after removing the intermediate association with x_{t+1}. All higher-order partials (e.g., $\rho_{14.23}, \rho_{15.234}$, etc.) will also vanish.

To summarize, if the process is AR(1), once we get to lag 2, all partial correlations are zero.

Now consider the AR(2) process. Using the expressions we derived for ρ_k in terms of a_1 and a_2, it can be shown with a bit of algebra that the partial correlation is not zero but that

$$\rho_{13.2} = \frac{\rho_2 - \rho_1^2}{1 - \rho_1^2} = a_2 \tag{13.2}$$

Thus, if the process is AR(2), lag 1 and lag 2 correlations, controlling for intermediate observation, are nonzero. Beyond lag 2, all partials are zero. Furthermore, the last nonzero partial, here the second one, takes the value

Table 13.1 *The duality between moving-average and autoregressive processes*

	Function	
Process	ACF	PACF
MA	Truncates	Dies out
AR	Dies out	Truncates

of precisely the autoregressive parameter of the order, here two, of the AR process. This result, we shall see, holds in general for AR(p) processes.

Usually, we do not know the order of the AR process, so we first fit an AR(1) and compute a_{11}. The Yule–Walker equations can be used for the fitting process. The second subscript denotes that this is our first approximation of the AR model. Then we fit an AR(2), computing a_{12} and a_{22}. Note that a_{12} need not equal a_{11}. *If we continue in this manner, a plot of* a_{kk} *versus the order* k *gives the partial autocorrelation function.*

What is the PACF at order k? It is easily interpreted as follows. Suppose that we have fit a process of order $(k - 1)$ and we wish to determine whether we gain anything by going back in the past of the series another time unit. This would assess the utility of increasing the order of the AR process from an AR($k - 1$) to an AR(k). We have the two functions (1) x_t, the data, and (2) $a_1x_{t-1} + a_2x_{t-2} + \cdots + a_{k-1}x_{t-k+1} = x_t'$, our AR($k - 1$) prediction of our data. The problem of whether we gain information by increasing the order of the AR process is equivalent to the question of whether a third function, (3) a_kx_{t-k}, correlated with (1) when we partial out (2). That is, to modify our earlier notation, we want to know if

$$\rho_{13.2} = 0$$

To summarize, the PACF can be computed as follows: First fit an AR(1) to the data, the coefficient a_1 is the lag-1 PACF, a_{11}; next fit an AR(2) to the data, the coefficient a_2 is the lag-2 PACF, a_{22}; for the kth order AR fit, the last AR coefficient is the PACF, a_{kk}. The Yule–Walker equations can be used for the AR fitting. Table 13.2 will illustrate this procedure (see also Appendix 14.C and Chapter 19).

Table 13.2. *Computation of the PACF by successive AR model fitting*

AR model	a_1	a_2	a_3	a_4	a_5	a_6	a_7	a_8	a_9
AR(1)	.8952								
AR(2)	1.4100	−.5752							
AR(3)	1.4830	−.7543	.1270						
AR(4)	1.4720	−.6855	−.0083	.0912					
AR(5)	1.4710	−.6854	−.0009	.0754	−.0107				
AR(6)	1.4710	−.6864	−.0009	.0841	−.0079	.0127			
AR(7)	1.4700	−.6862	−.0031	.0842	.0101	−.0260	.0263		
AR(8)	1.4700	−.6863	−.0030	.0847	.0101	−.0301	.0351	−.0060	
AR(9)	1.4710	−.6873	−.0022	.0844	.0078	−.0300	−.0541	−.0467	.0277

Note: Diagonals contain the PACF values, with lags increasing as the AR model's order increases.

Estimation using the PACF

Quenouille (1947) showed that if a time series is a realization of an AR(p) process, the estimated PACFs of order $p + 1$ and higher are approximately distributed with variance $1/N$, where N is the number of data points. Thus, despite the potential usefulness of the PACF, the reader should use it with caution. As an example of the potential problems with its orthodox use, a realization with $N = 200$ of an AR(4) process was simulated using the model

$$x_t = 1.820x_{t-1} - 1.320x_{t-2} + .455x_{t-3} - .063x_{t-4} + e_t$$

where the e_t was white noise normally distributed with mean zero and unit variance. The data are plotted in Figure 13.1. This model was selected so that its special density function would have two peaks. Its sample spectral density function, plotted in Figure 13.2a, shows evidence of two peaks, with less power for the second peak. The ACF, plotted in Figure 13.2b, appears more complicated than the damped sine wave of the periodic AR(2) processes.

Using Quenouille's estimate, $2/\sqrt{N} = 2/\sqrt{200} = 0.1414$. The Yule–Walker equations were then used to compute the coefficients of each AR(k) model fit and the residual variance of the fit. These coefficients are given through

Figure 13.1. Realization of an AR(4) process, with 200 observations.

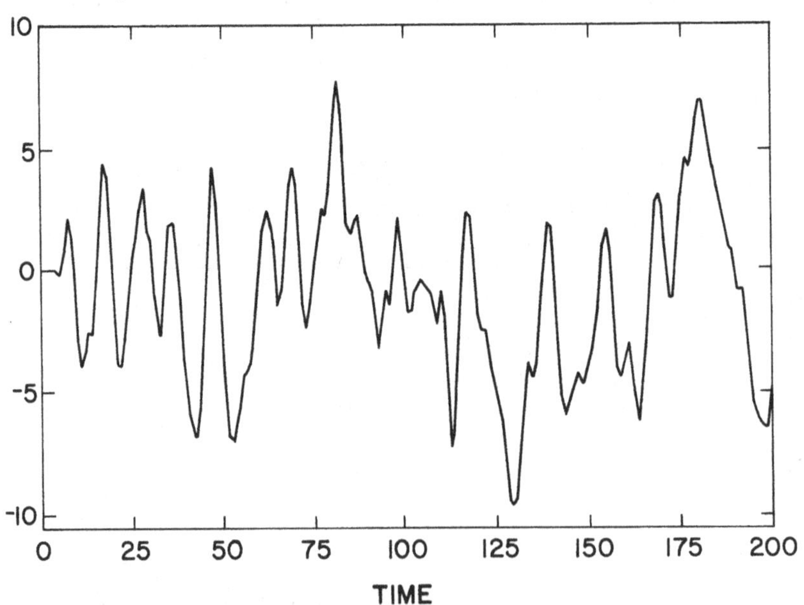

Figure 13.2. (a) Estimated spectral density function of the first 200 obervations of the AR(4) process plotted in Figure 13.1; (b) estimated autocorrelation function (ACF).

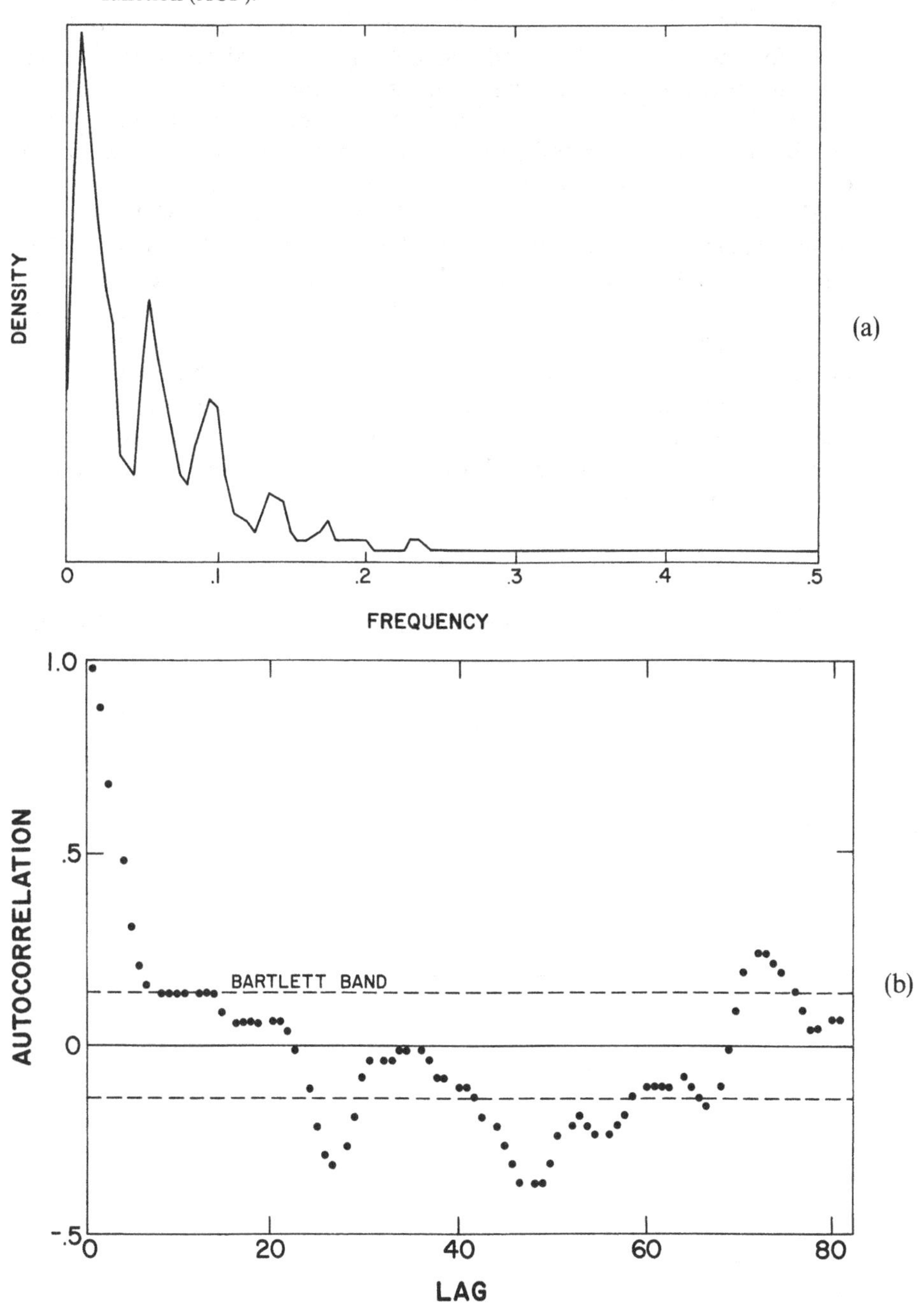

lag 9 in Table 13.2. *The PACF is represented by the diagonal numbers.* It would usually be plotted as a function of lag k, much as is the ACF (see Figure 13.3). The PACF would suggest an AR(2) model since the third and fourth partial correlations are not statistically significant. This model accounts for 86.71% of the original variance. However, a small and potentially interesting peak would remain in the spectral density in the residual from the AR(2) fit. To detect an AR(4), N would have to be such that $2/\sqrt{N}$ is less than $\hat{a}_{44} = .09122$, which gives the result $N \geq 481$. Another realization of this process was then generated with $N = 600$; it is plotted in Figure 13.4. Figure 13.5 is a plot of the PACF derived from Table 13.3. The PACF would suggest an AR(3) model, which will account for 92.40% of the variance. Also, note in Table 13.3 that the estimated coefficients of the AR process are closer to the process values used in the simulation than the values in Table 13.2.

This example points again out the need for large samples for time-series analysis. The large sample is needed to appropriately identify the internal structure; it would not be needed if we knew the order a priori and wanted only to estimate the coefficients.

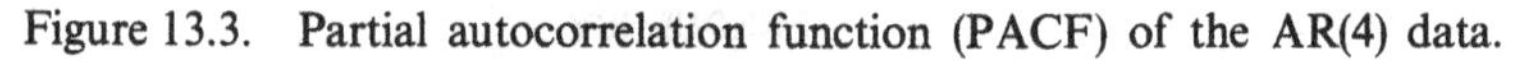

Figure 13.3. Partial autocorrelation function (PACF) of the AR(4) data.

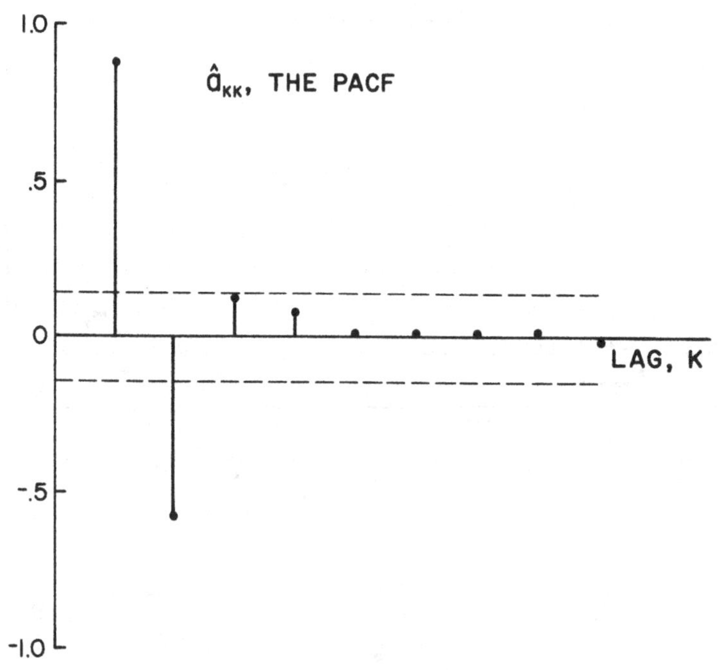

Figure 13.4. Another realization of the first 600 observations of the AR(4) process of Figure 13.3.

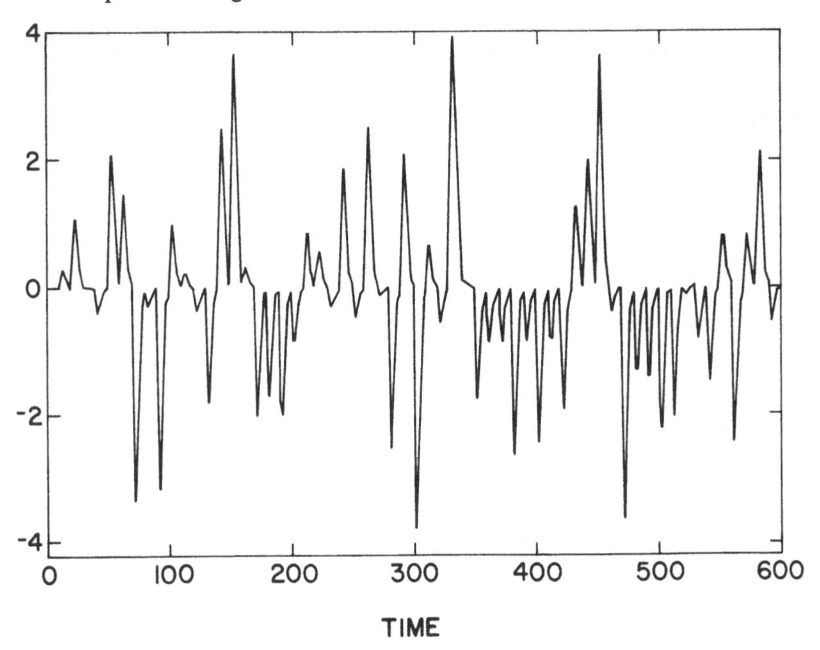

Figure 13.5. PACF of 600 observations of the AR(4) process.

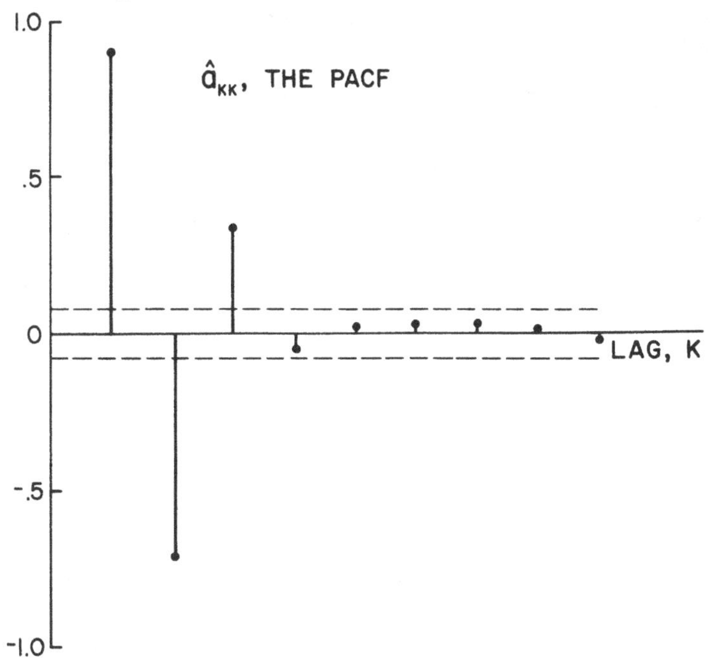

Box and Jenkins (1970) recommend using the ACF and PACF together as aids for model identification. For an AR(2) process, Figure 13.5a summarizes the theoretical ACF and PACF patterns that would be obtained for processes in each of the four regions of stationarity.

Table 13.3. *Computation of the PACF with 600 observations*

AR model	a_1	a_2	a_3	a_4	a_5	a_6	a_7	a_8	a_9
AR(1)	.9093								
AR(2)	1.5530	−.7077							
AR(3)	1.7990	−1.2470	.3472						
AR(4)	1.8160	−1.3100	.4377	−.0503					
AR(5)	1.8170	−1.3180	.4631	−.0856	.0194				
AR(6)	1.8160	−1.3160	.4506	−.0500	−.0298	−.0271			
AR(7)	1.8160	−1.3150	.4520	−.0629	.0082	−.0253	.0288		
AR(8)	1.8150	−1.3150	.4519	−.0618	.0002	−.0022	−.0031	.0176	
AR(9)	1.8150	−1.3150	.4519	−.0618	−.0004	.0021	−.0156	.0349	−.0095

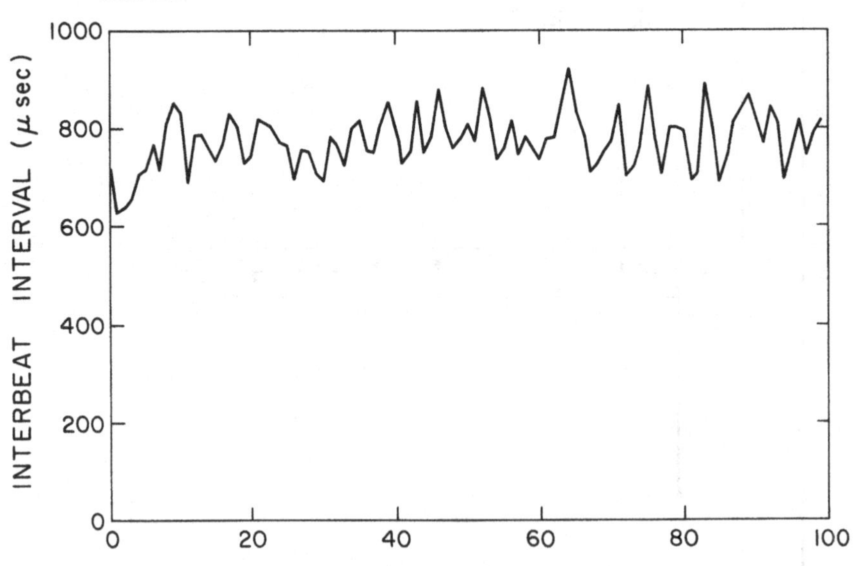

Figure 13.6. Heart rate of an adult male subject, measured as interbeat interval.

Examples: heart rate and blood velocity

The data analyzed in this section are from the laboratory of Robert Levenson. They are taken from physiological recordings of a male subject in a relaxed but vigilant state. Figure 13.6 is a graph of the subject's heart rate, measured as the average interval between beats, sampled 100 times in a 30-min period. The PACF of the heart-rate data is plotted in Figure 13.7. and it would probably be concluded from the PACF that the model is an AR(3) process. The estimated ACF and spectral density functions are plotted in Figure 13.8, and the two peaks in the spectral density function might suggest that the process is AR(4). There is not an adequate amount of power in the 100 observations to detect the peak of lower power.

Figure 13.9 presents the time between a contraction of the heart and the arrival of blood to the ear, a variable called ear pulse time. This variable is interesting because it could be a continuous measure that taps blood pressure, which cannot currently be assessed continuously in human beings. The PACF of these data are plotted in Figure 13.10. Using the PACF, only the first autocorrelation function would be considered significant, which suggests

Figure 13.7. PACF of the heart-rate data.

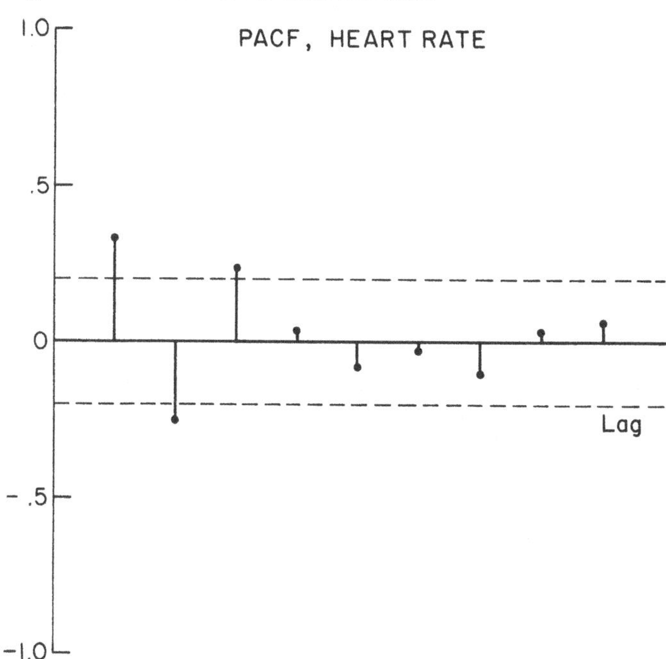

Figure 13.8. (a) ACF of the heart-rate data; (b) spectral density of the heart-rate data.

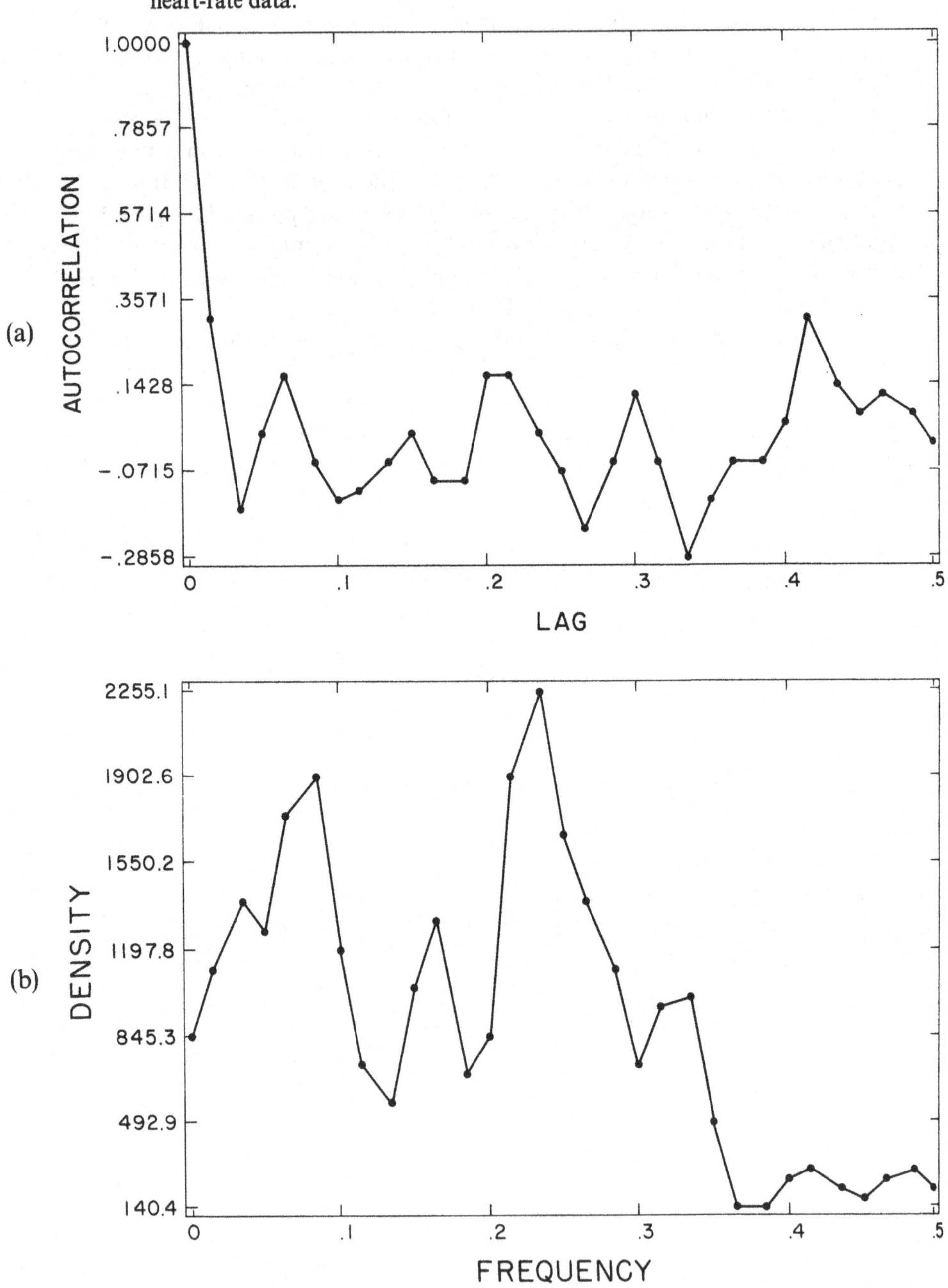

Figure 13.9. Ear pulse time of the subject in Figure 13.6.

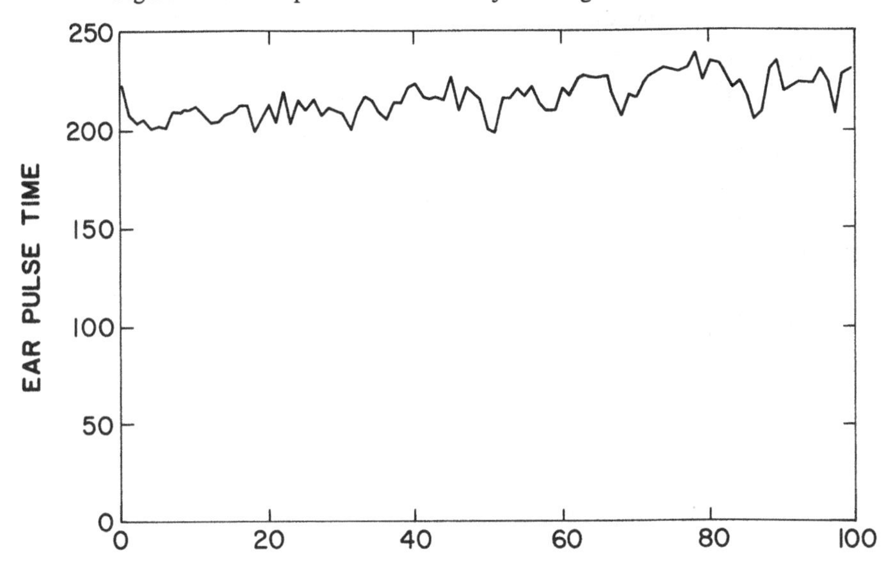

Figure 13.10. PACF of ear pulse time.

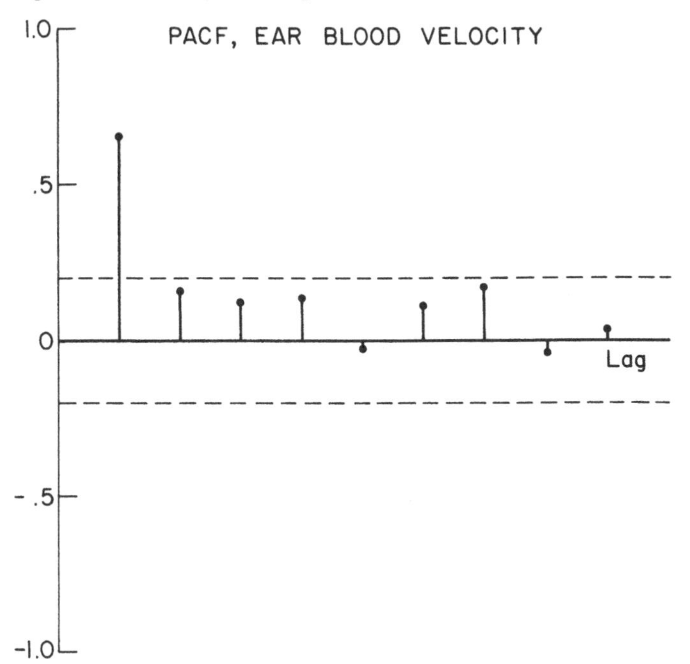

Figure 13.11. (a) ACF of the ear pulse time data; (b) spectral density of the ear pulse time data.

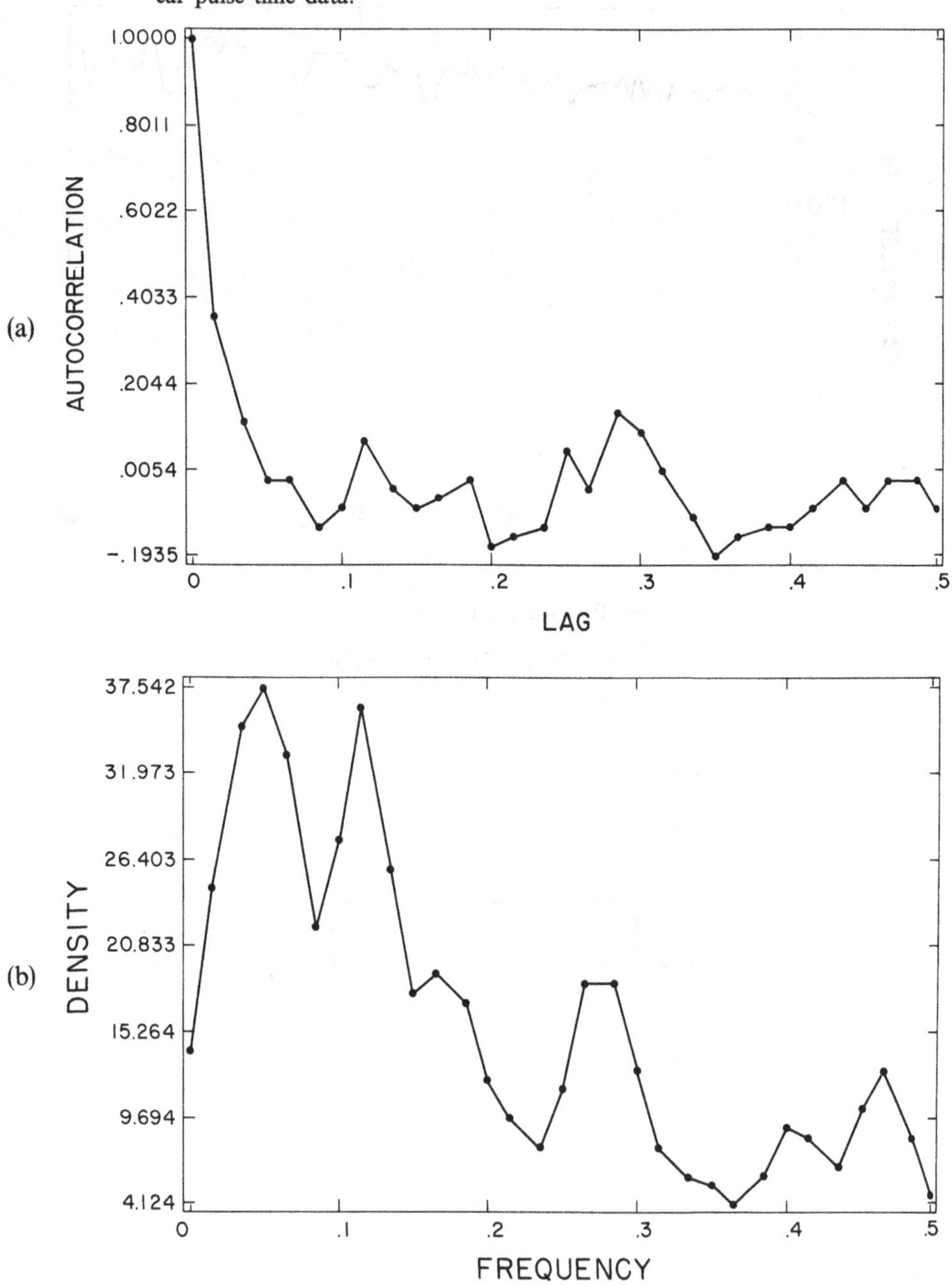

an AR(1) model. However, if the ACF and spectral density function estimates are examined (see Figure 13.11), two peaks would be suspected. It is thus likely that the first four PACF coefficients would be significant if more data points were available, so an AR(4) model might be preferred. These examples thus illustrate the usefulness of simultaneously employing both frequency- and time-domain methods, and interpreting significance levels loosely when we are identifying the form of the model. (More careful use of significance levels is appropriate once we wish to make statements about the parameters in an already identified model.)

Summary

This chapter introduced the PACF. Recall that previously it was shown that the ACF for an MA(q) process truncates after lag q. In this chapter we suggested that the PACF for an AR(p) process truncates after lag p.

What is the practical importance of these results? Many time-series analysts suggest using the PACF and ACF in tandem to identify AR and MA components of a stationary stochastic process. In this book we see that, although this may sometimes be sound advice, for many social science applications, AR model fitting is a useful and somewhat simpler alternative. In this chapter we also noted that the spectral density function of the process can also be helpful in suggesting a model.

14 The duality of MA and AR processes

This chapter introduces a useful notation (called operator notation), which can represent models of stationary stochastic components and explore the duality between moving-average and autoregressive processes. The new operator notation makes it possible to write AR and MA models succinctly and to discuss the autocorrelation generating function. The autocorrelation generating function will be useful later for deriving equations of the spectral densities of AR and MA processes (Chapter 17).

Appendices are included at the end of this chapter as a primer for the reader who is unfamiliar with matrix algebra.

There is a simple equation that we will eventually derive (Chapter 17) for the spectral density of moving-average and autoregressive models for the model parameters. The spectral density will also be useful in the model-identification stage of building a model.

To derive this equation we need to introduce some notation that will make the representation of the AR and MA models more compact and elegant. This notation will also make it easy to write the form of the autocorrelation and autocovariance functions in terms of the model parameters.

The backward-shift operator

The backward-shift operator, B, operates on a series to move it back one time unit, as follows:

t	x_t	Bx_t
1	2.1	—
2	5.6	2.1
3	.3	5.6
4	10.0	0.3

A new time series, Bx_t, is created by this transformation. In mathematical notation the backward shift is

$$Bx_t = x_{t-1} \tag{14.1}$$

The operator B is like a time machine that takes us back into the past. If we want to go back two time units, we apply B twice:

t	x_t	$B(Bx_t)$
1	2.1	—
2	5.6	—
3	.3	2.1
4	10.0	5.6

In mathematical notation,

$$B(Bx_t) = B(x_{t-1}) = x_{t-2} \quad \text{or} \quad B^2 x_t = x_{t-2} \tag{14.2}$$

Applying B twice (B^2) takes you back two units in time. In general, to shift back k time units,

$$B^k(x_t) = x_{t-k} \tag{14.3}$$

This holds for the whole series, $t = k, k + 1, \ldots, n$.

We can use powers of B to write AR and MA models. For example, to define the moving-average transformation, we could write

$$\begin{aligned} x_t' &= \tfrac{1}{2}x_t + \tfrac{1}{2}x_{t-1} = \tfrac{1}{2}(x_t + Bx_t) \\ &= \tfrac{1}{2}(1 + B)x_t \end{aligned}$$

In a similar fashion the difference operator is

$$\begin{aligned} (1 - B)x_t = x_t - Bx_t = x_t - x_{t-1} = \nabla x_t \\ \nabla x_t = (1 - B)x_t \end{aligned}$$

If we go back in time more than one time unit, we use powers of B, and this results in a *polynomial* in B. For example, for the AR(2) process,

$$\begin{aligned} \text{AR(2):}\quad x_t &= a_1 x_{t-1} + a_2 x_{t-2} + e_t \\ &= (a_1 B + a_2 B^2)x_t + e_t \end{aligned}$$

or

$$(1 - a_1 B - a_2 B^2)x_t = e_t \tag{14.4}$$

$$\text{AR}(p)\text{:}\quad x_t = \sum_{i=1}^{p} a_i x_{t-i} + e_t$$

$$(1 - a_1 B - a_2 B^2 - \cdots - a_p B^p)x_t = e_t \tag{14.5}$$

If we write the expression in the parentheses of equation (14.5) as a polynomial in B, AR(B), we have the simple representation for the AR(p) process:

$$\boxed{\text{AR}(p)\text{:}\quad \text{AR}(B)x_t = e_t} \tag{14.6}$$

The MA model has a similar representation:

$$\begin{aligned} \text{MA(2):}\quad x_t &= b_1 e_{t-1} + b_2 e_{t-2} + e_t \\ &= (1 + b_1 B + b_2 B^2)e_t \end{aligned} \tag{14.7}$$

$$\text{MA}(q)\text{:}\quad x_t = \sum_{s=0}^{q} b_s e_{t-s}$$

$$= (1 + b_1 B + b_2 B^2 + \cdots + b_q B^q)e_t \tag{14.7.1}$$

If we write the polynomial in the parentheses of equation (14.7.1) as a polynomial MA(B), we get

$$\boxed{\text{MA}(q)\text{:}\quad x_t = \text{MA}(B)e_t} \tag{14.8}$$

The AR/MA duality revisited

We will show that the drunkard's walk process

$$x_t = x_{t-1} + e_t \tag{14.9}$$

can be rewritten as an infinite moving average. Using the new notation, consider the moving average,

$$\begin{aligned} x_t &= e_t + e_{t-1} + e_{t-2} + \cdots \\ &= \sum_{i=0}^{\infty} e_{t-i} \\ &= (1 + B + B^2 + \cdots)e_t \\ &= \left(\sum_{i=0}^{\infty} B^i\right)e_t \end{aligned} \tag{14.10}$$

If we remember[1] that the series $\sum_0^\infty x^i = 1/(1 - x)$, we can write equation (14.10) as

$$x_t = \left(\frac{1}{1 - B}\right)e_t \tag{14.11}$$

Equivalently, equation (14.11) can be written as

$$(1 - B)x_t = e_t \tag{14.12}$$

or

$$x_t - x_{t-1} = e_t \tag{14.13}$$

or

$$x_t = x_{t-1} + e_t \tag{14.14}$$

which returns us to (14.9). Here we see that polynomial operators have inverses and that *infinite moving-average processes are finite autoregressive processes.* The reverse is also true; that is, *finite moving-average processes are infinite autoregressive processes.* As an example, consider a first-order moving-average process

$$x_t = e_t - b_1 e_{t-1} \tag{14.15}$$

In operator notation,

$$x_t = (1 - b_1 B)e_t$$

or equivalently,

$$\left(\frac{1}{1 - b_1 B}\right)x_t = e_t$$

$$(1 + b_1 B + b_1^2 B^2 + \cdots)x_t = e_t$$

$$x_t + b_1 Bx_t + b_1^2 B^2 x_t + \cdots = e_t$$

$$x_t + b_1 x_{t-1} + b_1^2 x_{t-2} + \cdots = e_t \qquad (14.16)$$

Compare these results using the B notation to those obtained in Chapter 11.

ARMA or "mixed" processes

Suppose that we have a set of data, x_t, that we fit with an AR(1) and obtain a residual n_t:

$$x_t = a_1 x_{t-1} + n_t$$

Suppose further that the ACF of the residual had only one nonzero value, so that we suspected that the residual was an MA(1) process, with parameter b. Then we could write

$$x_t = a_1 x_{t-1} + e_t + be_{t-1} \qquad (14.17)$$

where e_t is noise. Alternatively, we can write this as

$$x_t - ax_{t-1} = e_t + be_{t-1}$$

or

$$(1 - aB)x_t = (1 + bB)e_t \qquad (14.18)$$

$$\mathrm{AR}(B)x_t = \mathrm{MA}(B)e_t \qquad (14.19)$$

This is called a mixed or *autoregressive moving-average* (abbreviated ARMA) process.

Because the MA(1) process can also be written as an infinite AR process (if $|b| < 1$), equation (14.18) can be rewritten as follows:

$$(1 - aB)\left(\frac{1}{1 + bB}\right)x_t = e_t$$

or

$$(1 - aB)(1 - bB + b^2B^2 - b^3B^3 + \cdots)x_t = e_t$$

$$[1 - (a + b)B + (ab + b^2)B^2 + \cdots]x_t = e_t$$

This is an infinite-order AR process. Eventually, the powers of a and b will become small, so that it will be reasonable to approximate this process with a finite-order AR of reasonably low order.

These results are true more generally under some appropriate conditions. If we have two polynomials in B, $\mathrm{MA}(B)$ and $\mathrm{AR}(B)$, and an ARMA model

$$\mathrm{AR}(B)x_t = \mathrm{MA}(B)e_t$$

it is possible to write the model as an infinite AR process,

$$\left[\frac{\mathrm{AR}(B)}{\mathrm{MA}(B)}\right]x_t = e_t \tag{14.20}$$

or an infinite MA process,

$$x_t = \left[\frac{\mathrm{MA}(B)}{\mathrm{AR}(B)}\right]e_t \tag{14.21}$$

and approximate either by finite processes. However, the advantage of ARMA models lies in those (admittedly all too infrequent) cases where $\mathrm{AR}(B)$ and $\mathrm{MA}(B)$ are very simple, as in equation (14.18); however, the resulting AR or MA processes in equation (14.20) or (14.21) are complicated. Unfortunately, identifying such a model from the ACF and PACF may not be easy.

The condition necessary for dividing by $\mathrm{AR}(B)$ is that the AR process be stationary. However, since all finite MA processes are stationary, this condition will not be adequate for dividing by $\mathrm{MA}(B)$. The condition necessary for dividing by $\mathrm{MA}(B)$ is that the MA process be "invertible." Both conditions can be expressed in terms of the roots of the polynomials $\mathrm{AR}(B)$ and $\mathrm{MA}(B)$.

Roots of polynomials: stationarity and invertibility

The polynomial $x^2 - x - 2$ can also be written $(x - 2)(x + 1)$, and, in general, any polynomial of order n

$$f(x) = ax^n + bx^{n-1} + \cdots + cx^2 + dx + e$$

can be rewritten as a product

$$(x - s_1)(x - s_2)\cdots(x - s_n)$$

where the s_k are, in general, complex numbers. When x equals any of these numbers, the polynomial is zero, and the numbers s_k are called the *roots* of the polynomial. They are solutions to the equation $f(x) = 0$. For example, the polynomial $x^2 + 1$ can be written as $(x + i)(x - i)$, where $i = (-1)^{1/2}$.

The polynomials $\mathrm{AR}(B)$ and $\mathrm{MA}(B)$ also have roots determined entirely by their parameters. In general, every root s_k of the generating polynomials

AR(B) or MA(B) will be a complex number

$$s_k = u_k + iv_k$$

where u_k and v_k are real numbers and $i = (-1)^{1/2}$. The *modulus* of a complete number s_k is written $|s_k|$ and defined as

$$|s_k| = (u_k^2 + v_k^2)^{1/2}$$

In general, an AR process will be stationary if all the roots of AR(B) have modulus greater than 1.0. For the AR(1) case, AR(B) $= 1 - a_1 B$, and the root of the polynomial, $1 - a_1 x$, is simply $s_1 = 1/a_1$. The condition $|s_1| > 1.0$ is the same as the condition we are familiar with for the AR(1) process to be stationary, $|a_1| < 1$.

We can divide by the AR(B) polynomial to get equation (14.21) whenever the roots of AR(B) have modulus greater than 1.0; similarly, we can divide by the MA(B) polynomial to get equation (14.20) whenever the roots of the MA(B) polynomial have modulus greater than 1.0; in the latter case the MA(B) operator is called *invertible*, and the conditions on its roots are called *invertibility conditions.*

The Yule–Walker equations again

If we write the general AR(p) (pth-order autoregressive) process

$$x_t = a_1 x_{t-1} + a_2 x_{t-2} + \cdots + a_p x_{t-p} + e_t \tag{14.22}$$

we can multiply by x_{t-k}, take expectations, and obtain

$$\rho_k = a_1 \rho_{k-1} + a_2 \rho_{k-2} + \cdots + a_p \rho_{k-p} \tag{14.23}$$

In the operator notation we can write the AR(p) process as

$$\mathrm{AR}(B) x_t = e_t \tag{14.24}$$

where

$$\mathrm{AR}(B) = 1 - a_1 B - \cdots - a_p B^p \tag{14.25}$$

and the autocorrelation function satisfies the equation

$$\boxed{\mathrm{AR}(B)\rho_k = 0} \tag{14.26}$$

This is called the *autocorrelation generating function.*

For the AR(p) process, the autocorrelation generating function AR(B)$\rho_k =$ 0 gives the familiar Yule–Walker equations

$$\rho_k = a_1 \rho_{k-1} + \cdots + a_p \rho_{k-p}, \qquad k > 0 \tag{14.27}$$

Substituting $k = 1, 2, \ldots, p$ in this equation gives

$$\begin{aligned}
\rho_1 &= a_1 + a_2\rho_1 + \cdots + a_p\rho_{p-1} \\
\rho_2 &= a_1\rho_1 + a_2 + \cdots + a_p\rho_{p-2} \\
&\vdots \\
\rho_p &= a_1\rho_{p-1} + a_2\rho_{p-2} + \cdots + a_p
\end{aligned} \tag{14.28}$$

These equations are the *Yule–Walker equations.* They can be written in matrix form and we do so in Appendix 14B. Solving this linear system with r replacing ρ will give estimates $\hat{a}_i$ for the parameters, with computations similar to those used in familiar multiple regression.

Summary

For the past five chapters we have been discussing time-domain models for the stationary stochastic component of our data. In this brief introduction to models in the time domain we have demonstrated how patterning in the data is revealed by the autocorrelations and partial autocorrelations. We have also seen how the linear Yule–Walker equations can be used to estimate the parameters of AR models. Throughout our discussions in the time domain, occasional references have been made to the usefulness of simultaneous consideration of frequency-domain statistics. We will return to the time domain in Chapter 18, after we understand more about the spectral density function and its relationship to the autocovariance function.

APPENDIX 14A

Matrix algebra primer

The goal of this appendix is to review the *general linear model* and to point out why it needs to be used cautiously and perhaps modified in time-series data analysis. This appendix will not be very detailed, and I therefore refer the reader to other sources for a more complete introduction to matrix algebra (e.g., Reiner, 1971).

In elementary algebra if we have the equation

$$Ax = B \tag{14A.1}$$

we can divide by A and obtain

$$x = A^{-1}B \tag{14A.2}$$

where A^{-1} is the inverse, or reciprocal, of A. For example,

$$2x = 6$$

$$x = (2)^{-1}(6) = \frac{6}{2} = 3$$

These concepts can be generalized to solve systems of equations.

Linear equations

Suppose that we have two variables x and y, and what we know about x and y are the equations

$$a_{11}x + a_{12}y = b_1 \tag{14A.3}$$

$$a_{21}x + a_{22}y = b_2 \tag{14A.4}$$

where the a's and b's are known constants. We can solve these two equations for x and y if the constants satisfy some conditions that essentially mean that the two equations (14A.3) and (14A.4) are different (independent) equations. The solution proceeds by rewriting equation (14A.3) and solving for x:

$$x = \frac{1}{a_{11}}(b_1 - a_{12}y)$$

and then substituting this value of x into equation (14A.4) and solving for y:

$$a_{21}\left[\frac{1}{a_{11}}(b_1 - a_{12}y)\right] + a_{22}y = b_2$$

or

$$\frac{a_{21}}{a_{11}}b_1 - \frac{a_{21}a_{12}}{a_{11}}y + a_{22}y = b_2$$

or

$$y\left(a_{22} - \frac{a_{21}a_{12}}{a_{11}}\right) = b_2 - \frac{a_{21}}{a_{11}}b_1$$

or

$$y = \frac{b_2a_{11} - a_{21}b_1}{a_{22}a_{11} - a_{21}a_{12}} \tag{14A.5}$$

This much is straightforward high school algebra. We can also rewrite equations (14A.3) and (14A.4) in matrix form as

$$\begin{bmatrix} a_{11} & a_{12} \\ a_{21} & a_{22} \end{bmatrix} \begin{bmatrix} x \\ y \end{bmatrix} = \begin{bmatrix} b_1 \\ b_2 \end{bmatrix} \tag{14A.6}$$

Equation (14A.6) means exactly what equations (14A.3) and (14A.4) mean. For this to be true it means that the matrix

$$\mathbf{A} = \begin{bmatrix} a_{11} & a_{12} \\ a_{21} & a_{22} \end{bmatrix}$$

multiplies the vector $\mathbf{x}$ to get the vector $\mathbf{B}$, where

$$\mathbf{x} = \begin{bmatrix} x \\ y \end{bmatrix} \quad \text{and} \quad \mathbf{B} = \begin{bmatrix} b_1 \\ b_2 \end{bmatrix}$$

as follows:

$$\begin{bmatrix} a_{11} & a_{12} \\ a_{21} & a_{22} \end{bmatrix} \begin{bmatrix} x \\ y \end{bmatrix} = \begin{bmatrix} a_{11}x + a_{12}y \\ a_{21}x + a_{22}y \end{bmatrix}$$

In other words, a row of $\mathbf{A}$ multiplies a column of $\mathbf{x}$, as follows:

$$\begin{bmatrix} a_{11} & a_{12} \\ a_{21} & a_{22} \end{bmatrix} \begin{bmatrix} x \\ y \end{bmatrix}$$

The combination of these two arrows produces the linear combination $a_{11}x + a_{12}y$.

The advantage of all this shift in notation is that it is possible to write equation (14A.6) as

$$\mathbf{Ax} = \mathbf{B}$$

and then solve for $\mathbf{x}$ as

$$\mathbf{x} = \mathbf{A}^{-1}\mathbf{B}$$

Of course, we have to define what the inverse of a matrix is so that it gives us the kind of solution we expect from such equations as (14A.5).

General linear model

The advantage of matrix notation becomes apparent when we learn that we can extend much of simple algebra to matrices and vectors to obtain a system of *linear algebra*. The usefulness of linear algebra in statistics is illustrated by the ubiquitousness of the *general linear model*,

$$\mathbf{Y} = \mathbf{XB} + \mathbf{E} \tag{14A.7}$$

where

$$\mathbf{Y} = \begin{bmatrix} y_1 \\ y_2 \\ \vdots \\ y_N \end{bmatrix}$$

are the observations, $\mathbf{X}$ is an $N \times p$ matrix of known design factors and independent variables,

$$\mathbf{B} = \begin{bmatrix} B_1 \\ B_2 \\ \vdots \\ B_p \end{bmatrix}$$

are the unknown parameters, and $\mathbf{E}$ is a vector of random error. We will use the following notation. The transpose of a column vector (such as $\mathbf{x}$ above) is a row vector, written $\mathbf{x}'$ (pronounced "**x**-transpose"), where $\mathbf{x}' = (x, y)$. Recall that any time series can be represented (according to the Wold decomposition theorem) as a sum of deterministic and stochastic components, both of which may be functions of time. Here $\mathbf{XB}$ is the deterministic part and $\mathbf{E}$ the stochastic component series $\mathbf{Y}$. It is our goal to find least-squares estimates for a set of parameters in vector $\mathbf{B}$, when the matrices $\mathbf{X}$ and $\mathbf{Y}$ are given. The solution, used repeatedly in this book, is

$$\mathbf{B} = (\mathbf{X}'\mathbf{X})^{-1}\mathbf{X}'\mathbf{Y} \tag{14A.8}$$

where $\mathbf{X}'$, called the *transpose* of $\mathbf{X}$, is simply $\mathbf{X}$ with its rows and columns interchanged.

Under the conventional multiple regression or analysis-of-variance assumptions (independent, identically distributed normal random errors, E_i, a situation that may not be appropriate if the E_i form a time series), these elegant estimates are the best way to estimate $\mathbf{B}$. In Chapter 27 we learn why modifications are necessary when the E_i form a time series with autocovariance structure.

Setting up the design matrix

A few examples will illustrate how the general linear model [equation (14A.7)] and its solution [equation (14A.8)] can be set up from the actual observations and our model (assumptions) about the process that generated them. In this section we consider a special version of the *interrupted time-series experiment*. Suppose that there are N observations, with n_1 observations prior to intervention, and $n_2 = N - n_1$ observations after intervention. Let us begin by focusing on the deterministic component of the time series, which we will first represent with an equation and then in matrix form. Let us consider the simple case when prior to intervention, one trend line, and, after intervention, a different trend line, form the deterministic part. How can this be expressed mathematically?

We can represent all points *of the deterministic component* prior to intervention as

$$z(t) = B_1 + B_2(t-1), \qquad t \le n_1 \tag{14A.9}$$

and all points *after* intervention as

$$z(t) = (B_1 + B_3) + [B_2(t-1) + B_4(t - n_1 - 1)], \qquad t \ge n_1 + 1 \tag{14A.10}$$

B_3 represents the jump in level and B_4 represents the jump in slope as a result of the intervention (see Figure 14A.1).

Using matrix algebra

Continuing this discussion, the deterministic linear component can be represented in matrix form as follows:

$$\left[\begin{array}{c} z(1) \\ z(2) \\ z(3) \\ z(4) \\ \vdots \\ z(n_1) \\ \hline z(n_1+1) \\ z(n_1+2) \\ \vdots \\ z(N) \end{array}\right] = \left[\begin{array}{cccc} 1 & 0 & 0 & 0 \\ 1 & 1 & 0 & 0 \\ 1 & 2 & 0 & 0 \\ 1 & 3 & 0 & 0 \\ \vdots & \vdots & \vdots & \vdots \\ 1 & n_1 - 1 & 0 & 0 \\ \hline 1 & n_1 & 1 & 0 \\ 1 & n_1 + 1 & 1 & 1 \\ \vdots & \vdots & \vdots & \vdots \\ 1 & N-1 & 1 & N - n_1 - 1 \end{array}\right] \begin{bmatrix} B_1 \\ B_2 \\ B_3 \\ B_4 \end{bmatrix}$$

The solid line separates the preintervention from the postintervention points. The reader should multiply the matrices on the right-hand side together to verify that the

Figure 14A.1. Representation of the deterministic component of the effect of an intervention.

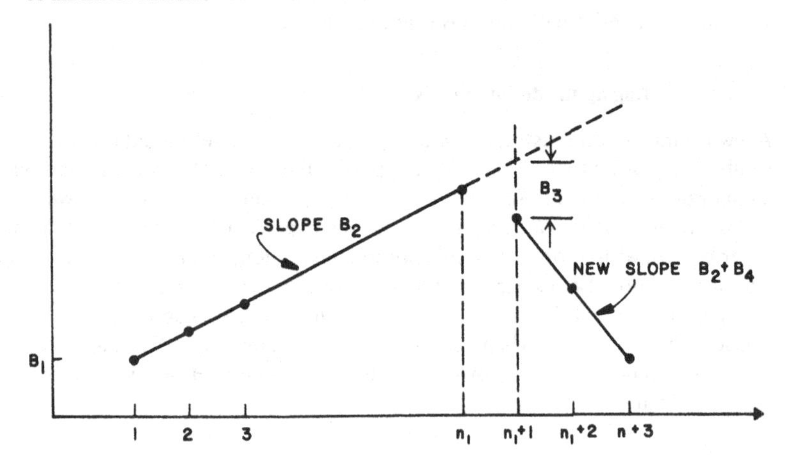

matrix equation does indeed represent equations (14A.9) and (14A.10). The matrices can be written briefly as

$$\mathbf{z} = \mathbf{XB}$$

where $\mathbf{X}$ is the design matrix and $\mathbf{B}$ is called the parameter vector.

An alternative representation

For more experience setting up matrix equations, consider another example, continuing the discussion of representing only the linear deterministic component of the interrupted time-series experiment.

We can rewrite the equations in terms of a separate preintervention level and slope and a separate postintervention level and slope. Prior to intervention ($1 \leq t \leq n_1$)

$$z(t) = B_1 + B_2[t - (n_1 + 1)/2]$$

and subsequent to intervention ($n_1 + 1 \leq t \leq N$),

$$z(t) = B_3 + B_4[t - (N + n_1 + 1)/2]$$

In this representation B_1 is the level in the middle of the preintervention period at $t = (n_1 + 1)/2$, B_2 is the preintervention slope, B_3 is the level in the middle of the post intervention period $[t = (N + n_1 + 1)/2]$, and B_4 is the postintervention slope.

In matrix form this can be written

$$\begin{bmatrix} z(1) \\ z(2) \\ z(3) \\ \vdots \\ z(n_1) \\ \hline z(n_1+1) \\ z(n_1+2) \\ z(n_1+3) \\ \vdots \\ z(N) \end{bmatrix} = \begin{bmatrix} 1 & \frac{1}{2}(1-n_1) & 0 & 0 \\ 1 & \frac{1}{2}(3-n_1) & 0 & 0 \\ 1 & \frac{1}{2}(5-n_1) & 0 & 0 \\ \vdots & \vdots & \vdots & \vdots \\ 1 & \frac{1}{2}(n_1-1) & 0 & 0 \\ \hline 0 & 0 & 1 & \frac{1}{2}(n_1-N+1) \\ 0 & 0 & 1 & \frac{1}{2}(n_1-N+3) \\ 0 & 0 & 1 & \frac{1}{2}(n_1-N+5) \\ \vdots & \vdots & \vdots & \vdots \\ 0 & 0 & 1 & \frac{1}{2}(N-n_1-1) \end{bmatrix} \begin{bmatrix} B_1 \\ B_2 \\ B_3 \\ B_4 \end{bmatrix}$$

This representation has the advantage that the matrix $\mathbf{X'X}$ is diagonal, so that it has a very simple inverse. All representations are, of course, mathematically equivalent. Now let us consider the stochastic component.

In the simplest case, the stochastic component will be pure white noise added to the deterministic component. In this case, the matrix equations become simply

$$\mathbf{z} = \mathbf{XB} + \mathbf{E}$$

where **E** is the white noise, written as a column vector:

$$\mathbf{E} = \begin{bmatrix} e_1 \\ e_2 \\ \vdots \\ e_N \end{bmatrix}$$

In the simplest case, the stochastic component will be distributed independently and normally with zero mean and constant, finite variance σ_e^2, which is independent of historical time. In this case the best estimates of the parameters are the least-squares estimates.

$$\hat{\mathbf{B}} = (\mathbf{X}'\mathbf{X})^{-1}\mathbf{X}'\mathbf{Y}$$

where, as before,

$\mathbf{X}$ = design matrix

$\mathbf{Y}$ = vector of time-series observations

$\mathbf{B}$ = vector of estimated parameters

The residual vector can be estimated from $\hat{\mathbf{Y}} = \mathbf{X}\hat{\mathbf{B}}$ as

$$\hat{\mathbf{e}} = \mathbf{Y} - \hat{\mathbf{Y}}$$

The residual variance is

$$s^2 = \frac{\hat{\mathbf{e}}'\hat{\mathbf{e}}}{N-k} = \frac{\sum_{i=1}^{N} \hat{e}_i^2}{N-k}$$

where N is the number of observations and k the number of estimated parameters.

To test whether any linear combination of the parameters $\mathbf{d}'\mathbf{B}$ differs from zero, we obtain the estimate $\mathbf{d}'\hat{\mathbf{B}}$ and an estimate of its standard error:

$$s(d'\hat{B}) = [s^2\mathbf{d}'(\mathbf{X}'\mathbf{X})^{-1}\mathbf{d}]^{1/2}$$

The ratio $\mathbf{d}'\hat{\mathbf{B}}/s(\mathbf{d}'\hat{\mathbf{B}})$ is distributed as a t statistic with $N - k$ degrees of freedom.

Departure from the classic case

When the stochastic component is not a white-noise process but the stochastic component can be represented as an AR process, the $\hat{\mathbf{B}}$ will remain unbiased, reasonable estimates. However, the estimated variance–covariance matrix of the parameters *will* be biased. The parameter estimates will no longer have the t distribution described above and will no longer be "best" in any sense. This case is discussed in detail in Chapter 26.

How serious are departures? Glass et al. (1975), Padia (1975), and Hibbs (1974) each discussed the consequences of violations of the assumption that residuals are uncorrelated (white noise). The evidence is that even moderate autocorrelation in the residuals can *seriously* bias the test of the significance of regression parameters. Thus, marked increases in precision can be obtained by fitting time-series models to the residual. Indeed, in some cases the validity of statistical conclusions can be questioned if time-series models are not at least considered.

APPENDIX 14B

The general linear model for regression

Review of classical linear least-squares regression

In linear regression we begin with a set of *fixed, known* points, X_i, at which we observe a random variable Y_i. This results in a set of pairs (X_i, Y_i). We translate the points so that X_i are expressed in terms of their deviations from the mean $\bar{X}$, x_i. This gives us a set of points (x_i, Y_i). Note that $\sum x_i = 0$. We then wish to find a least-squares straight line (i.e., find values for $\hat{a}$ and $\hat{b}$), as shown in Figure 14B.1,

$$\hat{Y}_i = \hat{a} + \hat{b}x_i \tag{14B.1}$$

that approximates the Y_i, such that the value

$$s(\hat{a}, \hat{b}) = \sum (Y_i - \hat{Y}_i)^2 \tag{14B.2}$$

is a minimum.

The solution is to select

$$\hat{a} = \bar{Y} = \frac{1}{N} \sum Y_i \tag{14B.3}$$

$$\hat{b} = \frac{\sum Y_i x_i}{\sum x_i^2} \tag{14B.4}$$

The problem can be repeated for an infinite number of samples of pairs of points, and we would then obtain a distribution of $\hat{a}$ and $\hat{b}$ around their population means a and b. The variances of $\hat{a}$ and $\hat{b}$, under the assumption that the Y_i are independent (i.e., have

Figure 14B.1. Classical linear least squares for two variables.

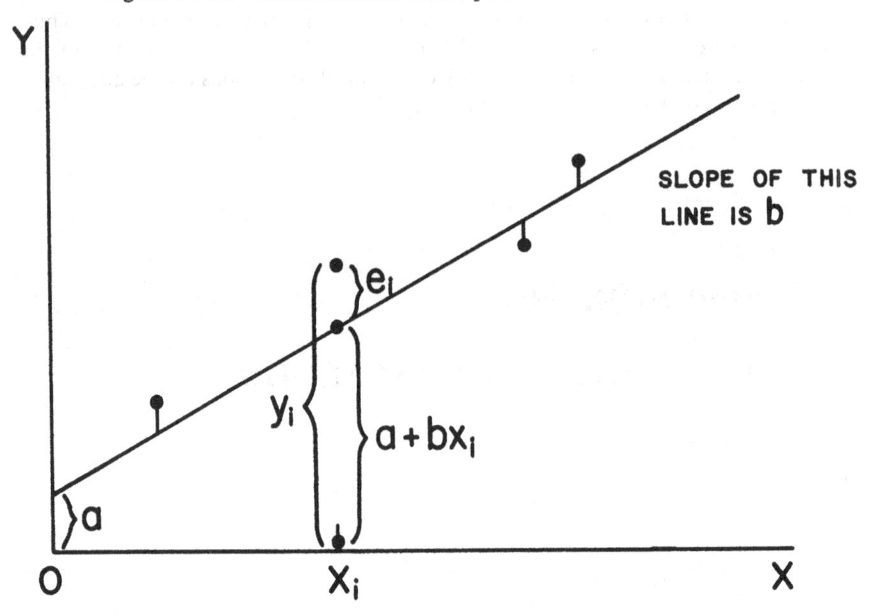

Figure 14B.2. Confidence interval around the regression line.

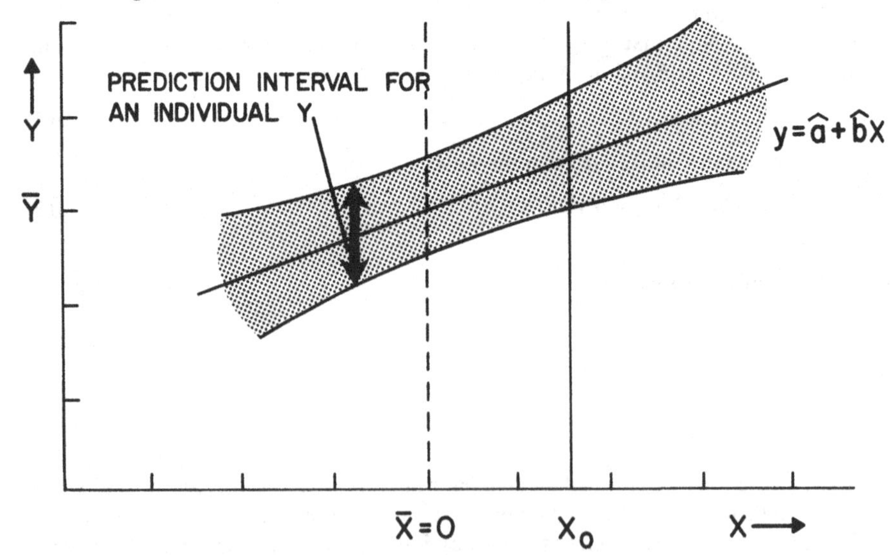

zero covariances), with variance σ^2, are

$$\operatorname{var}(\hat{a}) = \frac{\sigma^2}{N} \tag{14B.5}$$

$$\operatorname{var}(\hat{b}) = \frac{\sigma^2}{\sum x_i^2} \tag{14B.6}$$

and

$$\operatorname{cov}(\hat{a}, \hat{b}) = 0$$

and, furthermore, the least-squares estimators are also unbiased.

The Gauss–Markov theorem states that the least-squares estimators have minimum variance among the class of linear unbiased estimators. Furthermore, if the distribution of the Y_i is normal, the distribution of $\hat{b}$ is also normal. Confidence intervals around $\hat{b}$ can be obtained by estimating σ^2 as

$$s^2 = \frac{1}{N-2} \sum (Y_i - \hat{Y}_i)^2 \tag{14B.7}$$

which is unbiased. The 95% confidence interval for $\hat{b}$ is then

$$b = \hat{b} \pm t_{.025} \left[\frac{s}{(\sum x_i^2)^{1/2}} \right] \tag{14B.8}$$

and hypothesis tests are based on

$$t = \frac{\hat{b} - b}{\sqrt{s^2 / \sum x_i^2}} \tag{14B.9}$$

where t has a $t(N-2)$ distribution. The confidence interval for $\hat{a}$ is

$$a = \hat{a} \pm t_{.025} \left(\frac{s}{\sqrt{N}} \right) \tag{14B.10}$$

Note that this implies that the variance of a prediction $\hat{Y} = \hat{a} + \hat{b}x$ at a point $x = X - \bar{X}$ is

$$\operatorname{var}(\hat{Y}) = \operatorname{var}(\hat{a} + \hat{b}x) = \sigma^2 \left[\frac{1}{N} + \frac{(X - \bar{X})^2}{\sum x_i^2} \right] \tag{14B.11}$$

This means that the confidence interval around a regression line increases as one moves away from the mean $\bar{X}$ (see Figure 14B.2).

Classical linear regression in matrix form

Our linear estimate of Y is

$$\hat{Y} = \hat{a} + \hat{b}x \tag{14B.12}$$

If we denote the error of the estimate by $e = Y - \hat{Y}$, we have

$$Y = a + bx + e \tag{14B.13}$$

For each value $(x_1, x_2, \ldots)$ we have a corresponding set of values $(Y_1, Y_2, \ldots)$ and a corresponding set of errors $(e_1, e_2, \ldots)$. In fact, we can write

$$\begin{aligned} Y_1 &= a + bx_1 + e_1 \\ Y_2 &= a + bx_2 + e_2 \\ &\vdots \\ Y_N &= a + bx_N + e_N \end{aligned} \tag{14B.14}$$

or

$$\begin{bmatrix} Y_1 \\ Y_2 \\ \vdots \\ Y_N \end{bmatrix} = \begin{bmatrix} 1 & x_1 \\ 1 & x_2 \\ \vdots & \vdots \\ 1 & x_N \end{bmatrix} \begin{bmatrix} a \\ b \end{bmatrix} + \begin{bmatrix} e_1 \\ e_2 \\ \vdots \\ e_N \end{bmatrix} \tag{14B.15}$$

or, in matrix form,

$$\mathbf{Y} = \mathbf{XB} + \mathbf{E} \tag{14B.16}$$

The least-squares solution is

$$\boxed{\hat{\mathbf{B}} = (\mathbf{X'X})^{-1}\mathbf{X'Y}} \tag{14B.17}$$

Note that to check this solution, we write

$$\mathbf{X'X} = \begin{bmatrix} 1 & 1 & \cdots & 1 \\ x_1 & x_2 & \cdots & x_N \end{bmatrix} \begin{bmatrix} 1 & x_1 \\ 1 & x_2 \\ \vdots & \vdots \\ 1 & x_N \end{bmatrix} \tag{14B.18}$$

$$\mathbf{X'X} = \begin{bmatrix} N & \sum x_i \\ \sum x_i & \sum x_i^2 \end{bmatrix} = \begin{bmatrix} N & 0 \\ 0 & \sum x_i^2 \end{bmatrix} \tag{14B.19}$$

Multiply this matrix by the matrix $\hat{B}$:

$$(\mathbf{X'X})\hat{\mathbf{B}} = \begin{bmatrix} N & 0 \\ 0 & \sum x_i^2 \end{bmatrix} \begin{bmatrix} \hat{a} \\ \hat{b} \end{bmatrix} = \begin{bmatrix} Na \\ \sum x_i^2 \hat{b} \end{bmatrix} \tag{14B.20}$$

$$\mathbf{X'Y} = \begin{bmatrix} 1 & 1 & \cdots & 1 \\ x_1 & x_2 & \cdots & x_N \end{bmatrix} \begin{bmatrix} Y_1 \\ Y_2 \\ \vdots \\ Y_N \end{bmatrix} = \begin{bmatrix} \sum Y_i \\ \sum x_i Y_i \end{bmatrix} \tag{14B.21}$$

Therefore,

$$(\mathbf{X}'\mathbf{X})\hat{\mathbf{B}} = \mathbf{X}'\mathbf{Y} \tag{14B.22}$$

$$\begin{bmatrix} N\hat{a} \\ \sum x_i^2 \hat{b} \end{bmatrix} = \begin{bmatrix} Y_i \\ \sum x_i Y_i \end{bmatrix} \tag{14B.23}$$

which gives us our old solutions for $\hat{a}$ and $\hat{b}$:

$$\hat{a} = \frac{1}{N}\sum Y_i \tag{14B.24}$$

$$\hat{b} = \frac{\sum x_i Y_i}{\sum x_i^2} \tag{14B.25}$$

APPENDIX 14C

The PACF

The Yule–Walker equations [equations (14.23)] can be written more concisely in matrix form (see Appendix 14A):

$$\begin{bmatrix} 1 & \rho_1 & \rho_2 & \cdots & \rho_{p-1} \\ \rho_1 & 1 & \rho_1 & \cdots & \rho_{p-2} \\ \vdots & \vdots & \vdots & & \vdots \\ \rho_{p-1} & \rho_{p-2} & \rho_{p-3} & \cdots & 1 \end{bmatrix} \begin{bmatrix} a_1 \\ a_2 \\ \vdots \\ a_p \end{bmatrix} = \begin{bmatrix} \rho_1 \\ \rho_2 \\ \vdots \\ \rho_p \end{bmatrix} \tag{14C.1}$$

In matrix form

$$\mathbf{Pa} = \boldsymbol{\rho} \tag{14C.2}$$

where $\mathbf{P}$ is the matrix in equation (14C.1), $\mathbf{a}$ is the column vector it multiplies, and $\boldsymbol{\rho}$ is the vector of autocorrelations. This matrix equation has the solution

$$\mathbf{a} = \mathbf{P}^{-1}\boldsymbol{\rho} \tag{14C.3}$$

Applying the Yule–Walker equations

We can apply the Yule–Walker equations to our step-by-step process. Suppose we suspect that $p = 2$; that is, we suspect that we are dealing with an AR(2) process. The Yule–Walker equations are

$$\begin{aligned}\rho_1 &= a_1 + a_2\rho_1\\ \rho_2 &= a_1\rho_1 + a_2\end{aligned} \tag{14C.4}$$

Or, in matrix form, $\boldsymbol{\rho} = \mathbf{Pa}$:

$$\begin{bmatrix}\rho_1\\ \rho_2\end{bmatrix} = \begin{bmatrix}1 & \rho_1\\ \rho_1 & 1\end{bmatrix}\begin{bmatrix}a_1\\ a_2\end{bmatrix} \tag{14C.5}$$

The second-order partial autocorrelation coefficient is a_2, written a_{22}, which can be found by inverting the matrix $\mathbf{P}$, $\mathbf{a} = \mathbf{P}^{-1}\boldsymbol{\rho}$. Using Cramer's rule, this can be found as the ratio

$$a_{22} = \frac{\begin{vmatrix}1 & \rho_1\\ \rho_1 & \rho_2\end{vmatrix}}{\begin{vmatrix}1 & \rho_1\\ \rho_1 & 1\end{vmatrix}} = \frac{\rho_2 - \rho_1^2}{1 - \rho_1^2} \tag{14C.6}$$

For the next step, that is, the case in which we suspect (i.e., test for) that $p = 3$, the Yule–Walker equations are

$$\begin{aligned}\rho_1 &= a_1 + a_2\rho_1 + a_3\rho_2\\ \rho_2 &= a_1\rho_1 + a_2 + a_3\rho_1\\ \rho_3 &= a_1\rho_2 + a_2\rho_1 + a_3\end{aligned} \tag{14C.7}$$

In matrix form ($\boldsymbol{\rho} = \mathbf{Pa}$)

$$\begin{bmatrix}\rho_1\\ \rho_2\\ \rho_3\end{bmatrix} = \begin{bmatrix}1 & \rho_1 & \rho_2\\ \rho_1 & 1 & \rho_1\\ \rho_2 & \rho_1 & 1\end{bmatrix}\begin{bmatrix}a_1\\ a_2\\ a_3\end{bmatrix} \tag{14C.8}$$

Once again we are interested in the third-order partial autocorrelation coefficient, a_3, which can be shown to be (from $\mathbf{a} = \mathbf{P}^{-1}\boldsymbol{\rho}$)

$$a_{33} = a_3 = \frac{\begin{vmatrix}1 & \rho_1 & \rho_1\\ \rho_1 & 1 & \rho_2\\ \rho_2 & \rho_1 & \rho_3\end{vmatrix}}{\begin{vmatrix}1 & \rho_1 & \rho_2\\ \rho_1 & 1 & \rho_1\\ \rho_2 & \rho_1 & 1\end{vmatrix}} \tag{14C.9}$$

In general, in calculating a_{kk}, the PACF of any order, the determinant in the denominator is the determinant of the matrix $\mathbf{P}$, and the numerator has the same elements as $\mathbf{P}$, but the last column is replaced by $\boldsymbol{\rho}$.

In general, **P** matrix is striped:

$$\begin{bmatrix} 1 & \rho_1 & \rho_2 & \rho_3 & \cdots \\ \rho_1 & 1 & \rho_1 & \rho_2 & \cdots \\ \rho_2 & \rho_1 & 1 & \rho_1 & \cdots \\ \rho_3 & \rho_2 & \rho_1 & 1 & \cdots \\ \cdot & \cdot & \cdot & \cdot & \\ \cdot & \cdot & \cdot & \cdot & \\ \cdot & \cdot & & \cdot & \end{bmatrix} \tag{14C.10}$$

PACF for the MA(1) process

The ACF for the MA(1) process cuts off after lag 1; that is, we found

$$\rho_1 = \frac{b}{1+b^2}$$

$$\rho_k = 0, \qquad k > 1 \tag{14C.11}$$

If we let $\rho_1 = \rho$, we can calculate the PACF as follows. The striped **P** matrix is

$$\begin{bmatrix} 1 & \rho & 0 & 0 & 0 & \cdots & 0 \\ \rho & 1 & \rho & 0 & 0 & \cdots & 0 \\ 0 & \rho & 1 & \rho & 0 & \cdots & 0 \\ 0 & 0 & \rho & 1 & \rho & \cdots & 0 \\ & & & \cdot & \cdot & \cdot & \cdot \\ & & & & \cdot & \cdot & \cdot \\ & \textit{all zeros} & & & & \cdot & \cdot \\ & & & & \rho & 1 & \rho \\ & & & & 0 & \rho & 1 \end{bmatrix}$$

The determinant of this matrix is in the numerator of the PACF, and the denominator is the determinant of this matrix but with the last (right-hand side) column replaced by

$$\boldsymbol{\rho} = \begin{bmatrix} \rho_1 \\ \rho_2 \\ \\ \cdot \\ \cdot \\ \cdot \end{bmatrix} = \begin{bmatrix} \rho \\ 0 \\ \\ \cdot \\ \cdot \\ \cdot \end{bmatrix}$$

The solution can be shown to be

$$a_{kk} = \frac{-b^k(1-b^2)}{1-b^{2(k-1)}} \tag{14C.12}$$

The reader should check this for $k = 2$, the ambitious reader should check for $k = 3$, and the truly devoted should check it in general.

APPENDIX 14D

Stationarity and invertibility conditions for AR and MA processes

To state stationarity and invertibility conditions carefully, we must begin with some algebraic properties of complex numbers that may be unfamiliar. The goal of this appendix is to state, but not prove, these conditions.

Complex numbers and polar coordinates

When the quantity under the square-root sign is negative, it does not represent any real number. However, a new kind of number can be *defined* such that its square is negative. This kind of number is called an *imaginary number*. Imaginary numbers are real numbers times a number whose square is -1. The number whose square is -1 is usually written i:

$$(i)^2 = -1.0$$

Hence, all imaginary numbers are of the form $b(i)$, where b is a real number. A *complex number* S is a sum of a real number, a, and an imaginary number, bi:

$$S = a + bi$$

To represent a complex number we need to define a plane with two axes, a real axis and a complex axis (Figure 14D.1). The length of the line from the origin to the end point represented by S is denoted $|S|$, called the *modulus* of the complex number S; it is the length of the hypotenuse of the right triangle OaS, which by the Pythagorean theorem is

$$|S| = (a^2 + b^2)^{1/2}$$

The square of the modulus of S can also be written

$$S^2 = a^2 + b^2 = (a + ib)(a - ib)$$

because of the fact that $(i)^2 = -1.0$. The number $a - ib$ is called the *complex conjugate* of the number $S = a + ib$, and it will be denoted S^*. The equations can be rewritten

$$|S| = (S \cdot S^*)^{1/2}$$

By elementary trigonometry, recall that the *sine* of the angle θ is the length of the opposite side divided by the length of the hypotenuse, so that in (14D.1),

$$\sin\theta = \frac{b}{|S|}$$

or

$$b = |S| \sin\theta$$

Similarly, by the definition of the *cosine* as the length of the adjacent side divided by the length of the hypotenuse,

$$\cos\theta = \frac{a}{|S|}$$

or

$$a = |S| \cos\theta$$

Therefore, any complex number S can be written

$$S = |S| \cos\theta + i|S| \sin\theta$$

Figure 14D.1. The complex plane.

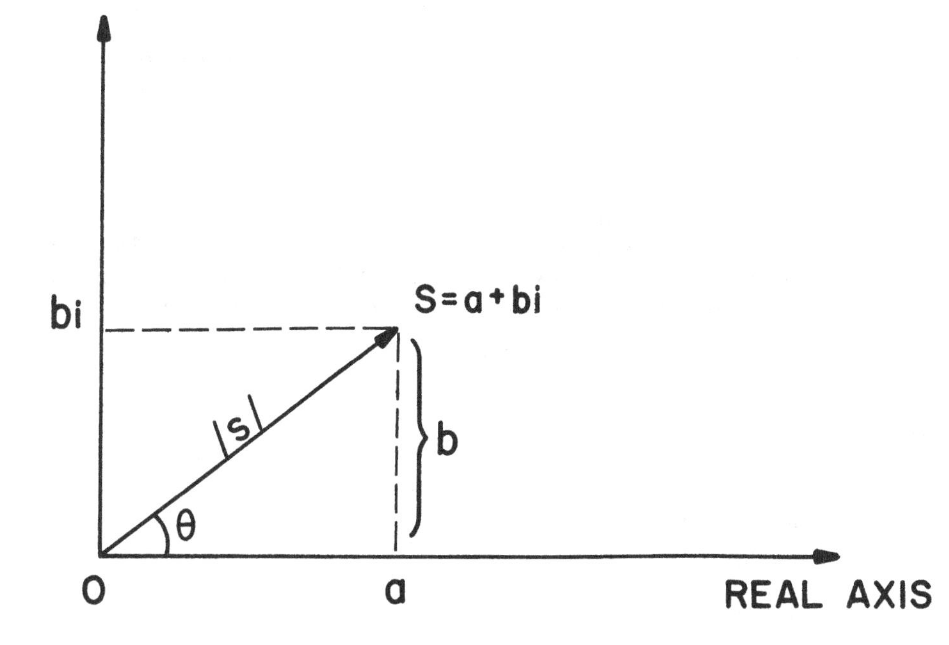

or

$$S = |S|(\cos\theta + i\sin\theta)$$

By a remarkable theorem called *deMoivre's theorem*, it can be shown that

$$\cos\theta + i\sin\theta = e^{i\theta} \tag{14D.1}$$

where e is the base of the natural logarithms and $e^{i\theta}$ satisfies all the usual algebraic properties of a real exponential. Thus, any complex number can be written in exponential form as

$$S = |S|e^{i\theta}$$

If equation (14D.1) is rewritten for negative θ, we obtain

$$\cos\theta - i\sin\theta = e^{-i\theta} \tag{14D.2}$$

because

$$\cos(-\theta) = \cos\theta \qquad \text{and} \qquad \sin(-\theta) = -\sin\theta$$

Using equations (14D.1) and (14D.2), it can be shown that

$$\cos\theta = \frac{e^{i\theta} + e^{-i\theta}}{2} \tag{14D.3}$$

and

$$\sin\theta = \frac{e^{i\theta} - e^{-i\theta}}{2i} \tag{14D.4}$$

These results will make it possible to write expressions involving the cosine and sine in terms of the exponential. This fact will be useful in Chapter 16.

Stationarity, invertibility, and the roots of polynomials

The second-order or quadratic polynomial $ax^2 + bx + c$ can be rewritten as $(x - S_1)\cdot(x - S_2)$, where S_1 and S_2 are solutions of the equation $ax^2 + bx + c = 0$. This is true for all values of a and b as long as we allow S_1 and S_2 to be complex numbers. In general, a polynomial $f(x)$ of order n has n (possibly complex) roots S_i and can be written

$$f(x) = (1 - G_1x)(1 - G_2x)\cdots(1 - G_nx) \tag{14D.5}$$

where

$$G_1 = \frac{1}{S_1}, \quad G_2 = \frac{1}{S_2}, \quad \ldots, \quad G_n = \frac{1}{S_n}$$

Any autoregressive model can be written

$$\mathrm{AR}(B)x_t = e_t$$

where

$$\mathrm{AR}(B) = 1 - a_1 B - a_2 B^2 - \cdots - a_p B^p$$

is a pth-order polynomial in B.

For each of these polynomials in B we can define corresponding "characteristic" polynomials in x, where x replaces B. The characteristic polynomials have a set of complex roots S_i, each with moduli $|S_i|$. It turns out to be a fact that AR(B) will be stationary if all the roots of its characteristic polynomial have moduli $|S_i|$ greater than 1.0, and also true that MA(B) will be invertible if all the roots of its characteristic polynomial have moduli greater than 1.0.

How nondeterministic periodicity arises in AR(2) models

We now have the notational machinery necessary for taking a close look at the stationarity region for the AR(2) model. In Chapter 12, regions 3 and 4 under the circle $a_1^2 + 4a_2 = 0$ were implicated as regions in which AR(2) models displayed nondeterministic periodicity, with one cycle. This section explains how this fact can be proved.

Using the autocorrelation generating function (for AR models also known as the Yule–Walker equations), we can write

$$\mathrm{AR}(B)\rho_k = 0$$

where

$$\mathrm{AR}(B) = 1 - a_1 B - a_2 B^2 - \cdots - a_p B^p$$

This, however, is a polynomial, with roots $S_1, S_2, \ldots, S_p$, so it can be factored as a product [see equation (14D.5)] as follows:

$$\mathrm{AR}(B) = (1 - G_1 B)(1 - G_2 B) \cdots (1 - G_p B)$$

when, once again, $G_i = 1/S_i$.

The general solution of the Yule–Walker equations is

$$\rho_k = A_1 G_1^k + A_2 G_2^k + \cdots + A_p G_p^k \qquad (14\mathrm{D}.6)$$

For stationarity, the roots S_i must satisfy $|S_i| > 1$ or lie outside the unit circle in the complex plane, which implies that $|G_i| < 1$.

There are two possibilities for the shape of the correlogram (ρ_k plotted against the lag, k) as given by equation (14D.6). The first is for a real root, S_i. In this case G_i is real and less than 1, and hence the term $A_i G_i^k$ geometrically decays to zero as k increases, much as is the case for an AR(1) model. This is the case in regions 1 and 2 for the AR(2) model discussed in Chapter 12.

The second possibility occurs if the root S_i is complex. Then G_i is also complex and less than 1 in modulus, and furthermore, S_i^* is also a root. In this case the pair G_i, G_i^* contributes a term of the form $Ad^k \sin(2\pi fk + F)$ to the autocorrelation function, which is a damped sine wave (see also Box and Jenkins, 1970, chap. 3.2).

For the AR(2) case, the roots will be complex-conjugate pairs when $a_1^2 + 4a_2 < 0$, which defines regions 3 and 4. The peak frequency of the oscillation can be shown to be

$$d = (-a_2)^{1/2}$$

and

$$\cos 2\pi f_0 = \frac{|a_1|}{2(-a_2)^{1/2}}$$

In Chapter 18 we return to this last equation.

Part IV

Stationary frequency-domain models

15
The spectral density function

This chapter, which deserves close reading, is important for conveying the conceptual basis of the mathematics and is central to understanding the frequency domain. The history of the spectral density function is discussed, as well as an old method for computing it by hand. The twentieth-century idea of a probabilistic basis for natural phenomena is also examined.

The history of frequency-domain approximation

In the eighteenth century Daniel Bernoulli suggested that a wide class of functions can be approximated using sums of trigonometric functions (Hawkins, 1970). The idea was initially rejected by the most famous mathematicians of Bernoulli's time (including Euler, d'Alembert, and Lagrange), so the concepts of trigonometric approximation are probably not initially intuitively obvious. It remained for Jean Baptiste Joseph Fourier, in his *The Analytical Theory of Heat* (1822), to examine the ability of sums of sines and cosines to approximate functions. The mathematical problems generated by what turned out to be an incorrect proof by Fourier took over 100 years to be straightened out; the process involved the greatest minds of the nineteenth and twentieth centuries. The process was also responsible for the creation of much of modern mathematics.

The success of Fourier approximation is based on simulating data using sums of sines and cosines, and estimating component frequencies, their phases, and their amplitudes. Figure 15.1 is a plot of part of a famous set of data gathered in 1924 by Whittaker and Robinson in their observations of the brightness of a variable star. They found that the data could be modeled very well using two sine waves summed together.

In determining this model, Whittaker and Robinson used an estimate of the spectral density function called the *periodogram*, a function that comes directly from Fourier analysis. This chapter will teach the reader an old method for computing the periodogram "by hand" (i.e., without a computer). The method was developed by Whittaker and Robinson. The periodogram estimates the spectral density function. For the Whittaker and Robinson data, the spectral density function would theoretically (with infinite N) have two peaks, because the data can be approximated well by the sum of two sine waves (see Figure 2.10). In Figure 2.10 the dashed line was constructed by the sum of two sine waves, each of which has a precise frequency and amplitude. The resulting spectrum would be called a *line spectrum*, because the component frequencies are precise.

Whittaker and Robinson were extending the work of Fourier. However, their work was still in the tradition of discovering deterministic patterns that were masked by noise. Following their work, a major conceptual revolution took place in which the work of Fourier was extended to include phenomena that were inherently statistical in nature. For such phenomena we see in this chapter that the periodogram (still a very useful function) is somewhat of a failure by itself. This chapter discusses this failure in detail, even though the problem is not without a solution. The reason that this chapter discusses the failure of the periodogram in detail is because the reader should be aware of the history of spectral time-series analysis.

Why is this important? The reason is that understanding the modern viewpoint is basic to our century's view of physical phenomena. It involves a shift from determinism to probabilism. By following this shift historically, we can see why it is needed and discover some of the basic pitfalls that may befall the reader just as they befell the original investigators. The two views are well expressed in Norbert Weiner's (1966) autobiography, in which he discussed the Charles River:

> The moods of the waters of the river were always delightful to watch. To me, as a mathematician and a physicist, they had another meaning as well. How could one bring to a mathematical regularity

Figure 15.1. Six hundred successive observations of the brightness of a variable star.

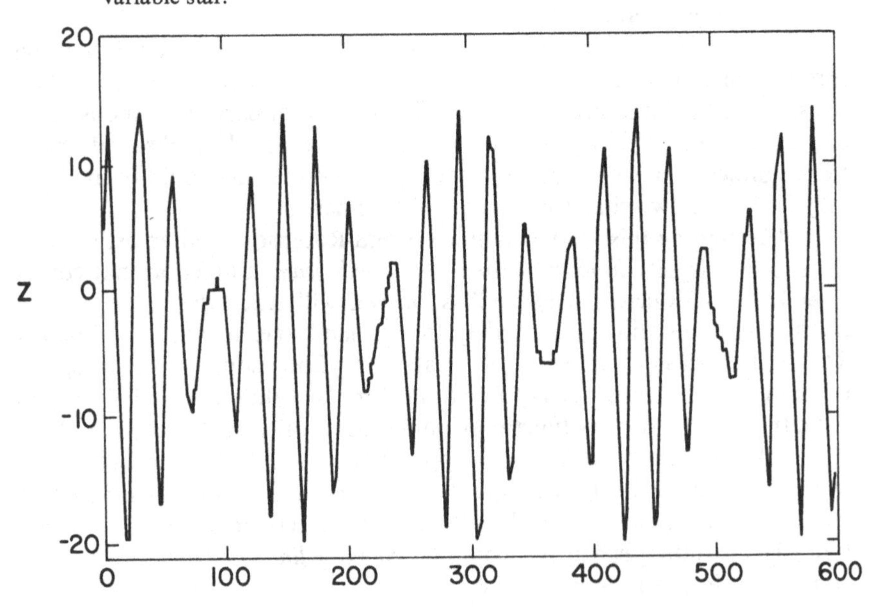

> the study of the mass of ever shifting ripples and waves, for was not the highest testing of mathematics the discovery of order among disorder? At one time the waves ran high, flecked with patches of foam, while at another they were barely noticeable ripples. Sometimes the lengths of the waves were to be measured in inches, and again they might be many yards long. What descriptive language could I use that would portray these clearly visible facts without involving me in the inextricable complexity of a complete description of the water surface? [P. 33]

Wiener wrote that the determinism of Newton would suggest that complete knowledge of the component forces would permit nearly perfect prediction of the motion of the waves. But because complete knowledge is unavailable, the physicist must work with probabilities. Wiener wrote:

> The problem of the waves was clearly one for averaging and statistics, and in this way closely related to the Lebesgue integral, which I was studying at the time. [P. 33]

These techniques made it possible for Wiener to generalize the concept of the periodogram for random processes. As Jenkins and Watts (1968) wrote:

> The basic reason why Fourier analysis breaks down when applied to time series is that it is based on the assumption of *fixed* amplitudes, frequencies and phases. Time series, on the other hand, are characterized by *random* changes of frequencies, amplitudes and phases. [P. 213]

In an indeterminate universe we seek knowledge *on the average*. We seek the estimate of the process of which an observed time series is but one realization, and in which order, by virtue of the Slutzky effect, may arise from the combination of an infinite collection of independent random events. Much like the Brownian motion of a gas molecule, prediction is possible *on the average*. Thus, the extension of the periodogram to the spectrum involves a shift away from deterministic thinking about time series to a statistical probabilism characteristic of modern physics.

The history of the search for hidden periodicities

This section presents a historical review of the attempt to detect hidden cycles in data. The goal of this section is to give the reader a sense of the conceptual revolution implicit in a probabilistic view of time series rather than a deterministic view.

Conceptual revolutions often seem trivial in retrospect. Consider, for example, the statements by Sir Ronald A. Fisher (1970) about the concept of statistics as the study of variation. He wrote:

> For until comparatively recent times, the vast majority of workers in this field appear to have had no other aim than to ascertain aggregate, or average, values. The variation itself was not an object of study, but was recognized rather as a troublesome circumstance which detracted from the value of the average. [P. 3]

The discussion of this section begins with a review of the practical, computational suggestions of Whittaker and Robinson (1924) for detecting hidden periodicities in a time series. Their work has great heuristic value even today. Their recommendations were based firmly on the earlier work of Sir Arthur Schuster (1898).

Testing for an assumed period

Suppose that we have a time series and we assume that it has a period of oscillation of p time units. This means that, within the limits of independent errors, the series more or less repeats itself every p time units. Whittaker and Robinson suggested writing out the values of the time series as shown in Table 15.1. This table was invented by Buys-Ballot in 1847.[1] Suppose that the standard deviation of the means M is σ_M and suppose that the standard deviation of the time series is σ_U. Then Whittaker and Robinson showed that the correlation ratio η, which is the ratio of σ_M/σ_U, is similar to Schuster's periodogram. They wrote: "Schuster's periodogram differs from that introduced above but the similarity of form and purpose is so great that is has seemed best to retain the name" (p. 346). They explained why the ratio would work to detect periodicities as follows:

> It is easy to see why the ratio of the standard deviation of the M's to the standard deviation of the U's is a suitable indicator of peri-

Table 15.1. *Buys-Ballot table for testing for an assumed period*

	1	2	3	$\cdots$	M	Variance
	x_1	x_2	x_3	$\cdots$	x_M	
	x_{M+1}	x_{M+2}	x_{M+3}	$\cdots$	x_{2M}	
	x_{2M+1}	x_{2M+2}	x_{2M+3}	$\cdots$	x_{3M}	
	$\vdots$	$\vdots$	$\vdots$		$\vdots$	
	x_{kM+1}	x_{kM+2}	x_{kM+3}	$\cdots$	$x_{(k+1)M}$	Variance of series $= \sigma_U^2$
Means	$\bar{x}_1$	$\bar{x}_2$	$\bar{x}_3$	$\cdots$	$\bar{x}_M$	Variance of means $= \sigma_M^2$

Note: Time-series data are $x_1, x_2, x_3, \ldots, x_N$ [N is here assumed, for convenience, equal to $(k+1)M$].

odicity. For in the course of one horizontal row of the above scheme, the part of the phenomenon which is of period p will pass through all the phases of one complete period, so that this periodic part is in the same phase at all terms which are above or below each other in the same vertical column.... Any accidental disturbance on the other hand, or any periodic disturbance of period different from p, will be enfeebled by the process of forming means, since positive and negative deviations will tend to annul each other; and therefore, when a periodicity of period p exists, the standard deviation of the M's has a value much larger than when a periodicity of this period does not exist in the phenomenon. [P. 346]

To illustrate why the Buys-Ballot table would work, consider a time series that is actually a sine wave, $\sin(2\pi/5)t$. In this case the frequency $f = 1/5 = .20$, and the period, $T = 1/f = 5$. The Buys-Ballot table for $p = 4$ and $p = 5$ would be:

	1	2	3	4	5
	.00	.95	.59	−.59	−.95
	.00	.95	.59	−.59	−.95
	.00	.95	.59	−.59	−.95
	.00	.95	.59	−.59	−.95
Means	.00	.95	.59	−.59	−.95

1	2	3	4
.00	.95	.59	−.59
−.95	.00	.95	.59
−.59	−.95	.00	.95
.59	−.59	−.95	.00
.95	.59	−.59	−.95
.00	.00	.00	.00

In this case, assuming that there were enough data, the ratio σ_M/σ_U would be 1.0 only for the true period, and zero elsewhere. The beauty of the method is that it works as well for nonsinusoidal repetitive patterns. For a proof that the proposed ratio is the periodogram, see Whittaker and Robinson (1924). See also note 2 for remarks on significance testing of the Whittaker periodogram.

Example of Whittaker and Robinson's method

Figure 15.1 is a graph of Whittaker and Robinson's (1924) observations of the brightness of a variable star at midnight on 600 successive days. The mean of these observations is 17 and the standard deviation is 8.63. There were 21 maxima and 21 minima in the 600 days, so one of the most important periods was suspected as being close to 600/21 = 28.6 days. Trial periods ranging from 20 to $32\frac{1}{2}$ days were therefore used for p; for example, with 17 means and $p = 24$, they obtained Table 15.2. For each of the p values selected,

Table 15.2. *Sample computation of eta*

25	28	31	32	33	33	32	31	28	25	22	18	14	10	7	4	2	0	0	0	2	4	8	11
15	19	23	26	29	32	33	34	33	32	30	27	24	20	17	13	10	7	5	3	3	3	4	5
7	10	13	16	19	22	24	26	27	28	29	28	27	25	24	21	19	17	15	13	12	11	11	10
10	11	12	12	13	14	15	16	17	18	19	19	19	19	20	20	20	20	20	20	20	20	21	20
20	20	20	19	18	17	16	15	13	12	11	10	9	9	10	10	11	12	14	16	19	21	24	25
27	28	29	29	28	27	25	23	20	17	14	11	8	5	4	2	2	2	4	6	9	12	16	19
23	27	30	32	33	34	33	32	30	27	24	20	16	12	9	5	3	1	0	0	1	3	6	9
13	17	21	24	27	30	32	33	33	32	31	28	25	22	19	15	12	9	7	5	4	4	5	5
7	9	12	14	17	20	22	24	25	26	27	27	26	25	24	22	20	18	17	15	14	13	13	12
12	12	13	13	13	14	14	15	15	16	17	17	17	17	18	18	19	19	20	20	21	21	22	22
22	22	22	21	20	19	17	16	14	12	11	9	8	7	8	8	9	10	12	14	17	20	23	25
27	29	30	30	30	29	27	25	22	19	16	12	9	6	4	2	1	1	2	4	7	10	14	17
21	25	29	31	33	34	34	33	31	29	26	22	19	14	11	7	4	2	1	0	1	2	5	7
11	15	19	22	25	28	30	32	32	32	31	29	26	23	21	17	14	11	9	7	6	5	6	6
7	9	11	13	15	18	20	22	23	24	25	25	25	24	24	22	21	19	18	17	16	15	15	14
14	14	14	14	14	14	14	14	14	14	15	15	15	15	16	16	17	18	19	20	21	22	23	23
24	24	24	23	22	21	19	17	15	13	11	9	7	6	6	6	7	8	10	12	15	18	22	24
285	319	353	371	389	406	407	408	392	376	359	326	294	259	242	208	191	174	173	172	188	204	238	254

Source: Whittaker and Robinson (1924).

Table 15.3. *Computations for the periodogram (mean values in columns)*

Period, p (days):

20	20½	21	21½	22	22½	23	23½	24	24½	25	25½	26	26½	27	27½	28	28½	29	29½	30	30½	31	31½	32	32½
292	274	285	282	285	283	251	211	285	359	333	295	294	290	301	281	227	282	437	442	337	291	315	324	300	293
291	278	290	285	285	287	252	227	319	373	330	297	297	295	310	283	240	318	454	431	320	290	315	323	296	295
297	281	295	286	293	288	254	247	353	381	326	298	304	300	315	284	261	350	472	419	307	284	315	318	287	296
292	284	297	288	295	291	258	272	371	384	317	299	304	305	318	285	272	379	471	396	284	282	314	312	287	295
291	288	301	291	305	296	264	297	389	379	308	301	307	310	321	287	289	405	471	373	270	277	311	307	282	296
293	293	305	292	309	296	273	321	406	371	299	300	306	312	322	287	306	424	452	344	249	274	308	299	277	295
297	297	305	295	308	299	284	344	407	354	291	300	307	314	319	288	328	437	432	313	234	273	304	291	279	296
293	299	307	296	311	302	296	362	408	337	281	301	303	315	319	290	338	445	412	280	222	271	300	284	270	294
292	303	308	299	315	302	307	379	392	313	272	297	303	314	315	290	351	443	375	248	216	271	294	275	267	296
293	306	304	301	314	301	317	386	376	291	266	294	296	316	308	289	361	434	339	220	206	270	289	268	266	294
298	306	306	305	315	302	326	390	359	267	263	294	297	312	305	292	374	420	303	193	208	273	285	253	271	292
292	306	304	306	312	301	333	385	326	245	260	291	290	310	298	292	371	395	265	172	205	272	279	257	266	290
292	305	298	305	311	298	338	375	294	228	259	287	290	307	289	293	371	371	229	153	210	278	275	253	270	289
290	304	297	303	305	298	340	358	259	214	262	286	285	303	284	294	367	342	194	144	219	279	272	252	271	289
296	300	293	300	303	296	338	337	242	209	267	284	288	300	278	296	365	310	161	139	236	286	269	252	281	287
287	297	287	298	292	292	332	314	208	205	273	281	282	295	270	298	349	278	143	146	247	289	267	253	280	287
287	295	287	295	287	291	324	287	191	212	279	280	286	291	267	298	336	246	127	155	268	295	266	258	285	286
284	287	282	291	276	288	313	261	174	220	286	278	283	286	263	301	320	215	112	173	282	299	267	263	291	285
289	285	279	288	274	285	302	237	173	237	295	279	288	283	259	303	308	189	115	195	301	304	268	269	302	285
282	281	279	283	266	285	289	215	172	254	302	280	287	279	260	302	284	168	118	222	319	309	271	278	301	284
		277	279	267	284	276	197	188	276	307	282	292	275	261	302	267	151	137	253	341	312	275	286	307	285
				262	282	263	189	204	298	310	283	291	272	261	303	249	141	157	285	352	316	280	295	311	286
						252	186	238	318	314	287	296	270	268	302	240	140	194	320	369	315	284	306	319	287
								254	337	314	287	295	268	272	299	222	144	231	349	373	318	289	312	316	288
										312	290	297	267	277	299	212	156	267	379	378	313	294	320	216	289
												296	266	285	297	206	175	301	400	379	313	299	327	315	290
														293	291	209	198	336	421	380	309	303	330	319	291
																205	225	369	431	363	304	308	332	309	292
																		403	442	361	297	310	334	304	294
																				343	292	313	331	299	294
																						314	327	298	296
																								286	296

these sums were calculated. For periods of $\frac{1}{2}$ day, entries are omitted occasionally. For example, for the period of $31\frac{1}{2}$ days the first row was days 1 to 31, the second row was 32 to 62, and 63 was left blank to bring the corresponding 64th day to the beginning of the third row. The table obtained is Table 15.3. This table produced a graph of η versus corresponding values of p shown in Figure 15.2. The figure shows two peaks, at $p = 24$ and $p = 29$, "with the side-peaks belonging to these as in a diffraction pattern in optics" (Whittaker and Robinson, 1924, p. 358).

Finer estimates could be obtained by taking p values near 24 and 29 and using more means (i.e., using a larger sample). This problem becomes serious if the two periods are very close together.

Phase estimates of the constituent oscillations can be determined by examining the residuals, or by least-squares estimates of the constants of

$$U_t = \alpha + \beta \cos\left(\frac{2\pi t}{29}\right) + \gamma \sin\left(\frac{2\pi t}{29}\right) + \delta \cos\left(\frac{2\pi t}{24}\right) + \varepsilon \sin\left(\frac{2\pi t}{24}\right)$$

which resulted in the fit

$$U_t = 17 + 10 \sin\left[\frac{2\pi(t+3)}{29}\right] + 7 \sin\left[\frac{2\pi(t-1)}{24}\right]$$

Figure 15.2. Whittaker and Robinson's computations of the periodogram of the variable star data; the abcissa represents cycles of different periods; the ordinate represents the importance of cycles of each period.

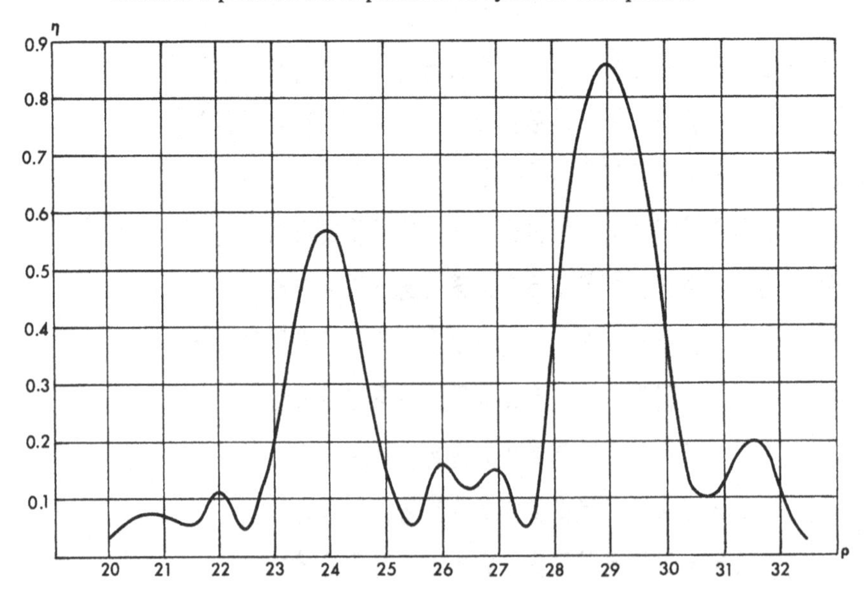

Whittaker and Robinson's example is valuable for providing an intuitive grasp of what the periodogram accomplishes in this example of line spectra.

A probabilistic foundation

In 1927, G. Udny Yule worked on sunspot numbers. He noted that the amplitudes and periods of his data *were irregular* and he constructed a model for the irregularity, which was well summarized by Kendall (1973) as follows:

> The illustration which he used to explain his fresh approach is classical: if we have a rigid pendulum swinging under gravity through a small arch, its motion is well known to be harmonic, that is to say, it can be represented by a sine or cosine wave, and the amplitudes are constant, as are the periods of a swing. But if a small boy now pelts the pendulum irregularly with peas the motion is disturbed. The pendulum will still swing, but with irregular amplitudes and irregular intervals. The peas, in fact, instead of leading to behavior in which any difference between theory and observation is attributable to an evanescent error, provide a series of shocks which *are incorporated into the future motion of the system.* [P. 4]

Yule presented this idea in a series of classic papers. In 1921, Yule wrote (Yule, 1971b):

> It is unfortunate that the word "periodic" implies rather too much as to the character of such more rapid movements; few of us, I suppose, now believe that they are strictly periodic in the proper sense of the term, and hence the occurrence in writings on the subject of such terms as "quasi-periodic" and "pseudo-periodic." They are wave-like movements, movements which can be readily represented with a fair degree of accuracy over a moderate number of years by a series of harmonic terms but which cannot be represented in the same way, for example, by a polynomial; movements in which the length of time from crest to crest of successive waves is not constant, and in which, it may be added, the amplitude is not constant either, but would probably, if we could continue our observations over a sufficient number of waves, exhibit a frequency distribution with a fairly definite mode. [P. 299]

Yule (1971d) analyzed Sir William Beveridge's (1921) data on wheat prices in western Europe from 1545 to 1844 and found that the autocorrelations of the data (and differences of the data) died out rather than remaining strongly sinusoidal. In 1927, Yule's paper on the Wolfer sunspot numbers (Yule,

1971a) proposed a contrast between the model designed for the periodogram analysis and a stochastic model. He wrote:

> When periodogram analysis is applied to data respecting any physical phenomenon in the expectation of eliciting one or more true periodicities, there is usually, as it seems to me, a tendency to start from the initial hypothesis that the periodicity or periodicities are masked solely by such more or less random *superposed fluctuations*—fluctuations which do not in any way disturb the steady course of the underlying periodic function or functions. [P. 390]

This model is a sine wave plus independent noise, and the presence of the pure harmonic may be masked to varying degrees, depending on the relative size of the noise. The *stochastic* model is the swinging pendulum whose motion is recorded by automatic apparatus, but

> unfortunately boys get into the room and start pelting the pendulum with peas, sometimes from one side and sometimes from the other. The motion is now affected, not by *superposed fluctuations* but by true disturbances, and the effect on the graph will be of an entirely different kind. The graph will remain suprisingly smooth, but amplitude and phase will vary continuously. [P. 390]

Herman Wold's 1938 thesis criticized the Schuster periodogram as a method for determining hidden periodicities on the grounds that a fixed sinusoidal pattern requires a correlogram that does not damp out with increasing lag. Yet Wold noted that in many cases the correlogram did just that. He cited air-pressure data from Port Darwin analyzed by Sir G. Walker (1931).[3] This condition that observations become more independent (and hence less predictable) as the lag between them increases has been called the "mixing" condition. Wold wrote:

> It follows from the above that in a descriptive analysis of a time series the serial coefficients of G. U. Yule are of functional importance.... In economics, the classical periodogram analysis has repeatedly been tried on business cycle material. The negative results support an opinion which has been maintained also on logical-theoretical grounds, and which now seems predominant, *viz.* that the hypothesis of hidden periodicities is inadequate in business cycle theory. [P. 27]

The search began for models that described data that were "almost periodic," or nondeterministically periodic. The work of Yule was extended simultaneously by Slutzky (1937), who examined the periodogram under a mixing and a normal assumption. During the same period, Wiener (1930) published a classic paper extending harmonic analysis to time series that included the models of Yule. Wiener's earlier work (see Wiener, 1948) on the Brownian motion of an individual molecule subjected to random shocks was an attempt to extend Einstein's (1906) paper on Brownian motion viewed statistically. The problem was similar to Yule's; it was also inspired by Wiener's interest in quantum mechanics (Wiener, 1930, 1933, 1949).

Thus, we see the historical progression in modeling repeating phenomena. First came purely deterministic (Newtonian) models. In the nineteenth and early twentieth centuries came models with deterministic, "true" cyclicities masked by the addition of random noise. Two decades into this century it became clear that the rhythmicity had to be *built into* the random error; it had to be internal to the system rather than something external; both deterministic and random components posited by the Wold decomposition theorem must play equal roles in describing observed phenomena.

How Yule invented autoregressive models

Yule's classic 1927 paper on the Wolfer sunspot numbers (Yule, 1971a) began by examining the kinds of data that were assumed by Schuster's periodogram – sinusoids plus noise. Figure 15.3 is a reproduction of his first figure; the top graph contains little noise and the bottom graph more noise. However, Yule's model of the pendulum pelted by peas from boys

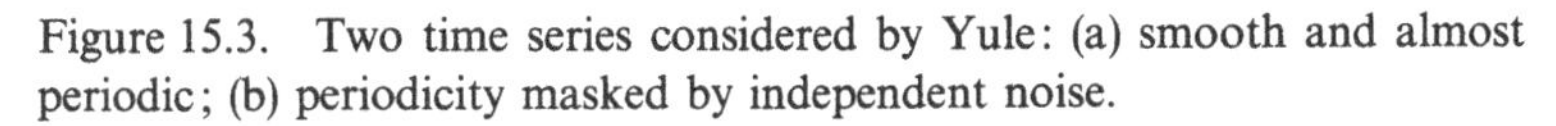
Figure 15.3. Two time series considered by Yule: (a) smooth and almost periodic; (b) periodicity masked by independent noise.

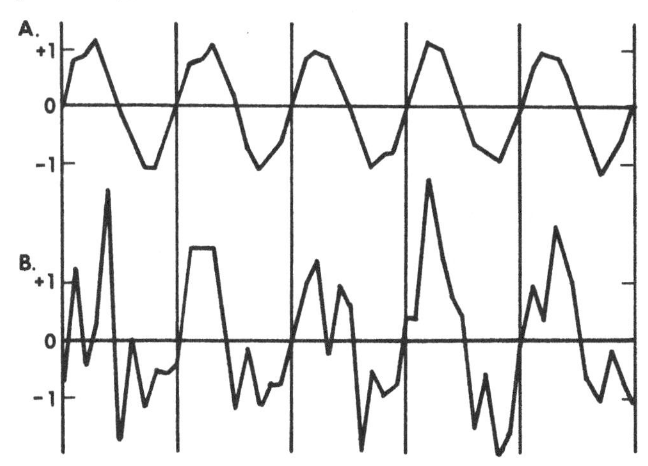

who "get into the room" gives a fundamentally different type of data because each random shock is absorbed into the system smoothly. The example Yule began with was a second-order process, a logical choice because harmonics arise from second-order differential equations. If a second-order autoregres-

Figure 15.4. Successive points of a second-order autoregressive process, which Yule used to represent the Wolfer sunspot numbers.

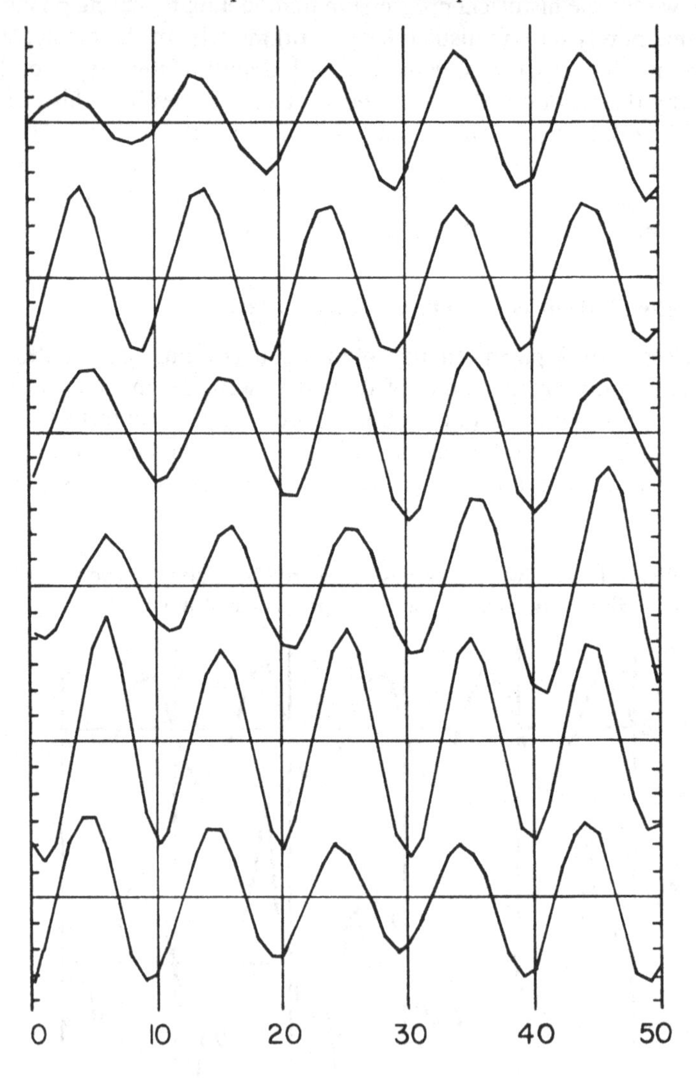

sive process is shocked at one time, the disturbance is felt in exponentially decreasing amounts in the future. The appearance of the data is thus smooth, as shown by Yule's second figure (see Figure 15.4), which is a plot of a realization of an AR(2) model.

Yule (1927) then examined the Wolfer sunspot data (see Figure 15.5), and wrote:

> Inspection of a graph of Wolfer's annual sunspot numbers . . . suggests quite definitely to my eye that we have to deal with a graph of the type of Fig. 2, not of the type of Fig. 1, at least as regards its principal features. [P. 395]

Yule began with fitting a second-order autoregressive process:

> Starting also, as I did, with the conception of periodogram analysis and harmonic periodicities in my mind, it was natural to assume an equation of the form (6) [an AR(2)], and an equation of corresponding form for two periodicities. This gave the first method tried.
>
> It only occurred to me later that the method started from an unnecessarily limited assumption; that it would be better simply to

Figure 15.5. Wolfer sunspot data.

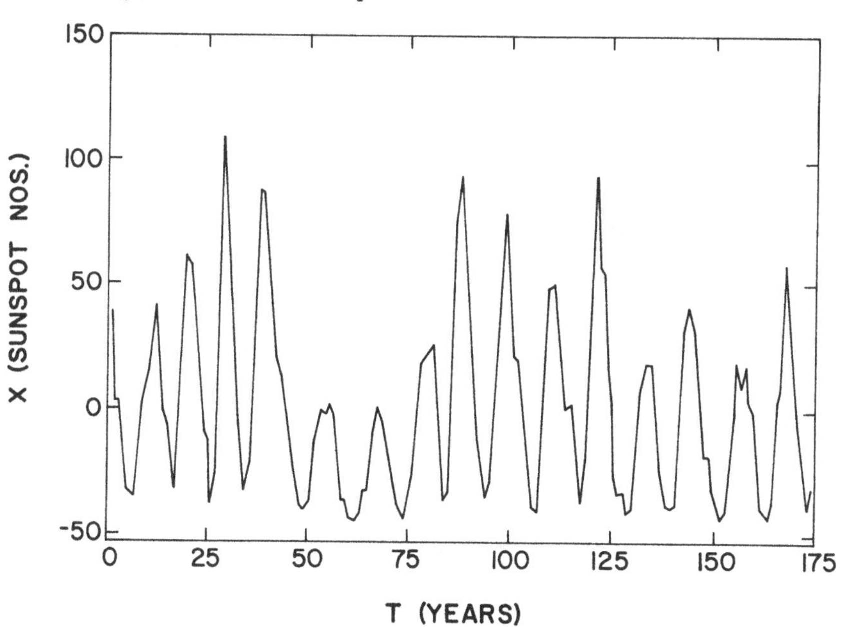

> find the linear regression equation of u_x on u_{x-1}, u_{x-2}, and more terms if necessary, and solve this as a finite difference equation. [P. 395]

This piece of history dramatically indicates the close linkage between the origins of time-domain time-series analysis, in the form of autoregressive models, from frequency-domain methods. (For a detailed discussion of various attempts to fit the Wolfer sunspot numbers, see Anderson, 1971, pp. 243–346.) Furthermore, it dramatizes the fact that time-domain models such as autoregressive models will be better models for nondeterministic cycles than simple deterministic models derived from Fourier approximation.

In short, the major modification of Fourier methods involves changing assumptions about the underlying process. It entails the conclusion that most processes have a nondeterministic component. They are thus not usually determinisms masked by measurement error or other noise. This was a major philosophical shift, and it has characterized most scientific thought in the twentieth century.

Introduction to spectral decomposition

We discovered that the AR(2) process was capable of representing series with one nondeterministic small band of cycles. This result is a logical beginning for us to introduce the mathematics of the frequency domain.

The most important result in the frequency domain is called the *spectral decomposition theorem*. This theorem states that the energy, or variance, of any time series can be broken down into the contribution of *statistically independent* oscillations of different frequencies. The graph of variance accounted for by all frequencies less than a given one is called the *spectral distribution function*. A time series with just one major frequency, such as a jittery sine wave or the right AR(2) process (in regions 3 or 4), will have one major peak in the *spectral density function*, which is the slope of the spectral distribution function. The relationship between the spectral density function and the spectral distribution function is analogous to the relationship between probability densities and cumulative distribution functions. The relationship is the same, although the spectral density function measures power on frequency bands, and the probability density function measures probability on intervals.

One of the major jobs in spectral time-series analysis is to estimate the spectral density function of the *process* from our sample of T observations. The major goal of this chapter is to discuss the central result in this estimation process, called the *Wiener–Khintchine theorem*.

The Wiener–Khintchine theorem tells us how to estimate the spectral density function from a sample of T observations using our usual estimates of the autocovariance functions. In fact, we will see that the Wiener–Khintchine estimate of the spectral density function is the *periodogram*, and that it can be derived by a remarkably simple and elegant formula, which is derived and explained in Chapter 16.

It is important for the reader to bear in mind the crucial fact that in estimating the spectral density function from the autocovariances, we are going from the time domain to the frequency domain.

Our estimate of the spectral density function, the *periodogram*, denoted $I(f_i)$, *is a function of the frequencies*, f_i, into which we try to decompose the energy of the series. The autocovariance function, c_k, is a function of the lag, k, computed in the time domain. The transformation is

$$\text{time domain} \qquad c_k \Longleftrightarrow I(f_i) \qquad \text{frequency domain} \tag{15.1}$$

The arrow in equation (15.1) goes two ways because the mathematics for the transformation is quite symmetric. In fact, the equations are extremely elegant. The transformation that accomplishes this marvelous estimate of the spectral density function is known as the *Fourier transform*.

The reader should be warned that an understanding of these results requires slogging through some trigonometry and algebra that may be painful. However, the ultimate results are worth the effort. No one can be a competent thinker with time-series analysis unless he or she can think in both the time and the frequency domains. The Wiener–Khintchine theorem is the shuttle between these domains.

Approximation and simulation

Recall that an advantage of trigonometric approximation is its ability to represent periodic phenomena. We could do very well approximating with polynomials, but beyond the range of the fit these polynomials would go to plus and minus infinity rather than repeat the same pattern.

Using the periodogram, I attempted to approximate data that describe the marital interaction of one couple. These data were obtained from scoring a videotape of the couple's interaction while they were discussing a sexual problem. The precise method of scoring is described in Gottman et al. (1977). Briefly, increases in the graph indicate positive interaction and decreases indicate negative interaction. The graph at time t is a cumulative total of the amount of positivity minus the amount of negativity in the interaction up to that time. Figure 15.6 shows how well four terms of the Fourier approximation are able to simulate the data. The periodogram is thus still a useful tool

Figure 15.6. Data obtained from marital interaction of one couple. Ordinate is the cumulative amount of positive minus negative interaction for (a) husband toward wife, and (b) wife toward husband.

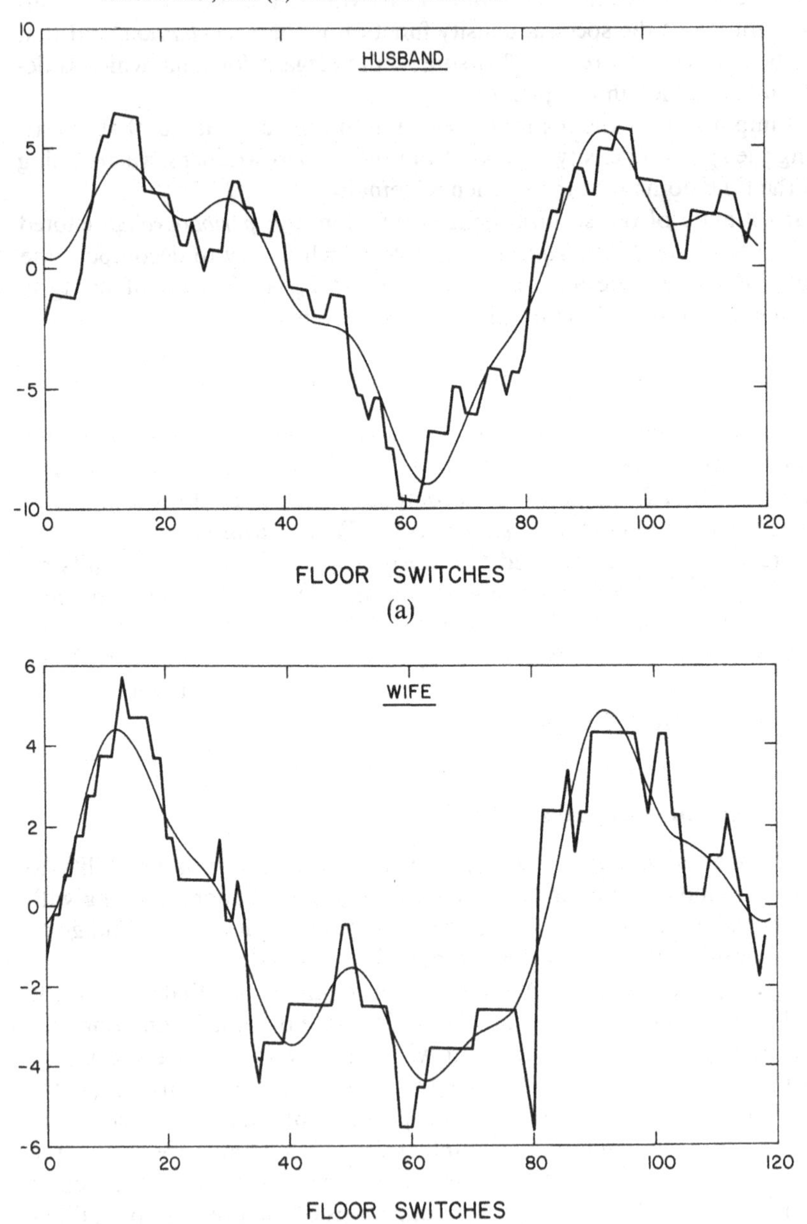

in approximating data, even data that do not come from physical phenomena as regular as the brightness of a variable star. Here, as with the variable star data, reasonable conclusions may be reached without building a model with random periodicities.

Chapter 16 develops the mathematics of the periodogram and its modern extensions. We will start, as here, with fixed (nonrandom) cyclicities. Appendix 16A is a primer on some of the basics of Fourier series. The reader is referred to this appendix for clarification of some of the more technical aspects of this chapter.

16
The periodogram

The Wiener–Khintchine theorem is introduced. This theorem makes it possible to go from the autocovariances of the time domain to the spectral density of the frequency domain. Many critical concepts for understanding both time- and frequency-domain time-series analysis are introduced.

The sine wave

A sine wave is a periodic function that can be obtained from the motion traced by the swinging pendulum shown in Figure 16.1 in a uniformly moving box of sand. In Figure 16.2 the more complex motion of two pendulums attached to one another describes the beat pattern discussed earlier in this book. By adding pendulums in this fashion, the resulting motion is a sum of sine waves. The amplitude of each wave is determined by how far the pendulum swings, which is determined by how much energy went into its initial displacement, much as how high a swing goes is determined by how hard it was shoved. The frequency of oscillation of a pendulum is determined by how fast it swings, which Galileo showed is determined entirely by its length. Shorter pendulums have shorter periods or faster frequencies.

To review periodic functions briefly, a function $f(t)$ is called a periodic function with period T if, for all t,

$$f(t) = f(T + t)$$

If a function has period T, it also has the periods $2T, 3T, \ldots$. A simple example of a periodic function is the sine wave:

$$x(t) = A\sin(wt + \phi)$$

The period of this function is $T = 2\pi/w$ because

$$x\left(t + \frac{2\pi}{w}\right) = A\sin\left[w\left(t + \frac{2\pi}{w}\right) + \phi\right]$$
$$= A\sin[(wt + \phi) + 2\pi] = A\sin(wt + \phi)$$

The period T is the time from peak to peak. The constant A is called the *amplitude*, the constant w is called the angular frequency, and the constant ϕ is called the initial phase. The angular frequency represents the number of complete cycles in 2π units of time. The frequency, in cycles per unit time, is given by

$$f = \frac{w}{2\pi}$$

Figure 16.1. Simple sine wave generated by the motion of one pendulum in a uniformly moving box of sand.

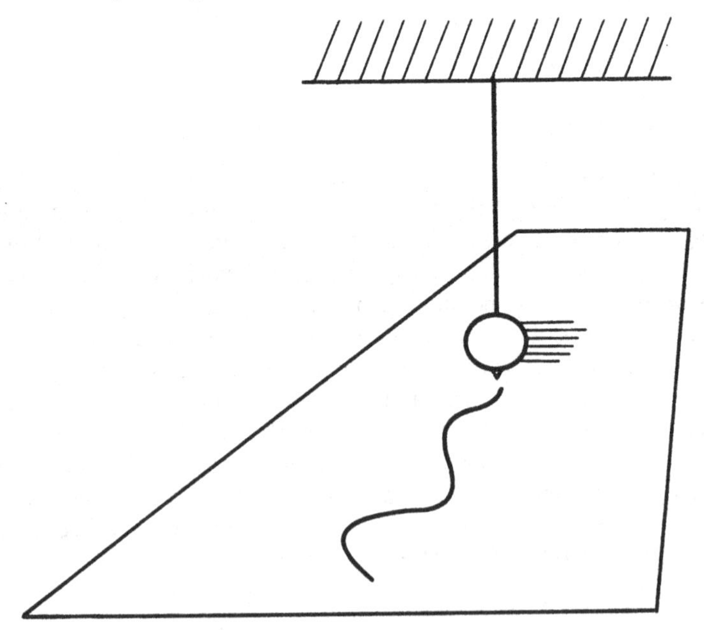

For example, if time is measured in minutes, the period is $2\pi/w$ and represents cycles that recur every $2\pi/w$ minutes. For example, if time is measured in months, what is the period of $\sin(2\pi/6)t$? The answer is

$$T = \frac{2\pi}{w} = \frac{2\pi}{2\pi/6} = 6 \text{ months}$$

Beats

Stated mathematically, if we add two sine waves (as in Figure 16.2), we obtain the equation

$$y_t = \sin(2\pi f t) + \sin(2\pi g t)$$

Figure 16.2. More complex patterns can be generated by the sum of two sine waves, or two pendula attached.

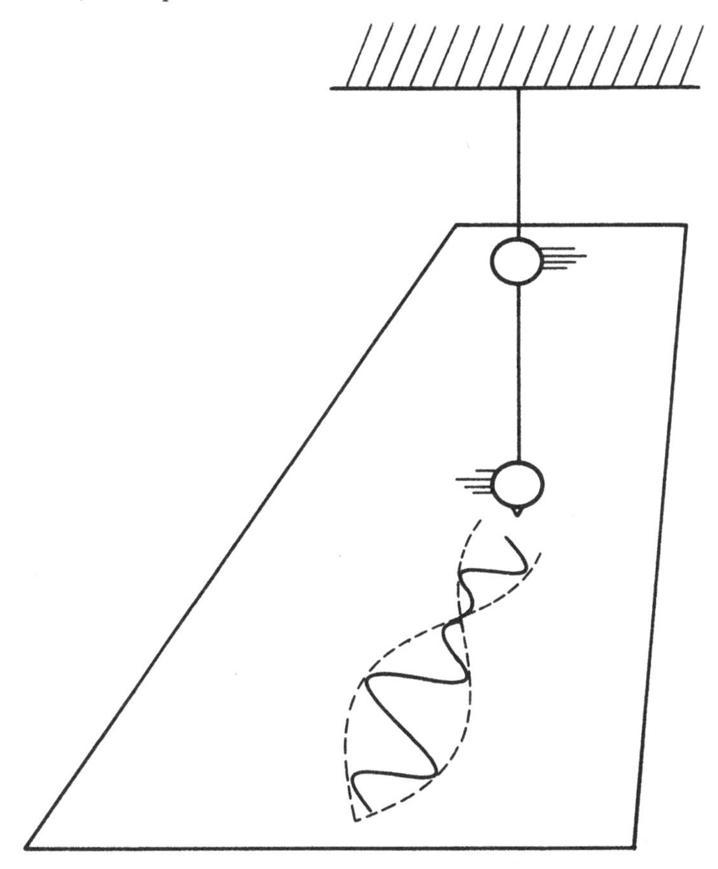

where the two different frequencies are g and f. A trigonometric identity can be used to show that we can also represent y_t as

$$y_t = 2\cos[\pi(f - g)t]\sin[\pi(f + g)t]$$
$$= A(t)\sin\pi[(f + g)t]$$

In this equation, $A(t)$ represents $2\cos[\pi(f - g)t]$, and if g is nearly equal to f, $f - g$ is nearly equal to zero, so that $A(t)$ is a slowly changing function. We can think of $A(t)$ as a variable amplitude, which is itself (slowly) periodic. The amplitude, $A(t)$, has a period equal to

$$T_1 = \frac{2\pi}{\pi(f - g)} = \frac{2}{f - g}$$

Suppose that $f = .5$ and $g = .4$; then $T_1 = 2/.1 = 20$. This is a long period. The sine wave that oscillates within this period has the short period of

$$T_2 = \frac{2\pi}{\pi(f + g)} = \frac{2}{.9} = 2.2$$

This is illustrated in Figure 16.3. As the reader may recall, this phenomenon of rhythmic changes in amplitude is called *beats*.

Obviously, in most cases we start with a time series, x_t, and wish to approximate it as a sum of sine waves. We wish to estimate the amplitudes,

Figure 16.3. The beat pattern.

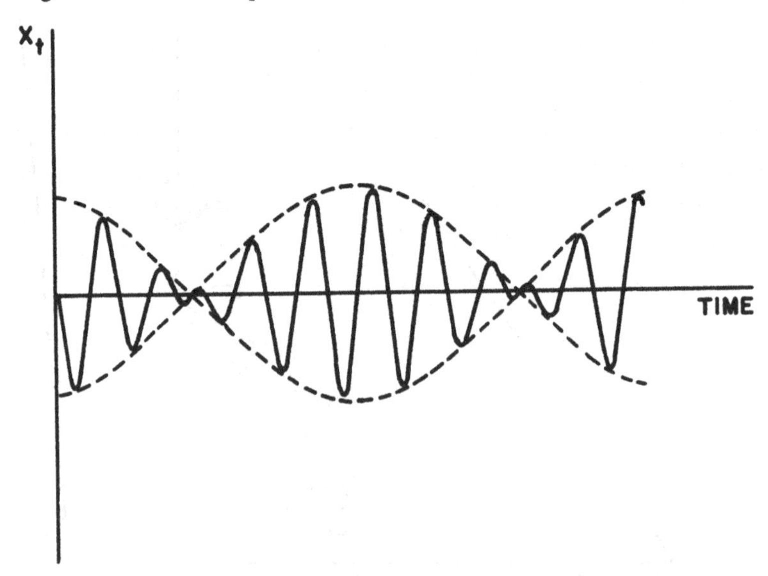

frequencies, and phases of the component cycles. In the deterministic case, Whittaker and Robinson (1924) were willing to assume that some noise might be added to this sum of sine waves, but they assumed that they were estimating cycles with fixed frequencies, amplitudes, and phases.

Sum of sine waves

Suppose that we begin with one sine wave, $A \sin(wt + \phi)$. Then, using a trigonometric identity it is possible to expand $A \sin(wt + \phi)$ as

$$A \sin(wt + \phi) = A(\cos wt \sin \phi + \sin wt \cos \phi)$$
$$= a \cos wt + b \sin wt$$

where

$$a = A \sin \phi$$

and

$$b = A \cos \phi$$

The objective of this mathematics is that $A \sin(wt + \phi)$ is nonlinear in the phase parameter ϕ, but $(a \cos wt + b \sin wt)$ is linear in the new parameters that describe phase. We can, if we wish, recover the original parameters, A and ϕ, because

$$A = (a^2 + b^2)^{1/2}$$

follows from

$$\sin^2 \phi + \cos^2 \phi = 1$$

Furthermore, the tangent of the phase angle ϕ is simply the ratio a/b, so we can compute the phase angle, ϕ, once we know a and b. If we knew w but not A and ϕ, it would be much easier to estimate a and b, by the usual regression methods (see Appendix 14A) rather than try to estimate A and ϕ directly.

We may write any trigonometric sum in two ways:

$$\sum_{j=0}^{\infty} A_j \sin(w_j t + \phi_j)$$

or

$$\sum_{j=0}^{\infty} (a_j \sin w_j t + b_j \cos w_j t)$$

and consider A_j and ϕ_j as functions of a_j and b_j. Obviously, in any application, this will be a finite and not an infinite sum. To discuss how we approximate the parameters of this trigonometric sum from the data, it will be illuminating to discuss the concept of *orthogonal functions*, which is a very beautiful notion because it turns the problem of statistical estimation into a problem of ordinary geometry.

Statistics as geometry: orthogonal functions

The idea of orthogonality comes from geometry. Two lines are orthogonal if they meet in a right angle. In the geometrical case there are thus two coordinates, x and y (see Figure 16.4). A line AB in the plane has a projection x_{AB} on the x coordinate and a projection y_{AB} on the y coordinate. These projections are like shadows cast by the line AB on each of the coordinates by lights from the far right or from above. The coordinates themselves cast no shadows on one another because they meet in a right angle. Now assume that the line AB is itself the shadow of a line in three-dimensional space. Then the three-dimensional line is represented by a sum in the plane, which can be written as a weighted sum:

$$\begin{aligned}\text{approximation of three-dimensional line} = AB &= x_{AB}(\text{unit of the } x \text{ axis}) \\ &\quad + y_{AB}(\text{unit of the } y \text{ axis})\end{aligned}$$

This entire metaphor can be generalized when the elements we wish to approximate are not lines in three dimensions, but functions.

Figure 16.4. Orthogonal projection in two dimensions; X_{AB} is the projection of the line AB on the X axis and Y_{AB} is the projection of the line AB on the Y axis.

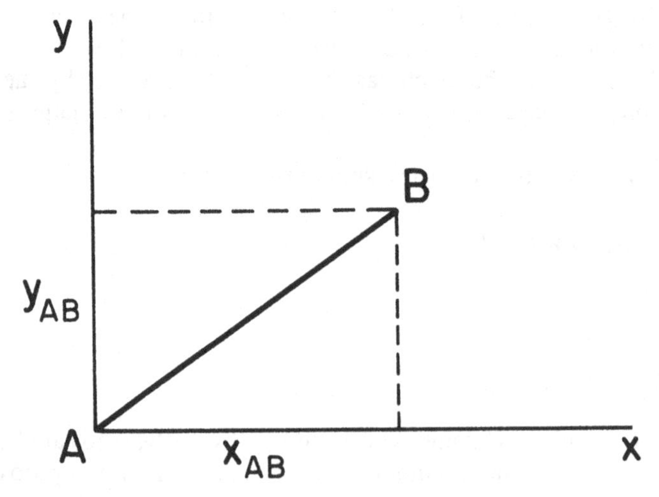

To accomplish the generalization, what are necessary are general definitions of distance and angle. The criterion of zero correlation is one example of how perpendicularity can be generalized. For normally distributed random variables, this criterion is equivalent to statistical independence.

In Fourier analysis, the orthogonal components are certain specially selected sine waves, called the *overtone series*, all of which are overtones or harmonics of one fundamental frequency. Furthermore, we are in a new "space," in which the axes are sine waves from the overtone series. The operation of "casting a shadow," or projection on one of these axes, is accomplished by using the generalized definition of angle – the correlation. The covariance of the function we are trying to approximate with one axis (say, $\sin 2\pi ft$) in the weighted sum (i.e., with one member of the overtone series) is the projection of the function on that axis. *If $f(t)$ is the function we are trying to approximate, this covariance can be thought of as how much of $\sin 2\pi ft$ is in $f(t)$.*

These are the basic concepts in Fourier analysis, and their generalization using geometrical notions is extremely elegant. It is the historical root of a field in mathematics called Hilbert space theory.

Fourier approximation

The actual formulas for this geometrical approximation appear complicated at first blush, but presenting them will make our discussion less glib, so that the reader can see that it is, in fact, possible to obtain actual numbers $\hat{a}_i$ and $\hat{b}_i$ from a real time series, x_t. To simplify matters slightly, let us assume that the time series has a zero mean (or that the mean has been subtracted). Let us also assume that there are an odd number of observations $x_0, x_1, \ldots, x_{T-1}$. If T is an even number, a correction term to the following formulas must be added. This term is given in Appendix 16A.

First, let us note that once we know how many observations there are, this determines the frequencies in our overtone series. Thus, given T, the angular frequencies w_j are determined as

$$w_j = \frac{2\pi j}{T}, \qquad j = 1, 2, 3, \ldots, \frac{(T-1)}{2}$$

To review our notation, these frequencies (in cycles/time unit) are

$$f_j = \frac{j}{T}, \qquad j = 1, 2, 3, \ldots, \frac{(T-1)}{2}$$

Note these correspond to cycles of period T, $T/2$, $T/3, \ldots, 2$ time units in length. For example, if $T = 31$, w_j would go from $2\pi(1/31) = .202$ in multiples up to $2\pi(15/31) = 3.040$. This is the grid of frequencies we have to work with for our approximation. *Note that this is an approximation.* The actual data might be cycling with some frequency that is trapped between the frequencies in our grid. The only simple way out of this problem is to make the grid finer, which can be done by increasing T, that is, by having more observations. Another solution, derived by Bloomfield (1976), is discussed in Chapter 22.

Notice that to get the projection of x_t on the axis $\cos w_j t$ we take the covariance of x_t with $\cos w_j t$. That is essentially all the formulas will entail. This elegant approximation is possible only because we have selected the overtone series of frequencies to use in our approximation, because the overtone series create an orthogonal set of functions that form the basis of a space that is Euclidean.

Now the problem is simply the estimation of a_j and b_j in the equation

$$x_t = \sum_j (a_j \cos w_j t + b_j \sin w_j t) + e_t \tag{16.1}$$

where e_t is noise. It can be shown (see Appendix 16A) that the least-squares estimates for a_j and b_j are just the sample covariances

$$\boxed{\begin{aligned} \hat{a}_j &= \frac{2}{T} \sum_{t=0}^{T-1} x_t \cos w_j t \\ \hat{b}_j &= \frac{2}{T} \sum_{t=0}^{T-1} x_t \sin w_j t \end{aligned}} \tag{16.2}$$

These are the basic equations of Fourier approximation.

Formal definition of the periodogram

The periodogram was invented in 1898 by Sir Arthur Schuster. He was fitting a model with one frequency, f, of the form

$$x_t = A \cos 2\pi f t + B \sin 2\pi f t + e_t \tag{16.3}$$

to the set of points $x_0, x_1, x_2, \ldots, x_{T-1}$, where A and B are constants and e_t is a set of zero mean, variance σ^2 normally distributed, independent errors. The least-squares estimates of A and B are

$$\begin{aligned} \hat{A} &= \frac{2}{T} \sum_{t=0}^{T-1} x_t \cos 2\pi f t \\ \hat{B} &= \frac{2}{T} \sum_{t=0}^{T-1} x_t \sin 2\pi f t \end{aligned} \tag{16.4}$$

Once again, recall that these can be viewed as the covariance of the series, x_t, with $\cos 2\pi ft$ and $\sin 2\pi ft$.

Now if we take the periodic part of equation (16.3), that is, that part of the right-hand side of the equation not including the e_t and take its sum of squares, using our estimates for A and B, we obtain

$$\sum_{t=0}^{T-1} (\hat{A} \cos 2\pi ft + \hat{B} \sin 2\pi ft)^2 \tag{16.5}$$

Then this quantity will be proportional to the *total energy of the series at the frequency* f, because it sums up the squares of the oscillations around our mean of zero.

Note that in testing $A = B = 0$, the quantity (16.5) is the "hypothesis sum of squares" (SSH) and the F test would be based on statistic $F = (\text{SSH}/2)/[\text{SSE}/(T-2)]$, where SSE is the error (residual) sum of squares:

$$\text{SSE} = \sum_{t=0}^{T-1} (x_t - \hat{A} \cos 2\pi ft - \hat{B} \sin 2\pi ft)^2$$

If we divide the quantity in (16.5) by 4π, this is the definition of the periodogram,[1] denoted $I(f)$:

$$I(f) = \frac{1}{4\pi} \sum_{t=0}^{T-1} (\hat{A} \cos 2\pi ft + \hat{B} \sin 2\pi ft)^2 \tag{16.6}$$

Note that the periodogram gives the *total energy* of the series at a frequency. As the number of observations increases, the energy is the sum of the energies contributed by each observation, so energy is proportional to the number of observations. Another useful concept is the *power*, which is the average energy, i.e., the energy divided by the number of observations or the average energy contributed by each observation.

Another form of the periodogram that is useful for calculations is

$$I(f) = \frac{1}{2\pi T} \left[\left(\sum_{t=0}^{T-1} x_t \cos 2\pi ft \right)^2 + \left(\sum_{t=0}^{T-1} x_t \sin 2\pi ft \right)^2 \right] \tag{16.7}$$

Once again, in Fourier approximation, the remarkable thing is that this result can be extended to the case when we assume a model wherein many frequencies compose x_t, with the frequencies estimated by

$$f_j = \frac{j}{T}, \qquad j = 1, 2, \ldots, \frac{(T-1)}{2} \tag{16.8}$$

In this case we have a set of A's and B's for each frequency f_j, estimated by $\hat{A}_j$ and $\hat{B}_j$. In this case at each frequency f_j, it is easy to show that

$$I(f_j) = \frac{T}{4\pi}\left[\frac{(\hat{A}_j)^2}{2} + \frac{(\hat{B}_j)^2}{2}\right] \tag{16.9}$$

This means that the total variability ($T\sigma^2$) accounted for by each frequency is proportional to the square of the amplitude at that frequency. If we have a line spectrum (i.e., only one frequency is present), $I(f_j)$ will be on the order of T. This means that for deterministic cycles, which give a line spectral density, the periodogram increases with the order of the number of observations; that is, if f_0 is a target cycle, $I(f_0)$ blows up as the number of observations, $T \to \infty$, and $I(f)$ goes to zero at nontarget frequencies.[2]

Furthermore, if we assume (as we did) that the e_t process was normal and uncorrelated in equation (16.3), we can test the significance of each periodogram value because the $\hat{A}$ and $\hat{B}$ at each frequency will be independent and normal. Hence, their sum of squares is a chi-square with two degrees of freedom, under the null hypothesis that $A_j = B_j = 0$ at that frequency.[3] In other words,

$$I(f) \sim \frac{\sigma^2}{2\pi}\frac{\chi_2^2}{2} \tag{16.10}$$

This result will hold asymptotically for large T.

In general, if x_t has theoretical spectral density function $p(f)$, then for each f, $I(f)$ has the distribution

$$I(f) \sim p(f)\frac{\chi_2^2}{2} \tag{16.11}$$

and at the frequencies k/T the periodogram estimates are asymptotically independent (i.e., become more independent as T, the sample size, increases).

Summary about the periodogram. In general, Schuster's periodogram was designed to fit a model to a time series that is the sum of deterministic sine waves plus noise. The parameters to be estimated are the amplitudes and phases of the component sine waves, or, alternatively, the A_j and B_j. The periodogram is *estimated* from the overtone series of frequencies $f_j = j/T, j = 1, 2, \ldots, (T-1)/2$, for which equation (16.9) holds. Once the number of observations, T, is given, the overtone series is determined. The greater T is, the finer is the mesh of estimated frequencies and the better the estimate of component cycles. The significance of each periodogram value for each frequency in the overtone series can be determined because the periodogram at each frequency is independently distributed as a chi-square with two degrees of freedom [equation (16.11)].

Thus, this section on the periodogram introduced the preprobabilistic theory for finding the significance of deterministic component frequencies, using the overtone series as estimates.

In the next two sections equations are derived that demonstrate the relationship between the autocovariance function of a time series and the periodogram. Then, the failure of the periodogram for probabilistic data is demonstrated. Finally, the extension of the periodogram is explained.

Exponential form of the periodogram

This section uses a few results from trigonometry previously introduced to obtain a convenient equation for the periodogram. In the next section that equation is used to derive a form of the Wiener–Khintchine theorem.

Using the square root of -1, recall that we can write sines and cosines in terms of exponentials. This is called deMoivre's equality, and it is written

$$e^{i\theta} = \cos\theta + i\sin\theta, \qquad \text{where } i = \sqrt{-1} \tag{16.12}$$

Using (16.12), we can write

$$\cos\theta = \frac{e^{i\theta} + e^{-i\theta}}{2} \tag{16.13}$$

$$\sin\theta = \frac{e^{i\theta} - e^{-i\theta}}{2i} \tag{16.14}$$

This much is simple algebra. If we use the common trigonometric notation for the square of the modulus of a complex number,

$$|a + ib|^2 = (a + ib)(a - ib) \tag{16.15}$$

then we can rewrite equation (16.7) for the periodogram very neatly and compactly as

$$\boxed{I(f) = \frac{1}{2\pi T}\left|\sum_{t=0}^{T-1} x_t e^{-i2\pi ft}\right|^2} \tag{16.16}$$

The Wiener–Khintchine theorem

If we use the trigonometry just introduced, we can show how the periodogram can be derived from the autocovariance function. Rewrite equation (16.16) as

$$I(f) = \frac{1}{2\pi T}\left|\sum x_t e^{-i2\pi ft}\right|^2$$

$$= \frac{1}{2\pi T}\left(\sum_t x_t e^{i2\pi ft}\right)\left(\sum_s x_s e^{-i2\pi fs}\right) \tag{16.17}$$

If we write out the two summations in the parentheses and group terms that multiply

$$e^{iM2\pi f} + e^{-iM2\pi f}$$

for M ranging between zero and $(T - 1)$, we get

$$\begin{aligned}
x_1^2 + x_2^2 + \cdots + x_T^2 &= c_0 T && \text{multiplies } e^0 = 1\\
x_1x_2 + x_2x_3 + \cdots + x_{T-1}x_T &= c_1 T && \text{multiplies } e^{i2\pi f} + e^{-i2\pi f} = 2\cos 2\pi f\\
x_1x_3 + x_2x_4 + \cdots + x_{T-2}x_T &= c_2 T && \text{multiplies } e^{2(i2\pi f)} + e^{-2(i2\pi f)}\\
& && = 2\cos 4\pi f\\
&\vdots\\
x_1x_t + x_tx_1 &= c_{T-1}T && \text{multiplies } e^{(T-1)(i2\pi f)} + e^{-(T-1)(i2\pi f)}\\
& && = 2\cos(T-1)2\pi f
\end{aligned} \tag{16.18}$$

where $c_0, \ldots, c_{T-1}$ are just estimates of the autocovariance.[4] If we regroup all these messy algebraic terms, we have the result of the Wiener–Khintchine theorem:

$$\boxed{I(f) = \frac{1}{2\pi}\left[c_0 + 2\sum_{k=1}^{T-1} c_k \cos k(2\pi f)\right]} \tag{16.19}$$

This equation transforms c_k into $I(f)$ and is called the Fourier transform of the c_k.

Illustrative computations of the periodogram and the fast Fourier transform

This section is designed to make equation (16.19) less abstract. The computations for the periodogram in equation (16.19) are conceptually not very difficult. Let us pretend, for pedagogical purposes only, that $T = 11$. Then since $(T-1)/2 = 5$, the overtone series of frequencies f_j is

$$f_1 = \tfrac{1}{11}$$
$$f_2 = \tfrac{2}{11}$$
$$f_3 = \tfrac{3}{11}$$
$$f_4 = \tfrac{4}{11}$$
$$f_5 = \tfrac{5}{11}$$

Using equation (16.19), we can write the first value of the periodogram as

$$\begin{aligned} I\left(\frac{1}{11}\right) = \frac{1}{2\pi}\bigg[c_0 &+ 2c_1 \cos 2\pi\left(\frac{1}{11}\right) + 2c_2 \cos 2\pi\left(\frac{2}{11}\right) \\ &+ 2c_3 \cos 2\pi\left(\frac{3}{11}\right) + 2c_4 \cos 2\pi\left(\frac{4}{11}\right) \\ &+ 2c_5 \cos 2\pi\left(\frac{5}{11}\right) + 2c_6 \cos 2\pi\left(\frac{6}{11}\right) \\ &+ 2c_7 \cos 2\pi\left(\frac{7}{11}\right) + 2c_8 \cos 2\pi\left(\frac{8}{11}\right) \\ &+ 2c_9 \cos 2\pi\left(\frac{9}{11}\right) + 2c_{10} \cos 2\pi\left(\frac{10}{11}\right)\bigg] \end{aligned}$$

Note that this is a lot of computation for just one periodogram value. However, many of the same cosine terms will appear in the equation for the next periodogram value, so a system of accounting that keeps track of all the relevant cosines will reduce the computations. The next periodogram value is

$$\begin{aligned} I\left(\frac{2}{11}\right) = \frac{1}{2\pi}\bigg[c_0 &+ 2c_1 \cos 2\pi\left(\frac{2}{11}\right) + 2c_2 \cos 2\pi\left(\frac{4}{11}\right) \\ &+ 2c_3 \cos 2\pi\left(\frac{6}{11}\right) + 2c_4 \cos 2\pi\left(\frac{8}{11}\right) \\ &+ 2c_5 \cos 2\pi\left(\frac{10}{11}\right) + 2c_6 \cos 2\pi\left(\frac{12}{11}\right) \\ &+ 2c_7 \cos 2\pi\left(\frac{14}{11}\right) + 2c_8 \cos 2\pi\left(\frac{16}{11}\right) \\ &+ 2c_9 \cos 2\pi\left(\frac{18}{11}\right) + 2c_{10} \cos 2\pi\left(\frac{20}{11}\right)\bigg] \end{aligned}$$

In general, the matrix of cosines we need to compute is

$$\begin{bmatrix} \cos 2\pi\left(\frac{1}{T}\right) & \cos 2\pi\left(\frac{2}{T}\right) & \cdots & \cos 2\pi\left[\frac{(T-1)/2}{T}\right] \\ \cos 4\pi\left(\frac{1}{T}\right) & \cos 4\pi\left(\frac{2}{T}\right) & \cdots & \cos 4\pi\left[\frac{(T-1)/2}{T}\right] \\ \cos 6\pi\left(\frac{1}{T}\right) & \cos 6\pi\left(\frac{2}{T}\right) & \cdots & \cos 6\pi\left[\frac{(T-1)/2}{T}\right] \\ \vdots & \vdots & & \vdots \\ \cos(T-1)\pi\left(\frac{1}{T}\right) & \cos(T-1)\pi\left(\frac{2}{T}\right) & \cdots & \cos(T-1)\pi\left[\frac{(T-1)/2}{T}\right] \end{bmatrix}$$

Although the matrix of cosines may seem forbiddingly complicated to compute, because of the properties of the cosine function, many of these computations are redundant. For example, the matrix is symmetric, so automatically only half the values need to be computed. Because the cosine is periodic, the work can be reduced further. A careful accounting system is necessary to keep track of the computations. Nonetheless, if T is large, the number of computations will be extremely large, and even machine computations will be sluggish. However, it turns out that it is possible to reduce the work dramatically if the number of observations is not a prime number but can be factored. The more composite T is (e.g., if T is a power of 2), the greater the reduction in computations. This elegant, and somewhat subtle, elimination of redundant computation is called the fast Fourier transform; it will not be discussed in this book (see Brigham, 1976).

The failure of the periodogram

Because (except at zero and π) $I(f)$ is distributed as $(\chi_2^2/2)p(f)$ [equation (16.11)], and because we know that a χ_{DF}^2 with DF degrees of freedom has variance $2DF$, it follows that the variance of the periodogram estimate of the spectral density function is

$$\begin{aligned} \operatorname{var}[I(f)] &= \frac{p^2(f)}{4} \operatorname{var}(\chi_2^2) \\ &= p^2(f) \end{aligned} \tag{16.20}$$

This means that the variance of the periodogram estimate does not go down as the sample size, T, increases, and since estimates of frequencies near one another have independent distributions, the plot of $I(f)$ would be unstable.

Demonstration of the failure and the solution of the problem

To dramatize the failure of the periodogram as an estimate of the spectral density, compute the periodogram of realizations of white noise of different sample sizes. We know that the spectral density of white noise is theoretically a straight line, so if the periodogram were a straight line, its peaks should smooth out with increasing sample size. Figure 16.5 shows that this is not what happens for the periodogram. With increasing sample size, the periodogram stays as wild as ever.

Figure 16.5. Spectral density function of white noise estimated from the periodogram. The flat dashed line is the theoretical spectral density of white noise. The failure of the periodogram lies in the fact that its variance does not decrease as the number of observations increases.

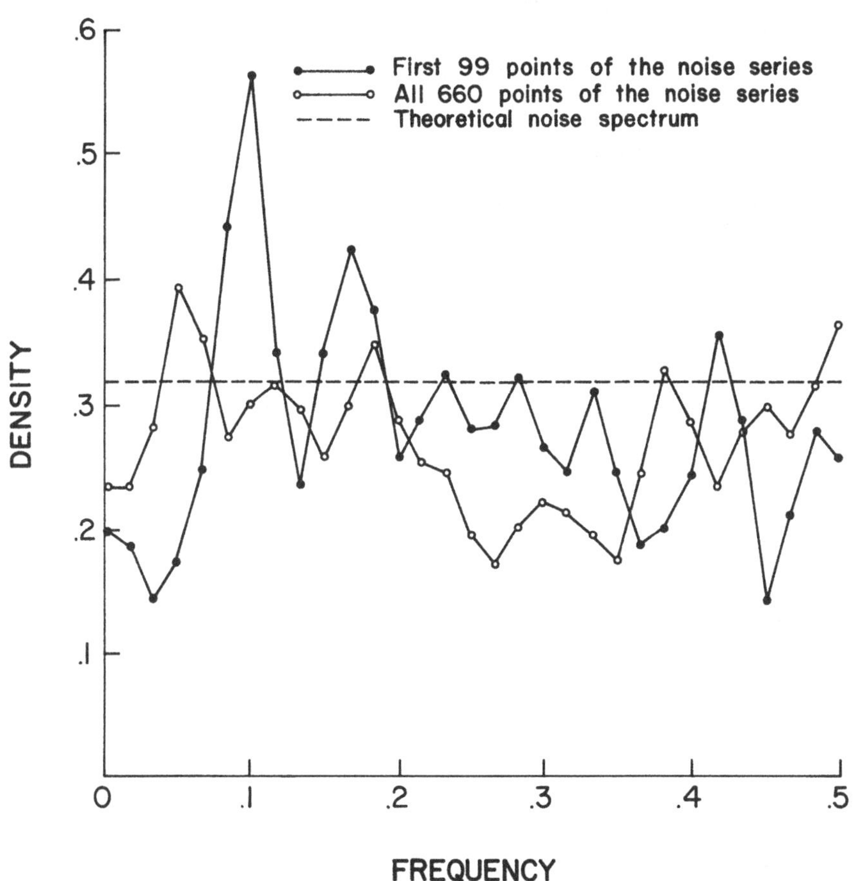

However, the periodogram can still be salvaged by noticing that the peaks of the periodogram of noise are randomly placed, so if we were to overlay all these graphs together and average them, we would obtain a closer estimate to a straight line. If we had a great deal of data we could divide them into separate chunks and average different chunks. This solution is computationally unfeasible, but fortunately, there are two equivalent alternatives. The first is to average the autocovariances, c_k, in equation (16.19) with a set of weights, w_k, called a *time-domain window*. The second alternative is to smooth $I(f)$ itself with a *frequency-domain window*, $W(f)$. Both alternatives are discussed in detail in Chapter 17.

To summarize, this chapter showed, mathematically, how models had to change from deterministic sinusoids plus noise to accommodate intrinsically probabilistic structures. Chapter 17 discusses the concepts involved in selecting a useful spectral window.

APPENDIX 16A

Orthogonal functions

If we have T observations (T assumed odd) in a time series, the slowest cycle we can fit to the data completes one cycle in the whole set of data points. This means that $2\pi/w = T$, and thus the lowest angular frequency is $2\pi/T$. If the w_j are limited to integer multiples of this frequency

$$w_j = \frac{2\pi j}{T}, \qquad j = 1, 2, \ldots, \frac{(T-1)}{2}$$

the frequencies will range from the slowest frequency up to the Nyquist frequency, the fastest frequency that can be fit to the data (see Chapter 2).

In this case the functions $(1, \cos w_j t, \sin w_j t)$ are called orthogonal functions because

$$\sum_t \cos w_n t \cos w_m t = 0 \qquad \text{if } n \neq m \tag{16A.1}$$

$$\sum_t \sin w_n t \sin w_m t = 0 \qquad \text{if } n \neq m \tag{16A.2}$$

$$\sum_t \sin w_m t \cos w_n t = 0 \qquad \text{for all } n \text{ and } m \tag{16A.3}$$

These relationships can be proved by using the identities

$$\cos xt \cos yt = \tfrac{1}{2}[\cos(x+y)t + \cos(x-y)t] \tag{16A.4}$$

$$\sin xt \sin yt = \tfrac{1}{2}[\cos(x-y)t - \cos(x+y)t] \tag{16A.5}$$

$$\sin xt \cos yt = \tfrac{1}{2}[\sin(x+y)t + \sin(x-y)t] \tag{16A.6}$$

These equations make it possible to solve for the constants in a Fourier series,

$$x(t) = \sum_{n=0}^{\infty} (a_n \sin w_n t + b_n \cos w_n t) \tag{16A.7}$$

by multiplying through first by $\cos w_m t$ and summing:

$$\sum_t x(t) \cos w_m t = b_m \sum_t \cos w_m t \cos w_m t = \frac{T}{2} b_m \tag{16A.8}$$

which proves that

$$b_m = \frac{2}{T} \sum_t x(t) \cos w_m t \tag{16A.9}$$

Similarly, it can be shown that

$$a_m = \frac{2}{T} \sum_t x(t) \sin w_m t \tag{16A.10}$$

If the number of observations is even, an additional frequency must be added:

$$a_{T/2} = \frac{1}{T} \sum_{t=0}^{T-1} x_t(-1)^t \tag{16A.11}$$

The $(-1)^t$ arises from $\cos(w_{T/2})t$, which is $\cos \pi t = (-1)^t$.

APPENDIX 16B

Removing seasonal components

Matrix form (see Appendix 14A)

$$z_t = \mu + A \cos \theta t + B \sin \theta t + e_t \tag{16B.1}$$

$$\begin{bmatrix} z_1 \\ z_2 \\ \vdots \\ z_T \end{bmatrix} = \begin{bmatrix} 1 & \cos\theta & \sin\theta \\ 1 & \cos 2\theta & \sin 2\theta \\ \vdots & \vdots & \vdots \\ 1 & \cos T\theta & \sin T\theta \end{bmatrix} \begin{bmatrix} \mu \\ A \\ B \end{bmatrix} + \begin{bmatrix} e_1 \\ e_2 \\ \vdots \\ e_T \end{bmatrix} \tag{16B.2}$$

or

$$\mathbf{z} = \mathbf{X}\boldsymbol{\beta} + \mathbf{e}, \tag{16B.3}$$

which has the least-squares solution (assume that T is even)

$$\hat{\boldsymbol{\beta}} = (\mathbf{X}'\mathbf{X})^{-1}\mathbf{X}'\mathbf{z} \tag{16B.4}$$

$\mathbf{X}'\mathbf{X}$ is a diagonal matrix and

$$\hat{\mu} = \sum_t \frac{z_t}{T}$$

$$\hat{A} = \frac{2}{T}\sum_t z_t \cos\theta t$$

$$\hat{B} = \frac{2}{T}\sum_t z_t \sin\theta t \tag{16B.5}$$

This generalizes to multiple frequencies θ_k:

$$\theta_k = \frac{2\pi k}{T}, \qquad k = 1, \ldots, \frac{T}{2} \tag{16B.6}$$

in which A is replaced by A_k and B is replaced by B_k; that is, we can simultaneously estimate μ and get the same A and B.

APPENDIX 16C

Detecting a deterministic cycle

To detect a deterministic cycle, we examine the behavior of the periodogram when x_t is a pure cycle:

$$x_t = A\sin(\lambda t + \theta), \qquad \lambda = 2\pi f \tag{16C.1}$$

In this case it can be shown that if we assume that $f_k = k/T$, the sample estimate of the covariance at lag k is, asymptotically,

$$c_k = \frac{A^2}{2}(\cos\lambda k)\left(1 - \frac{k}{T}\right) \tag{16C.2}$$

The periodogram is

$$I(\lambda) = \frac{1}{2\pi}\left(c_0 + 2\sum_{k=1}^{T-1} c_k \cos\lambda k\right) \tag{16C.3}$$

and by substituting equation (16C.2) into (16C.3), we obtain

$$I(\lambda) = \frac{A^2}{4\pi}\left[1 + 2\sum_{k=1}^{T-1}\left(\cos^2 \lambda k\right)\left(1 - \frac{k}{T}\right)\right] \tag{16C.4}$$

By using the trigonometric identity that

$$\cos^2 \theta = \tfrac{1}{2}(1 + \cos 2\theta) \tag{16C.5}$$

it is not difficult to show with some algebra that, to within a constant term,

$$I(\lambda) = \left(\frac{A^2}{8\pi}\right)T \tag{16C.6}$$

which implies that $I(\lambda)$ blows up as the sample size, T, increases. In Chapter 22 we will use this as a test of a time series that is a sine wave plus a stationary stochastic time series, by computing the periodogram for subsamples of points that increase to the entire sample size. It can be seen that the periodogram can be used to detect a deterministic cycle. Appendix 16B contained the matrix equations for removing a cycle of a particular frequency with least-squares estimates of the amplitude.

To derive equation (16C.2), in which the var $A = \sigma^2$, θ is uniformly distributed on the interval $(0, 2\pi)$, and A and θ are independent, write

$$\begin{aligned}\gamma_k &= E[A\sin(\lambda t + \theta)A\sin(\lambda t + \theta + \lambda k)] \\ &= \tfrac{1}{2}E(A^2)E[\cos(\lambda k) + \cos(2\lambda t + 2\theta + \lambda k)]\end{aligned}$$

Since A and θ are independent,

$$\gamma_k = \tfrac{1}{2}\sigma^2(\cos \lambda k + E\{\cos[\lambda(2t + k) + 2\theta]\})$$

But since θ is uniformly distributed,

$$E[\cos(\text{anything} + 2\theta)] = 0$$

so we have

$$\gamma_k = \frac{\sigma^2}{2}\cos \lambda k$$

This does not imply that the sample estimator, c_k, will be exactly $(\sigma^2/2)\cos \lambda k$. We can write, from the definition of the sample estimate,

$$\begin{aligned}c_k &= \frac{1}{T}\sum_{t=1}^{T-k} x_t x_{t+k} \\ &= \frac{1}{T}\sum_{t=1}^{T-k} A^2 \sin(\lambda t + \theta)\sin[\lambda(t + k) + \theta] \\ &= \frac{1}{T}\sum_{t=1}^{T-k} A^2 \tfrac{1}{2}[\cos \lambda k + \cos(2\lambda t + 2\theta + \lambda k)] \\ &= \frac{A^2}{2T}\left[(T - k)\cos \lambda k + \sum_{t=1}^{T-k} \cos(2\lambda t + \theta^*)\right]\end{aligned} \tag{16C.7}$$

where $\theta^* = 2\theta + \lambda k$. The sum $\sum_{t=1}^{T-k} \cos(2\lambda t + \theta^*)$ is bounded by $\sum_{t=1}^{T^*} \cos 2\lambda t$, where T^* is such that $2\lambda T^* \approx \pi$ (i.e., $T^* \approx \pi/2\lambda$). When we rewrite [equation 16C.7)], we obtain the result

$$c_k = \frac{A^2}{2}\left(\frac{T-k}{T}\right)\cos \lambda k + \frac{A^2}{2T}\left[\sum \cos(2\lambda t + \theta^*)\right]$$

$$= \frac{A^2}{2}\left(\frac{T-k}{T}\right)\cos \lambda k + \text{terms of order}\left(\frac{1}{T}\right)$$

The terms of order $1/T$ go to zero as T increases.

17 Spectral windows and window carpentry

This chapter discusses various spectral windows. It also reviews Satterthwaite's approximation and an approximate F test that make it possible to compare two (or more) spectral densities across a range of frequencies.

The idea of a spectral window

Recall that the sample periodogram estimates, $\hat{p}(f)$, are asymptotically independent (i.e., they become independent as the number of observations increases), and they are distributed as $p(f)\chi_2^2/2$. Also, recall that, unfortunately, the variance of the estimate does not decrease as the number of observations increases. One solution to this problem is to average neighboring values of the periodogram using a moving average. If independent estimates are averaged, the variance of the average is less than the variance of each of the values in the average. This fact led Daniell to propose the following solution.

Suppose that the periodogram estimates for a series with T observations are $\hat{p}(f_1), \hat{p}(f_2), \ldots, \hat{p}(f_R)$, when $f_k = k/T$, is, as usual, the overtone series of frequencies and R is either $T/2$ or $(T-1)/2$, depending on whether T is even or odd.

The general symmetrical moving average would average points to the left and right of a frequency, and the averaged value would be the new estimate. For example, suppose that a simple five-point average, with equal weights for each frequency were used. Then, at a particular frequency f_j, the periodogram values $\hat{p}(f_{j-2})$, $\hat{p}(f_{j-1})$, $\hat{p}(f_j)$, $\hat{p}(f_{j+1})$, and $\hat{p}(f_{j+2})$ would be averaged to give a new estimate $\hat{p}^*(f_j)$. As is usual with moving averages, the procedure moves along with a five-point average. More generally, if we employ an average that includes m values to the right and m values to the left of a target frequency, $\hat{p}(f_k)$, the new estimate is

$$\hat{p}^*_{\text{Daniell}}(f_k) = \frac{1}{2m+1} \sum_{j=-m}^{m} \hat{p}(f_{j+k}) \tag{17.1}$$

This method of smoothing (remember, every moving weighted average smoothes) is called the Daniell window. The expected value of the new estimate is unbiased as T increases, as long as the spectral density function is continuous. Also, the variance of the new estimate is

$$\begin{aligned} \text{var}[\hat{p}^*(f_k)] &= \frac{1}{(2m+1)^2} \sum_{j=-m}^{m} \text{var}[\hat{p}(f_{k+j})] \\ &\approx \frac{1}{(2m+1)^2} \hat{p}^2(f_k)(2m+1) \\ &= \frac{1}{2m+1} \hat{p}^2(f_k) \end{aligned} \tag{17.2}$$

As m increases the variance of this estimate decreases, which overcomes the problem with the periodogram, because, as T increases, m can also increase, as we have more and more frequencies to work with in the overtone series. However, note that this estimate will be biased unless $p(f)$ is reasonably flat within the $(2m + 1)$-step window.

The width of the averaged values in the Daniell window is called the *bandwidth* of the window. For the five-point window we are averaging the frequencies $f - 2/T, f - 1/T, f, f + 1/T$, and $f + 2/T$, and the width of this band of frequencies is $4/T$. In general, the bandwidth is $2m/T$.

It thus makes good theoretical sense to take m large, but small relative to T (e.g., $m = \log T$), but it makes good practical sense not to do this, because more and more values are averaged as m increases and relatively adjacent peaks will be melded into one peak. We lose our ability to resolve peaks within the bandwidth. Herein lies a basic dilemma in employing spectral windows.

The Daniell window in the time domain and the Bartlett 1 window

An alternative procedure for smoothing the periodogram estimates is to smooth neighboring autocovariances with a time-domain moving average instead of smoothing neighboring frequencies with a frequency-domain moving average. In fact, every frequency-domain window (moving average) has a time-domain representation and, conversely, every time-domain window has a frequency-domain representation. In fact, once again the Fourier transform is the shuttle between time- and frequency-domain representations. To summarize, and simultaneously introduce some new notation, a frequency-domain window is a set of weights $W(f_k)$ that creates a new estimate

$$\hat{p}^*(f_k) = \sum W(f_{k+j})\hat{p}(f_{k+j}) \tag{17.3}$$

and the autocovariance estimates c_k are weighted by a time-domain set of weights w_k.

The Daniell window can be described, although not particularly elegantly, in the time domain. Recall that the periodogram is

$$I(f) = \frac{1}{2\pi}\left[c_0 + 2\sum_{k=1}^{T-1} c_k \cos k(2\pi f)\right] \tag{17.4}$$

The Daniell window can be written as

$$\hat{p}^*_{\mathrm{Daniell}}(f) = \frac{1}{2\pi}\left[c_0 + 2\sum_{k=1}^{T-1} w_k c_k \cos k(2\pi f)\right] \tag{17.5}$$

where the w_k follow the dashed line in the left-hand side of Figure 17.1. Note

Figure 17.1. The Daniell window is square in the frequency domain (dashed line, right) and wiggly in the time domain in weighting the autocovariances (dashed line, left). The Bartlett window is square in the time domain (solid line, left) and wiggly in the frequency domain in weighting the periodogram (solid line, right).

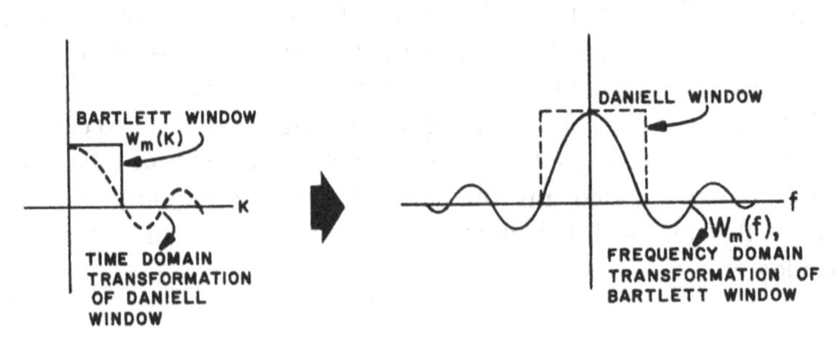

that the weights $W(f_k)$ are symmetric around zero in the frequency domain, but the w_k are not symmetric in the time domain, because we are not working with negative lags for c_k [which is the reason for the 2 multiplying the sum in equations (17.4) and (17.5)].

Note that what this time-domain weighting accomplishes is to give less credence to each estimate $\hat{p}^*(f)$ of autocovariances at large lags. This has the somewhat appealing feature that w_k is large for small k, where c_k, being an average of many terms, is a good estimate of ρ_k, and w_k is small for large k, where c_k is both unreliable and, as suggested by the time-domain models of the preceding section, possibly rather small. However, the transition from large to small weights as k increases is extremely irregular for the time-domain equivalent of the Daniell window, and, indeed, some of the weights are negative.

An alternative is the dual notion of the Daniell window – a moving average of a few neighboring *autocovariances* that truncate to zero abruptly. This is illustrated by the solid step function on the left-hand side of Figure 17.1, and it is called the Bartlett 1 window.

The Bartlett 1 window thus follows a conceptual direction opposite to that of the Daniell window. The Bartlett 1 window uses only a fixed number of c_k's and removes c_k's for large k, as illustrated in Figure 17.1. Thus,

$$\hat{p}^*_{\text{Bartlett 1}}(f) = \frac{1}{2\pi}\left[c_0 + 2\sum_{k=1}^{m} c_k \cos k(2\pi f)\right] \tag{17.6}$$

where m is the number of c_k's included in the estimate.

An additional problem with the Daniell window is that its computation, which would usually proceed using a variant of equation (17.5), requires computation and repeated calculations (for each f) involving *all* the covariances. Thus, a nice feature of the Bartlett 1 window is that it is computationally more feasible than the Daniell window. The Bartlett 1 window in the time domain gives an equivalent spectral window $W_m(f)$, shown as the solid line on the right-hand side of Figure 17.1. However, this has created another problem.

The wiggly form of $W_m(f)$ in Figure 17.1 dramatizes another issue that must be considered in the design of spectral windows, that of *side-lobe distortion.* For example, assume that the true spectral density is that of noise, a straight line. The raw periodogram will be rather irregular and the negative values of some side lobes can give negative spectral estimates, which should never happen. The positive values where they do not belong can produce *ghost peaks.* These peaks are actually not in the data; they are introduced by the window's side lobes. If adjacent peaks were spaced so that when the window centered halfway between each peak was on one of the negative side lobes, the result would be a large *negative* peak!

The Daniell and Bartlett 1 windows thus each look good in one setting (spectral domain and time domain, respectively) but poor in the other setting. Each has been used in practice, although other windows, which combine the best features of each, have been devised. The general idea of these windows is to avoid the sharp corners of the Daniell and Bartlett 1 windows, which, when Fourier-transformed,[1] produce undesirably large side lobes.

One simple solution is to alter the Bartlett 1 window so that its shape is triangular (see top of Figure 17.2), called the Bartlett 2 window:

$$w_k = 1 - \frac{k}{m} \tag{17.7}$$

which has the frequency-domain effect illustrated in the bottom part of

Figure 17.2. A triangular version of the Bartlett 1 window has less side-lobe distortion and no negative values in the frequency domain.

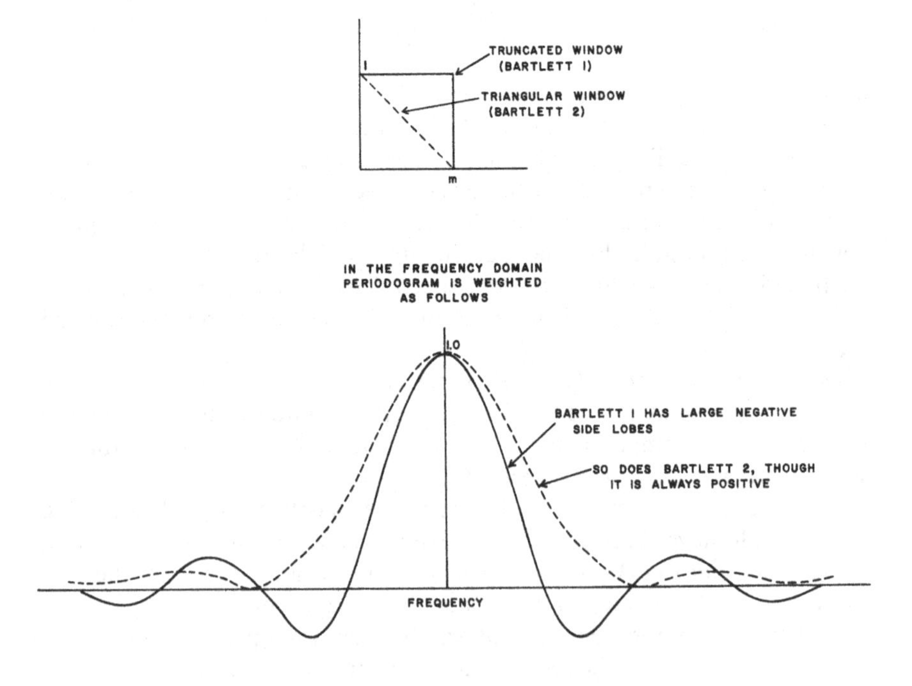

Figure 17.2. The figure also shows that the situation is somewhat improved with less side-lobe distortion and nonnegative values.

A host of other windows have been designed. Figure 17.3 compares two windows proposed by Tukey, the Tukey–Hamming and the Tukey–Hanning windows. The Tukey–Hanning window has very nice properties, one of which, ease in computation, is not as important now as when Tukey invented it. Its time-domain form is

$$w_k = \frac{1}{2}\left[1 + \cos \pi\left(\frac{k}{m}\right)\right] \tag{17.8}$$

which can also be rewritten using a trigonometric identity as

$$w_k = \cos^2\left(\frac{\pi}{2}\right)\left(\frac{k}{m}\right) \tag{17.9}$$

Figure 17.3. Two spectral windows proposed by Tukey.

IN THE TIME DOMAIN
ACVF's ARE WEIGHTED
AS FOLLOWS

1
HANNING WINDOW
HAMMING WINDOW
m

IN THE FREQUENCY DOMAIN
PERIODOGRAM IS WEIGHTED
AS FOLLOWS

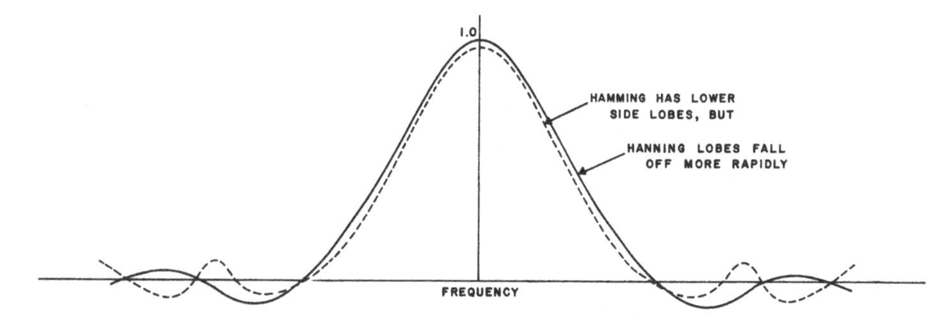

where m, as usual, would be specified in practice by the user.[2] For most applications in the social sciences, it will not make too much difference which of the latter two windows is employed.

Significance testing

Recall from introductory statistics how a 95% confidence interval was constructed for the variance. If we have N independent observations on a normally distributed variable, the sample estimate, $\hat{S}^2$, for the population variance, σ^2, satisfies the equation

$$\text{prob}\left[\chi^2_{(N-1;.975)} \leq \frac{N-1}{\sigma^2}\hat{S}^2 \leq \chi^2_{(N-1;.025)}\right] = .95$$

This is not a useful form for the confidence interval, so we can rewrite it as

$$\text{prob}\left[\frac{(N-1)\hat{S}^2}{\chi^2_{(N-1;.025)}} \leq \sigma^2 \leq \frac{(N-1)\hat{S}^2}{\chi^2_{(N-1;.975)}}\right] = .95$$

A similar technique is used for obtaining a confidence interval for the sample spectral density estimate. Recall that for a chi-square distribution with degrees of freedom, DF,

$$E(\chi^2_{\text{DF}}) = \text{DF}$$
$$\text{var}(\chi^2_{\text{DF}}) = 2(\text{DF})$$

Because windows smooth the periodogram, the individual estimates of the windowed spectral density estimate are no longer independent. For the periodogram, each estimate, $\hat{p}(f)$, is distributed as

$$\hat{p}(f) \sim \frac{p(f)(\chi^2_{\text{DF}})}{\text{DF}} \tag{17.10}$$

where DF = 2. When a spectral window is used, however, each spectral estimate has an "equivalent degrees of freedom," usually abbreviated EDF. In the Daniell case this is just $(4m + 2)$, the degrees of freedom of the sum of $(2m + 1)$, $\chi^2(2)$ random variables. In other cases we essentially are taking a weighted sum of adjacent frequencies, which causes a need for adjustment. The exact distribution of $\hat{p}(f)$ is quite complicated; however, we approximate it by

$$\hat{p}(f) \sim \frac{p(f)(\chi^2_{\text{EDF}})}{\text{EDF}}$$

(see Appendix 17B). This creates the following 95% confidence interval:

$$\frac{p(f)\chi^2_{\text{EDF}}(.975)}{\text{EDF}} \leq \hat{p}(f) \leq \frac{p(f)\chi^2_{\text{EDF}}(.025)}{\text{EDF}}$$

or

$$\frac{\text{EDF}\ \hat{p}(f)}{\chi^2_{\text{EDF}}(.025)} \leq p(f) \leq \frac{\text{EDF}\ \hat{p}(f)}{\chi^2_{\text{EDF}}(.975)} \tag{17.11}$$

For the Daniell window, then, since the EDF $= 4m + 2$, a $(1 - \alpha) \times 100\%$ confidence interval for $p(f)$ is

$$\frac{4m+2}{\chi^2_{4m+2}(\alpha/2)}\hat{p}(f) \leq p(f) \leq \frac{4m+2}{\chi^2_{4m+2}[1-(\alpha/2)]}\hat{p}(f)$$

If $m = 3$ and $\alpha = .05$, this is

$$\frac{14}{26.1}\hat{p}(f) \leq p(f) \leq \frac{14}{5.63}\hat{p}(f)$$

$$.54\hat{p}(f) \leq p(f) \leq 2.49\hat{p}(f)$$

If $\hat{p} = 1.0$, the 95% confidence interval is

$$.54 \leq p(f) \leq 2.49$$

This example illustrates the fact that significance tests of the spectral density estimate are often not very powerful, unless we take large bandwidths, giving a large EDF, and requiring a large number of observations.

The confidence interval can be used to test the significance of the spectral density function at each frequency by determining if the value of the spectral density for white noise falls outside the interval. Theoretically, the density for white noise is $\sigma_x^2/2\pi$ because all the autocovariances are zero for nonzero lags. Tests are also available for the density at more than one frequency value. For example, the Kolmogorov–Smirnov test for the integrated spectrum (see Hannan, 1960).

Comparing spectral density estimates

Two independent spectral density estimates can be compared using a procedure similar to testing for homogeneity of variance in analysis of variance. Since $\hat{p}(f_1) \sim p(f_1)\chi_r^2/r$ and $\hat{p}(f_2) \sim p(f_2)\chi_r^2/r$, under the null hypothesis that $p(f_1) = p(f_2)$, the ratio

$$\frac{\hat{p}(f_1)/r}{\hat{p}(f_2)/r}$$

will be distributed asymptotically as an F ratio with degrees of freedom (r, r). The approximation is only appropriate if $\hat{p}(f_1)$ and $\hat{p}(f_2)$ are independent, that is, farther apart than the bandwidth of the spectral window. The reader should be careful because, although some computer programs report only independent frequencies (i.e., a wider bandwidth implies a sparcer spectral density), some programs give back *all* the original frequencies in the overtone series, even though they are not independent. The use of the procedure recommended in this section on that kind of nonindependent $\hat{p}(f_1)$ and $\hat{p}(f_2)$ would produce spurious significance. In the case of the Daniell window, for example, two *windowed* spectral density values $\hat{p}(k/T)$ and $\hat{p}(k'/T)$ are asymptotically independent if $|k - k'| > 2m$, which is the size of the window from its center.

To approximate the distribution of a *sum* of the independent spectral density estimates, $\hat{p}(f_1) + \hat{p}(f_2)$, we can use a general method known as Sattertwaite's approximation (see Appendix 17A). If r is the equivalent degrees of freedom of a single spectral density estimate, the sum $\hat{p}(f_1) + \hat{p}(f_2)$ is approximately distributed as $C\chi_q^2$, where

$$C = \frac{\hat{p}^2(f_1) + \hat{p}^2(f_2)}{r[\hat{p}(f_1) + \hat{p}(f_2)]} \tag{17.12}$$

$$q = \frac{r[\hat{p}(f_1) + \hat{p}(f_2)]^2}{\hat{p}^2(f_1) + \hat{p}^2(f_2)} \tag{17.13}$$

Once again, the approximation for the distribution of the sum is only appropriate if $\hat{p}(f_1)$ and $\hat{p}(f_2)$ are independent. For the case of the data from a schizophrenic patient discussed in Chapter 1 (see Figures 1.6–1.8), before intervention the sum of the three spectral density estimates at $f = .075, .100$, and $.125$ is 35.8, with EDF 1 for the sum equal to 17.5, whereas after intervention, the sum is 16.1, with EDF 2 for the sum equal to 17.5. The F ratio is 2.22, $p < .05$, so that we can tentatively reject the null hypothesis. However, this approximation will be useful only for large EDFs, because the distribution of the sums as chi-squares is not exact and it is asymptotic distribution theory.

The importance of using a range of bandwidths

You will notice that for each window an additional parameter must be specified. For the Daniell window, it was $(2m + 1)$, the number of frequencies averaged, whereas for the time-domain windows it was the effective maximum lag of the autocovariances beyond which they were truncated. These parameters will alter the equivalent degrees of freedom for the spectral density estimate and the degree of independence of neighboring values of the estimate. These parameters determine a band of frequencies whose

Figure 17.4. The spectral density estimate varies with the bandwidth selected. In (a) a finer grid makes it possible to detect the low-frequency peaks, whereas in (b) there is no evidence of low-frequency peaks.

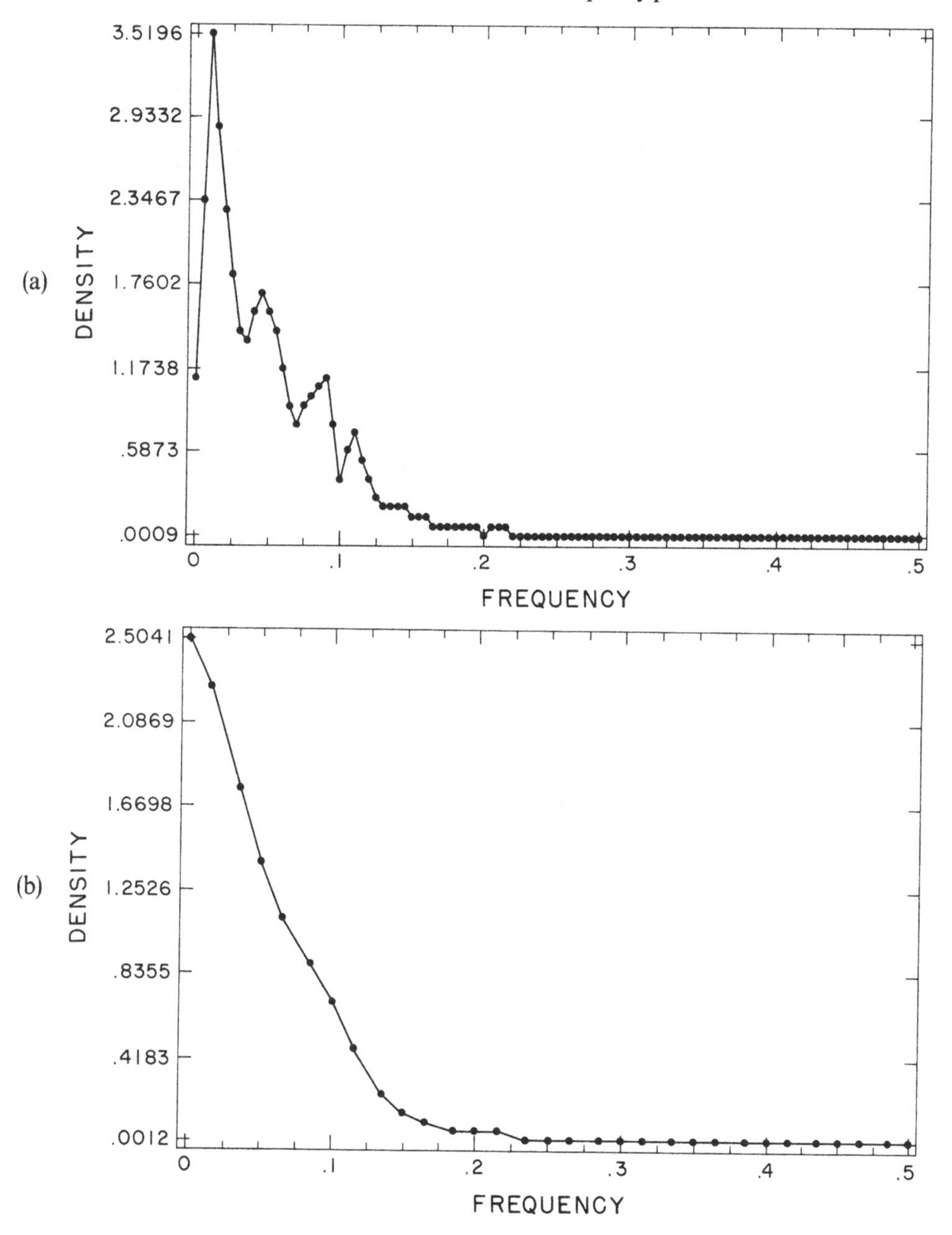

periodogram values are smoothed together, and the width of this band of frequencies was dubbed the bandwidth.

How is the researcher to choose an appropriate bandwidth? In most applied problems there is no adequate answer to this question, and the best practical advice is to try a selected set. The wider the bandwidth, the larger the EDF will be, and thus the narrower the confidence intervals on the spectral density estimate, but at a price in resolution. I always recommend computing the periodogram for a close look at each frequency, and then also trying several other bandwidth values. To illustrate the art of selecting an appropriate bandwidth, the 600 data points generated from the AR(4) model displayed in Figure 13.4 were used to compute two spectral density estimates, with varying bandwidths.

With 600 observations, there are 299 values for the periodogram, so that the frequency grid of the overtone series has width $\frac{1}{299} = .00334$. If we compute the spectral density estimate using the Tukey–Hanning window with $m = 3$ and with $m = 10$, we would obtain two different estimates of the spectral density function. Figure 17.4 is a plot of these two estimates of the spectral density function. Note that with the larger bandwidth, the low-frequency peak cannot be detected.

APPENDIX 17A

Equivalent degrees of freedom for a windowed spectral density estimate

Sattertwaite's approximation can be used to find the equivalent degrees of freedom (EDF) of a sum of chi-squares, approximated by

$$C\chi_r^2$$

where C and r are chosen to make the mean and variance of the sum and the mean and variance of this distribution the same. In general, we have a windowed periodogram

$I_w(f)$:

$$E[I_w(f)] \approx p(f)$$

$$\text{var}[I_w(f)] \approx p^2(f)\frac{2\pi}{T}[\text{area under the curve } W^2(f)]$$

$$\approx p^2(f)\frac{2\pi}{T}[\text{area of } W^2(f)]$$

and

$$E(C\chi_r^2) = C(r)$$

$$\text{var}(C\chi_r^2) = C^2(2r)$$

so

$$r = \frac{T}{\pi[\text{area of } W^2(f)]}$$

$$c = \frac{p(f)}{r}$$

and hence

$$I_w(f) \sim \frac{p(f)\chi_r^2}{r}$$

APPENDIX 17B

Equivalent degrees of freedom for a sum of spectral density estimates

If $\hat{p}(f) \sim p(f)\chi_r^2/r$, where $r = \text{EDF}$, we want to find the distribution of

$$\text{sum} = \hat{p}(f_1) + \hat{p}(f_2) \sim \frac{p(f_1)(\chi_r^2)}{r} + \frac{p(f_2)(\chi_r^2)}{r}$$

The variance of the sum is

$$\text{var(sum)} = p^2(f_1)\frac{2r}{r^2} + p^2(f_2)\frac{2r}{r^2}$$

$$= \frac{2}{r}[p^2(f_1) + p^2(f_2)]$$

The expected value (mean) of the sum is

$$E(\text{sum}) = p(f_1) + p(f_2)$$

If the new equivalent degrees of freedom is q, we estimate our sum as being distributed as

$$\text{sum} \sim C\chi_q^2$$

We now choose C and q so that the mean and variance of $C\chi_q^2$ are the same as the sum. Recall that the mean of a chi-square with M degrees of freedom is M and its variance is $2M$. Hence, set

$$Cq = p(f_1) + p(f_2)$$

$$2C^2q = \frac{2}{r}[p^2(f_1) + p^2(f_2)]$$

Solving for C and q gives

$$C = \frac{p^2(f_1) + p^2(f_2)}{r[p(f_1) + p(f_2)]}$$

$$q = \frac{r[p(f_1) + p(f_2)]^2}{p^2(f_1) + p^2(f_2)}$$

The result extends to more than two terms in the sum.

18
Explanation of the Slutzky effect

This chapter discusses a fundamental theorem that makes it possible to obtain the new spectral density function of a transformed series. The theorem is applied to derive the spectral density function of moving average and autoregressive processes. The examples discussed include a useful formula for the peak frequency of a pseudoperiodic AR(2) process.

This chapter brings to full circle our discussion of the fact that moving averages of noise processes can be cyclic, which the reader will recall was Slutzky's discovery. The fundamental theorem will be quoted that describes how a moving-average transformation affects noise. The theorem is proved in Appendix 18A.

The fundamental theorem is not actually very practical, but it is a conceptual device for understanding how the spectral densities of moving-average models arise. This chapter also derives equations for the spectral densities of AR models. Also, the notation used in this chapter will be helpful later when the bivariate case is discussed.

The theoretical spectral density of noise is a flat line. The moving-average model is a transformation of noise that weights a set of noise values and slides along the noise series. The noise can be thought of as input to a moving-average transformation. The fundamental theorem is quite general. Any linear transformation, called a *linear filter*, of a time series, x_t, can be derived from the spectral density of x_t. The input series, x_t, need not be noise.

The fundamental theorem

Suppose that a time series, x_t, with a spectral density function, $p_x(f)$, is transformed by a linear filter as follows:

$$y_t = \sum_s a_s x_{t-s} = A(B)x_t \tag{18.1}$$

where $A(B)$ can be written in operator notation as

$$A(B) = \sum_s a_s B^s \tag{18.2}$$

and $A(B)$ defines a characteristic polynomial $A(z)$ of any number z as follows:

$$A(z) = \sum_s a_s z^s \tag{18.3}$$

Then the question is: What is the spectral density of the y_t time series, $p_y(f)$? The fundamental theorem states that $p_y(f)$ can be derived by substituting $e^{i2\pi f}$ for z in equation (18.3), taking the square of the modulus of A, and multiplying it by $p_x(f)$. Stated mathematically, this is

$$\boxed{p_y(f) = |A(e^{i2\pi f})|^2 p_x(f)} \tag{18.4}$$

This is a new use of the operator notation, and it is probably unfamiliar to the reader. To explain the notation of equation (18.4), the following section will derive the spectral density of an MA(1) model and then the spectral density of an MA(q) model. Recall that the MA model can be considered a linear filter just like equation (18.1), in which the input process, x_t, is white noise, e_t.

Spectral density for an MA(q) process

The MA(q) process is a linear filter of a white noise process e_t of the form, using operator notation,

$$x_t = \text{MA}(B)e_t \tag{18.5}$$

where

$$\text{MA}(B) = 1 + b_1B + b_2B^2 + \cdots + b_qB^q \tag{18.6}$$

We know[1] that the spectral density function of white noise, $p_e(f)$, is a constant $\sigma_e^2/2\pi$. Therefore, applying equation (18.4), we have

$$p_x(f) = |\text{MA}(e^{i2\pi f})|^2\frac{\sigma_e^2}{2\pi} \tag{18.7}$$

For an MA(1) process,

$$x_t = e_t + be_{t-1} \tag{18.8}$$

which can be written

$$x_t = (1 + bB)e_t$$

and hence

$$\text{MA}(B) = 1 + bB$$
$$\text{MA}(z) = 1 + bz$$

If we substitute $z = e^{i2\pi f}$, we obtain

$$\text{MA}(e^{i2\pi f}) = 1 + be^{i2\pi f}$$

which is a complex function of the frequency, f. Its modulus squared is simply the product of it and its complex conjugate. The complex conjugate is $\text{MA}(e^{-i2\pi f})$, which is simply $1 + \text{b}e^{-i2\pi f}$. Hence, its modulus squared is

$$(1 + be^{i2\pi f})(1 + be^{-i2\pi f}) = |\text{MA}(e^{i2\pi f})|^2$$

When the left-hand side of this equation is expanded and deMoivre's equality

$$e^{i\theta} = \cos\theta + i\sin\theta$$

is employed, it can easily be shown that

$$|\text{MA}(e^{i2\pi f})|^2 = 1 + b^2 + 2b\cos 2\pi f$$

Therefore, the spectral density of the MA(1) process, $p_x(f)$, must be the quantity above times the spectral density of noise, which is a constant σ_e^2/π

and we have shown that

$$p_x(f) = (1 + b^2 + 2b\cos 2\pi f)\,\frac{\sigma_e^2}{2\pi} \tag{18.9}$$

for an MA(1) process.

Consider two examples, one with $b = .7$ and one with $b = -.7$. Figure 18.1 plots the value of the spectral density. The positively weighted moving average $b = .7$ positively weights lower frequencies, which is sensible because it smooths the noise spectrum. The $b = -.7$ spectrum is exactly opposite.

For the MA(q) process, we have ($\lambda = 2\pi f$)

$$p_x(f) = \frac{\sigma_e^2}{2\pi}(1 + b_1 e^{i\lambda} + b_2 e^{2i\lambda} + \cdots + b_q e^{qi\lambda})(1 + b_1 e^{-i\lambda} + b_2 e^{-2i\lambda} + \cdots + b_q e^{-qi\lambda}) \tag{18.10}$$

Expanding this density, we can compute $p_x(f)$ as

$$\begin{aligned} p_x(f) = \frac{\sigma_e^2}{2\pi}\,[&(1 + b_1^2 + b_2^2 + \cdots + b_q^2) \\ &+ (b_1 + b_1 b_2 + b_2 b_3 + \cdots + b_{q-1} b_q) 2\cos\lambda \\ &+ (b_2 + b_1 b_3 + b_2 b_4 + \cdots + b_{q-2} b_q) 2\cos 2\lambda \\ &+ \cdots + (b_{q-1} + b_1 b_q) 2\cos(q-1)\lambda] \end{aligned} \tag{18.11}$$

Figure 18.1. Theoretical spectral density functions for two processes, an MA(1) with b = −.7 (dashed line) and an MA(1) with b = +.7 (solid line).

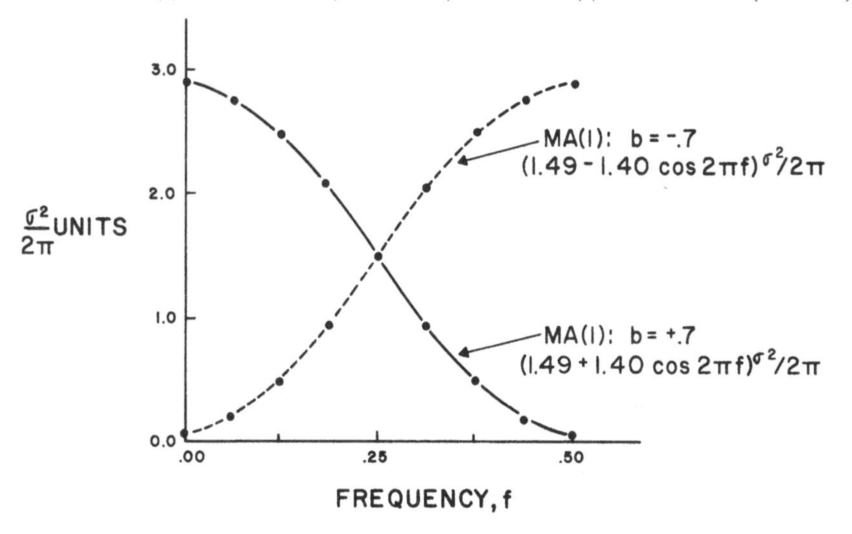

Equation (18.11) can be used to derive the spectral density of any finite-order moving-average model.

Spectral density for an AR(p) process

Consider any stationary pth-order autoregressive process, AR(p), which can be written

$$x_t = \sum_{s=1}^{p} a_s x_{t-s} + e_t \tag{18.12}$$

This can be rewritten

$$\mathrm{AR}(B)x_t = e_t \tag{18.13}$$

where

$$\mathrm{AR}(B) = 1 - a_1 B - a_2 B^2 - \cdots - a_p B^p \tag{18.14}$$

Since the AR process is stationary, this can be expressed as a moving-average model with generating function MA(B) as

$$x_t = \mathrm{MA}(B)e_t \tag{18.15}$$

Figure 18.2. Theoretical spectral density functions for two processes, an AR(1) with a = −.7 (dashed line) and an AR(1) with a = +.7 (solid line).

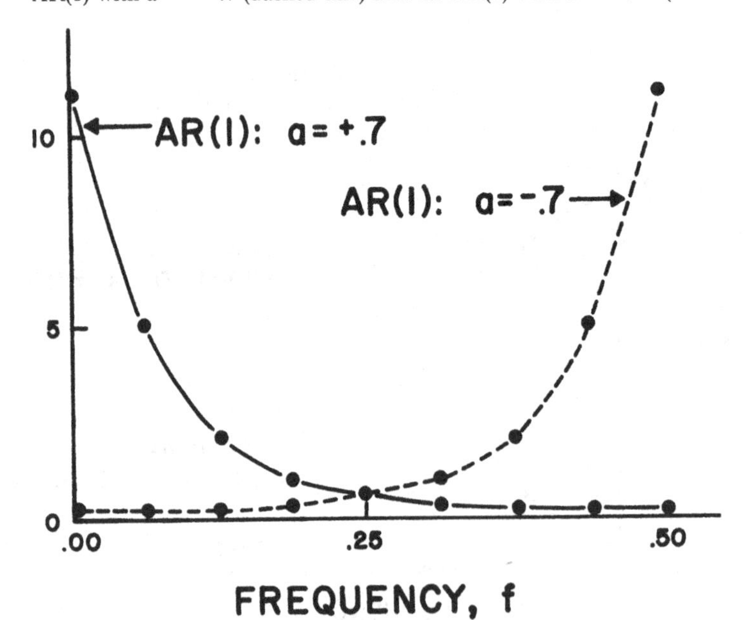

where

$$\mathrm{MA}(B) = \frac{1}{\mathrm{AR}(B)} \tag{18.16}$$

Once we have written the AR(p) model as a moving average, we can use the result in equation (18.4) to derive the result that the spectral density function of an AR(p) process is

$$p_x(f) = \frac{\sigma_e^2}{2\pi} \frac{1}{|\mathrm{AR}(e^{i\lambda})|^2} \tag{18.17}$$

For the AR(1) process this reduces to

$$p_x(f) = \frac{\sigma_e^2}{2\pi} \frac{1}{(1 - ae^{i\lambda})(1 - ae^{-i\lambda})}$$

$$= \frac{\sigma^2}{2\pi} \frac{1}{1 + a^2 - 2a\cos\lambda} \tag{18.18}$$

The shape of this function will be extremely similar to the spectrum of the MA(1) process but much steeper. It is displayed in Figure 18.2. The differences in shape between Figures 18.1 and 18.2 may at times be useful in deciding whether to use an AR or an MA model.

Spectral density for an AR(2) model

The spectral density of the AR(2) model can be computed from equation (18.17) as

$$p(f) = \frac{\sigma_\varepsilon^2}{2\pi} \frac{1}{(1 + a_1^2 + a_2^2) - 2a_1(1 - a_2)\cos 2\pi f - 2a_2 \cos 4\pi f} \tag{18.19}$$

The spectral density function, when it has a peak, will have one[2] at

$$f_0 = \frac{1}{2\pi} \arccos\left[\frac{|a_1|}{2(-a_2)^{1/2}}\right] \tag{18.20}$$

and the average peak-to-peak period will approximate $1/f_0$. Figure 18.3 displays the spectral density of a realization of the process

$$z_t = -.63z_{t-1} - .52z_{t-2} + e_t \tag{18.21}$$

This spectral density peaks at $f_0 = .35$, which is close to the value calculated using equation (18.20):

$$f_0 = \frac{1}{2\pi} \arccos(-.44) = \frac{2.25}{2\pi} = .36 \tag{18.22}$$

Figure 18.3. Estimated spectral density function and autocovariance function of an AR(2) model, showing nondeterministic periodicity.

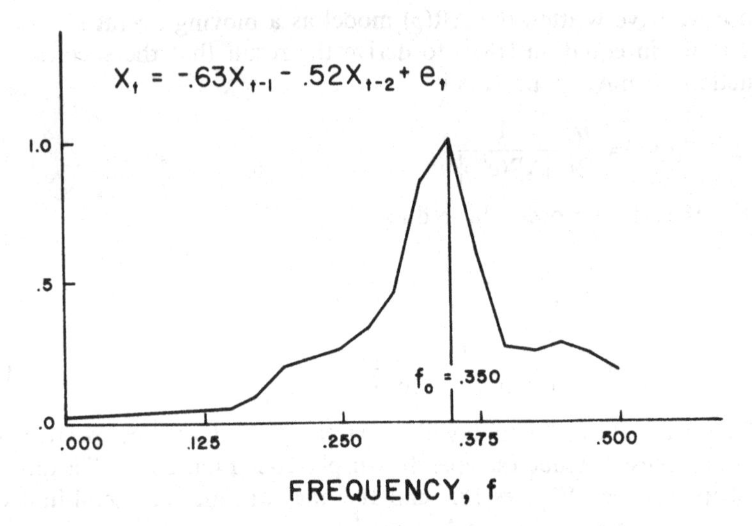

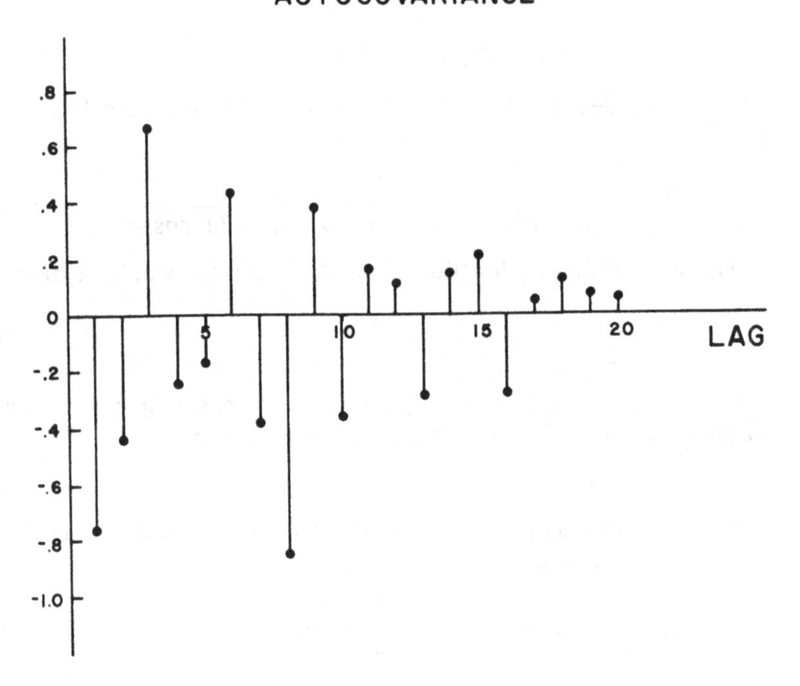

The general form of the spectral density function for an AR(p) model is easily derived by expanding $|\text{AR}(e^{i\lambda})|^2$. For $p = 3$, some simple algebra will show that

$$\hat{p}(f) = \frac{\sigma_e^2}{2\pi} \frac{1}{k_0 + 2k_1 \cos\lambda + 2k_2 \cos 2\lambda + 2k_3 \cos 3\lambda} \tag{18.23}$$

where $\lambda = 2\pi f$ and

$$\begin{aligned} k_0 &= 1 + \hat{a}_1^2 + \hat{a}_2^2 + \hat{a}_3^2 \\ k_1 &= -\hat{a}_1 + \hat{a}_1\hat{a}_2 + \hat{a}_2\hat{a}_3 \\ k_2 &= -\hat{a}_2 + \hat{a}_1\hat{a}_3 \\ k_3 &= -\hat{a}_3 \end{aligned} \tag{18.24}$$

In general, for an AR(p) model,

$$\hat{p}(f) = \frac{\hat{\sigma}_e^2}{2\pi} \frac{1}{k_0 - 2\sum_{j=1}^{p} k_j \cos j\lambda} \tag{18.25}$$

where $\lambda = 2\pi f$ and

$$\begin{aligned} k_0 &= 1 + \hat{a}_1^2 + \hat{a}_2^2 + \cdots + \hat{a}_p^2 \\ k_1 &= -\hat{a}_1 + \hat{a}_1\hat{a}_2 + \hat{a}_2\hat{a}_3 + \cdots + \hat{a}_{p-1}\hat{a}_p \\ k_2 &= -\hat{a}_2 + \hat{a}_1\hat{a}_3 + \hat{a}_2\hat{a}_4 + \cdots + \hat{a}_{p-2}\hat{a}_p \\ k_3 &= -\hat{a}_3 + \hat{a}_1\hat{a}_4 + \hat{a}_2\hat{a}_5 + \cdots + \hat{a}_{p-3}\hat{a}_p \\ &\vdots \\ k_{p-1} &= -\hat{a}_{p-1} + \hat{a}_1\hat{a}_p \\ k_p &= -\hat{a}_p \end{aligned} \tag{18.26}$$

Part V

Estimation in the time domain

19
AR model fitting and estimation: Mann–Wald procedure

The reader who explores the matter in depth will discover in the time-series literature a host of recommendations for time-domain models of the stationary stochastic component. This chapter refers to three approaches: Box–Jenkins, Wu–Pandit, and Mann–Wald. These approaches differ in a variety of ways. They are compared in this chapter and a recommendation for Mann–Wald is made. Table 19.1 presents an integrated set of suggestions for model fitting that considers trend, deterministic cycles, the stationary stochastic component, and the residual. An outline of the Mann–Wald procedure is given in two appendices.

Despite these recommendations, no modern book on time-series analysis would be complete without a discussion of the Box–Jenkins and related procedures. This is deferred to Chapter 20.

The next two chapters focus on two major alternatives currently in use for time-domain model fitting. One alternative, which forms the subject matter of this chapter, is to employ autoregressive (AR) models. This book has discussed the fact that AR model parameters can be fit by solving the *linear* Yule–Walker equations. This chapter will introduce a stepwise procedure that is computationally simple; furthermore, this chapter will show how to derive asymptotically normal tests of significance of the AR coefficients. These tests, derived by Anderson (1971), make use of the key work by both Mann and Wald (1943) and Quenouille (1947). Using normal distribution theory, the Anderson procedure derives the least-squares estimates of the coefficients of an AR(p) model and tests the hypothesis that coefficients of higher-order models are zero. In other words, this procedure estimates the order of the AR model and estimates the AR parameters as well. Thus, this first alternative fuses the model identification and estimation stages of finding an appropriate time-domain model.

In Chapter 20 the second alternative is discussed. It employs autoregressive moving-average (ARMA) models. These models, whose use was developed by Box and Jenkins (1970), employ a class of models that arise – by fitting an AR model and obtaining an MA residual, or by fitting an MA model and obtaining an AR residual. How does employing this class of models compare with the first alternative? Recall that if a low-order AR model is fit to a set of data and an MA residual is obtained, AR fitting can continue until eventually the residual is white noise. However, at times more parameters may be required in the AR models than in the ARMA model. Thus, ARMA models may be more parsimonious than AR models. However, the advantage of AR model fitting is that the estimation procedure requires solving linear rather than nonlinear equations.

A third alternative is reviewed briefly in Chapter 20. This third procedure is a variant of the Box–Jenkins method, developed by Pandit and Wu (Wu and Pandit, in preparation). The Wu–Pandit method uses ARMA$(2n, 2n-1)$ models and employs a stepwise model-fitting method that uses techniques of nonlinear estimation. The Wu–Pandit method has some advantages over the Box–Jenkins method, which are discussed briefly in Chapter 20.

Fitting an autoregressive scheme of any order

This section will continue the discussion of fitting AR models. Anderson (1971) presented a general method for least-squares estimation of the order of the AR process, assuming a normal distribution for the residuals. The parameters of a trend term can also be estimated, but for our purposes we will assume that this term is zero, and that the mean of the series is zero. Since Anderson's presentation is summarized in its *general* form in Appendices 19A and 19B, this chapter will simplify the presentation.

Let us begin with the familiar Yule–Walker equations (see Chapter 11), in which sample autocovariances replace process parameters. The equations are

$$c_k = a_1 c_{k-1} + a_2 c_{k-2} + \cdots + a_p c_{k-p}, \qquad k = 1, 2, \ldots, p \tag{19.1}$$

where the c_k are the sample autocovariances and the a_j are the autoregressive parameters.

From these equations it is possible to obtain a stepwise procedure for computing the autoregressive coefficients. If the process is AR(2),

$$X_t - a_1 X_{t-1} - a_2 X_{t-2} = e_t \tag{19.2}$$

Then the a_i are given by

$$\hat{a}_1 = \frac{r_1(1 - r_2)}{1 - r_1^2}$$

$$\hat{a}_2 = \frac{r_2 - r_1^2}{1 - r_1^2} \tag{19.3}$$

The next step is an AR(3) process in which we wish to fit three parameters, say b_1, b_2, and b_3. They are given in terms of the a_1 and a_2 above by

$$\hat{b}_3 = \frac{r_3 - r_2 a_1 - r_1 a_2}{1 - r_2 a_2 - r_1 a_1}$$

$$\hat{b}_1 = a_1 - a_2 b_3$$

$$\hat{b}_2 = a_2 - a_1 b_3 \tag{19.4}$$

The next step is an AR(4) process in which we wish to fit four parameters, c_1, c_2, c_3, and c_4, in terms of the b_1, b_2, and b_3:

$$\hat{c}_4 = \frac{r_4 - r_3 b_1 - r_2 b_2 - r_1 b_3}{1 - r_3 b_3 - r_2 b_2 - r_1 b_1}$$

$$\hat{c}_1 = b_1 - b_3 c_4$$

$$\hat{c}_2 = b_2 - b_2 c_4$$

$$\hat{c}_3 = b_3 - b_1 c_4 \tag{19.5}$$

This general form can be continued to any desired order.[1] Alternatively, equations (19.1), the Yule–Walker equations, can be written (as usual) in terms of the sample estimates of the autocorrelations. These equations are

$$r_k = \hat{a}_1 r_{k-1} + \hat{a}_2 r_{k-2} + \cdots + \hat{a}_p r_{k-p}, \qquad k = 1, \ldots, p \tag{19.6}$$

For example, if we are fitting an AR(3) to the headache data (see Chapter 12), we would first compute the autocorrelations r_1, r_2, and r_3, which equal .546, .148, and $-.185$, respectively.

The Yule–Walker equations are

$$\begin{aligned}
k = 1:&\quad r_1 = \hat{a}_1 r_0 + \hat{a}_2 r_{-1} + \hat{a}_3 r_{-2}\\
&\quad r_1 = \hat{a}_1 + \hat{a}_2 r_1 + \hat{a}_3 r_2\\
k = 2:&\quad r_2 = \hat{a}_1 r_1 + \hat{a}_2 r_0 + \hat{a}_3 r_{-1}\\
&\quad r_2 = \hat{a}_1 r_1 + \hat{a}_2 + \hat{a}_3 r_1\\
k = 3:&\quad r_3 = \hat{a}_1 r_2 + \hat{a}_2 r_1 + \hat{a}_3 r_0\\
&\quad r_3 = \hat{a}_1 r_2 + \hat{a}_2 r_1 + \hat{a}_3
\end{aligned}$$

These three equations can be solved for $\hat{a}_1$, $\hat{a}_2$, and $\hat{a}_3$ to obtain .609, $-.048$, and $-.249$, respectively. The stepwise procedure would yield the following estimates: (1) for the AR(1) fit, $\hat{a}_1 = .546$; (2) for the AR(2) fit, $\hat{a}_1 = .662$, $\hat{a}_2 = -.213$; (3) for the AR(3) fit, $\hat{a}_1 = .609$, $\hat{a}_2 = -.048$, $\hat{a}_3 = -.249$. It should be recalled that *the last coefficient of each of these successive models is the estimated PACF.* Hence, an added benefit of the stepwise procedure is that it produces the PACF estimates as a convenient by-product.

Significance testing: three alternatives

How is one to know when to stop fitting? There are three alternatives. The first alternative is to employ asymptotic distribution statistics for the estimates of the autoregressive model parameters. The second alternative is a procedure developed by Quenouille, which is an approximation procedure that decides to stop fitting when the PACF is no longer significantly different from zero, and remains nonsignificant for higher lags. The third alternative is to continue fitting until the residuals resemble white noise.

Alternative 1

Mann and Wald (1943) proved that as long as the residual series after AR model fitting is stationary with zero mean, finite variance, and finite fourth moment, the estimates of the autoregressive parameters will be asymptotically normal. This means that normality is approached as the number of observations increases. Thus, *we do not have to assume that either the process or the residual is normal.* Hannan (1960), in reviewing Mann and Wald's work, wrote:

> Our results . . . provide us with a large sample theory for testing the significance of the coefficients in an hypothesized autoregression when the residuals are independent random variables with finite fourth moment. Indeed these results show that we may proceed in the case of an autoregression exactly as we should for the classic case of a regression of one variable on a set of variables entirely distinct from it, when the residuals from that regression are normally and independently distributed, and our test procedures will be, at least asymptotically, valid. [P. 83]

In this first alternative, the striped covariance matrix plays an important role in estimating the covariances of the model parameters. Precise computations are presented in Appendices 19A and 19B. At this point approximate computations (that ignore end effects) will be reviewed. Suppose that σ^2 is the residual variance after the AR(p) model has been fitted and γ_k is the theoretical lag k autocovariance of the original series. Assume further that the original data have been detrended. Actually, this assumption is not necessary and we may simultaneously estimate these deterministic model parameters for any order of trend at the same time that we obtain least-squares estimates of the autoregressive parameters (see Appendix 19A). But for now, it will simplify the presentation if we make the assumption that trend has been removed from the data before AR(p) model fitting. This will become a practical recommendation later in this chapter (see Table 18.1). In this case, the covariance matrix for the difference between our model parameter estimates, $\hat{a}_i$, and values given by any hypothesis, a_i, is asymptotically $(\sigma^2/T)\mathbf{A}^{-1}$, where $\mathbf{A}^{-1}$ is the inverse of the striped covariance matrix, $\mathbf{A}$, whose elements are

$$\mathbf{A} = \begin{bmatrix} \gamma_0 & \gamma_1 & \gamma_2 & \gamma_3 & \gamma_4 & \cdots & \gamma_{p-1} \\ \gamma_1 & \gamma_0 & \gamma_1 & \gamma_2 & \gamma_3 & \cdots & \gamma_{p-2} \\ \gamma_2 & \gamma_1 & \gamma_0 & \gamma_1 & \gamma_2 & \cdots & \gamma_{p-3} \\ \vdots & \vdots & \vdots & \vdots & \vdots & & \vdots \\ & & & & & & \gamma_1 \\ \gamma_{p-1} & \gamma_{p-2} & \gamma_{p-3} & \gamma_{p-4} & \gamma_{p-5} & \cdots & \gamma_0 \end{bmatrix} \tag{19.7}$$

The matrix is the familiar striped matrix, striped as shown above. For an AR(2) process, for example, this means that the covariance matrix for $(\hat{a}_1, \hat{a}_2)$ is

$$\frac{\sigma^2}{T}\begin{bmatrix} \gamma_0 & \gamma_1 \\ \gamma_1 & \gamma_0 \end{bmatrix}^{-1} = \frac{\sigma^2}{T(\gamma_0^2 - \gamma_1^2)}\begin{bmatrix} \gamma_0 & \gamma_1 \\ -\gamma_1 & \gamma_0 \end{bmatrix}$$

This can be estimated by replacing σ^2 by $S^2/(T - p)$, where S^2 is the residual sum of squares and T is the number of observations, and replacing γ_0 and γ_1 by c_0 and c_1 as given in Chapter 11. Recall that $c_0 = \hat{\sigma}_x^2$ and $c_1 = r_1\hat{\sigma}_x^2$.

$$\operatorname{cov}(\hat{a}_1, \hat{a}_2) \approx \frac{\hat{\sigma}^2}{T(\hat{\sigma}_x^4 - r_1^2\hat{\sigma}_x^4)}\begin{bmatrix} \hat{\sigma}_x^2 & -\hat{r}_1\hat{\sigma}_x^2 \\ -\hat{r}_1\hat{\sigma}_x^2 & \hat{\sigma}_x^2 \end{bmatrix} = \frac{\hat{\sigma}^2}{T\hat{\sigma}_x^2(1 - r_1^2)}\begin{bmatrix} 1 & -r_1 \\ -r_1 & 1 \end{bmatrix}$$

Thus, for the AR(2) process, the estimated variance of both $\hat{a}_1$ and $\hat{a}_2$ is

$$\frac{S^2/(T - p)}{T\hat{\sigma}_x^2(1 - r_1^2)} \tag{19.8}$$

where $\hat{r}_1$ is, as usual, the lag 1 sample autocorrelation coefficient. If the null hypothesis is that the coefficients are zero, then the $\hat{a}_i/\mathrm{SD}(\hat{a}_i)$ are approximately standard normal deviates. To be cautious, the t distribution with $(T - p)$ degrees of freedom can be used instead of the normal for significance testing.

This distributional result can be used in several ways. One approach, called a step-up procedure, is to build successively models of order 1, 2, 3, . . . , p until a_p is not significant. This approach will, however, lead to difficulty if, say, a_1 is large, a_2 and a_3 are small, but a_4 is very large, indicating a delayed reaction. It is possible to fit an AR(1) and miss the additional delayed response. To be careful, then, it might be best not to stop at the first small a_i, but overshoot a bit. Alternatively, one could use a step-down approach. Fit too large a model and then successively remove small a_k's from the top down. However, this approach may lead to orders that are larger than necessary. The only suggestion is for the experimenter to be flexible; in picking the order of a model there are no hard-and-fast rules.

Alternative 2

The second alternative is a procedure developed by Quenouille (1947). Quenouille found that if one is fitting a $(p + 1)$st parameter model when in reality the process is of order p, the autoregressive estimate, a_{p+1}, will be asymptotically normal with variance $1/T$. This coefficient is the same as the partial autocorrelation coefficient. Thus, a *rough* asymptotic test of the order of an AR process, under the null hypothesis that the pth autoregressive coefficient is zero, is to compare the pth partial autocorrelation coefficient to a normal distribution with mean zero and variance $1/T$. Recall that the pth coefficient of the AR(p) process is the partial correlation between x_t and x_{t+p}, holding $x_{t+1}, \ldots, x_{t+p-1}$ constant. In general, the first alternative is probably superior to Quenouille's method, since in the first procedure we are estimating the variance instead of assuming it. Asymptotically (i.e., for large T), they will be the same.

We must be careful not to cut off the order of estimation too soon. Anderson (1971, pp. 242ff.) gives an example of sunspot data that Yule fit to an AR(2) but which Anderson recommends could be fit profitably (in terms of accounting for residual variation) by either an AR(9) or AR(18). An examination of the spectral density may help suggest a reasonable order. Anderson's computation (p. 661) shows one major peak, which would suggest an AR(2) process. In fact, the coefficients of AR processes up to order 5 and the approximate normal deviate [as given by Quenouille's (1947) estimate] are $b_1 = -.81, \sqrt{T}b_1 = -.10.76; b_2 = .66, \sqrt{T}b_2 = 8.69; b_3 = .10, \sqrt{T}b_3 = 1.34$; $b_4 = -.01$, $\sqrt{T}b_4 = -.18$; and $b_5 = .05$, $\sqrt{T}b_5 = .66$. It does not seem like a bad approximation to assume that the data can be fit by an AR(2) process, which was Yule's original fit.[2] Much depends on the amount of residual variation that an investigator believes is reasonable for purposes of model building.

Alternative 3

The third alternative is appropriate for the step-up procedure previously described. It requires that fitting continue until the residuals look like a realization of a white-noise process. One test of whether the residual is white noise, the Box–Pierce test, will be discussed here.

The Box–Pierce test. The third alternative is to continue fitting until the residual is white noise. Since white noise has no autocorrelation, we might be inclined simply to wish to test the autocorrelations of the residual using our version of Bartlett's $(2/\sqrt{N})$ bands. However, that procedure is not correct for the following reasons.

If we *knew* the true parameters of the model, the residuals would have estimated autocorrelations distributed approximately independently with mean zero and variance $1/N$ (see Box and Jenkins 1970, section 8.2), but in practice we do not know the model parameters. In this case it would be misleading to use $1/N$ as the variance. Durbin (1970) showed that for an AR(1) with parameter a, the variance of r_1 is a^2/N, which can be much less than $1/N$. With ARMA models, this problem was shown to be minimal by Box and Pierce (1970) for r_k at longer lags.

Box and Pierce (1970) suggest caution in examining the correlogram of the residual series. For small values of k, r_k may be nonzero. For example, for an AR(1) process it can be shown that

$$\operatorname{cov}(r_i, r_j) \simeq \frac{1}{N}\left[\delta_{ij} - a^{i+j-2}(1 - a^2)\right] \tag{19.9}$$

where δ_{ij} is the Kroencker delta ($\delta_{ij} = 1$ if $i = j$, and 0 otherwise). If $i = j$, this gives

$$\operatorname{var}(r_k) \simeq \frac{1}{N}\left[1 - a^{2k-2}(1 - a^2)\right] \tag{19.10}$$

Hence, the $\operatorname{var}(r_k)$ will always be smaller than $1/N$. For example, for $k = 1$,

$$\operatorname{var}(r_1) \simeq \frac{a^2}{N} \tag{19.11}$$

For large k however, var r_k will approach $1/N$, although how large a k is needed to make $1/N$ a good approximation depends on a. Thus, it is best to examine the autocorrelations of the residual series beyond the first few lags. (For a discussion of the derivation of these results, see Anderson, 1975, pp. 78ff.)

The test suggested by Box and Pierce (1970) for the independence of the residuals is given by the statistic

$$R = n \sum_{j=1}^{k} r_j^2 \tag{19.12}$$

R is distributed as χ^2_{k-p-q}, where k is usually at least 20. The r_j's are calculated on the residual series. If the residual series has N observations and the original series was fit by an ARIMA(p, d, q) model (3), then $n = N - d$. There are several worked examples in this book (Chapter 22). A significant chi-square indicates model inadequacy (Box and Pierce, 1970). Several other tests for white noise are possible, including the Kolmogorov–Smirnov test and the integrated periodogram (see Box and Jenkins, 1970, sec. 8.2.3). In practice, it is always wise to check diagnostically the residuals, for example, by using both the Box–Pierce test and by examining the spectral density of the residual series. This method, too, is appropriate for a step-up procedure, and when used in combination with the "overshooting" procedure described earlier will give good results in most cases. This recommendation is incorporated in the next section.

Critique of the Box–Pierce statistic. The Box–Pierce criterion is based on the overall size of the autocorrelations of the residual series, independent of the pattern of these autocorrelations with lag. The spectral density of the residuals may still show a clear peak; even if the power of this peak is not great, it may be theoretically interesting. The Box–Pierce test could be insensitive to this situation. Note that the Box–Pierce criterion is different from the one outlined by Anderson (1971), which tests the significance of autoregressive coefficients of higher-order models, and Quenouille's (1947) test of the PACF, and it is not well known how the criteria compare. Chatfield (1975) raised doubts about the Box–Pierce test's ability to detect systematic deviations from white noise. Wu and Pandit (in preparation), as part of their stepwise ARMA model-fitting procedure (see Chapter 20), use a criterion that is distributed as an F ratio under some assumptions that may be satisfied asymptotically. Their F ratio is a check of whether the next-higher-order model in their series of models provides any useful reduction in residual variance.

Recommendations for model fitting

Table 19.1 is a suggestion for model fitting and estimation. This table integrates both AR(p) least-squares and Box–Jenkins procedures. Basically, this table suggests that the possible components of the model be examined in the order given and that each component be handled separately. First check for and remove trends and deterministic cycles, then for the residuals build a stochastic model (this chapter suggests ARs) by repeated fitting and diagnostic checking until the terms in the model are significant (or *nearly* so – note that strict adherence to 10% or 5% levels is not appropriate when you are still trying to diagnose a model) and until the residual looks reasonably white.

One potential goal of model fitting is to understand the process that may have generated the observed series. This is not meant to be facetious. At times accounting for variance is not the same as understanding the process that may have generated the observed series. The "best" model in a scientific sense is the one in which we can attach some interesting interpretation to the parameters of the model. This point is explored in a practical example in Chapter 22.

For the stochastic component, the processes of autoregressive model fitting are summarized by the flow chart in Figure 19.1. Note in step I the phrase "Estimation and identification are combined." This means that the order of the AR model is determined at the same time as the model parameters

Table 19.1. *Recommendations for model fitting*

Model component	Indicators	Solutions
Trend	Zero frequency in spectral density Correlogram does not decrease faster than linearly $\sum r_{k,d}^2$ has a minimum at some degree of differencing (see Chapter 20 for a discussion)	Least-squares detrending Differencing
Deterministic cycle	Periodogram peak increases with increasing sample length	Least-squares de-sining Check residuals
Stationary stochastic component	Parsimony not an issue: spectral density function has peaks in a band of frequencies, or resembles AR or MA classical spectral densities	Least-squares AR fitting
	Parsimony an issue: correlogram and partial autocorrelation show patterns as described in text	Box–Jenkins ARIMA fitting procedure Pandit–Wu models Frequency-domain, spectral decomposition (significance testing)
Residual	Examine spectral density of residuals Examine ACF and PACF	Test significance of peaks White-noise test

are estimated and their significance tested. The reader should compare this with the flowchart of the Box–Jenkins procedure in Chapter 20, in which identification and estimation phases of model fitting are separate. This point makes AR fitting considerably more streamlined than Box–Jenkins ARIMA model fitting.

One caution bears repeating. If a step-up procedure is used, it is possible to decide mistakenly that the process is of a particular order, when, for example, a higher-order PACF coefficient will be significant, suggesting the appropriateness of a higher-order model.

Figure 19.1. Summary flowchart for autoregressive model fitting of the stationary stochastic component of the time series.

Step I. Fit AR(p) models varying p, using stepwise procedure. The last autoregressive coefficient at each order of the fit is the PACF. Estimation and identification are combined. It is important to compute higher estimates and step-down as well as to step up. (See text and the example in Chapter 20.)

Step II. Diagnostic Checking. Three alternatives:

1. Mann and Wald's work showed distributions of the estimates of the AR coefficients are asymtotically normal, with variance estimates given by the diagonal $\sigma^2_T A^{-1}$, where A is the striped covariance matrix.
2. Quenouille's procedure is a rough version of #1; uses the PACF, comparing $\sqrt{T}$ times the PACF to a standard normal distribution; i.e., at 95% confidence, ask is $\sqrt{T}$ times the PACF $>$ 1.96?
3. Box-Pierce test of the residuals from the AR fit. Should be used in conjunction with the spectral density estimate of the residual to check the whiteness of the residual. Also, consider the prewhitening section of this chapter.

Fussy model fitting using prewhitening

This section introduces Tukey's concept of *prewhitening*, which has many uses (see, e.g., Chapter 27). For a definition of prewhitening, we return to the frequency domain.

Recall that a linear filter applied to a time series x_t can be designed so that the filtered series

$$y_t = A(B)x_t \tag{19.13}$$

has a smoother spectral density function. An example of such a filter is an autoregressive model AR(B) fit to x_t such that the filter is white noise e_t:

$$e_t = \mathrm{AR}(B)x_t \tag{19.14}$$

By the fundamental theorem, the spectral density of the process can be computed from

$$\hat{p}_y(f) = \hat{p}_x(f)|A(e^{i2\pi f})|^2 \tag{19.15}$$

Now, if the AR model is a good fit, the residual series will theoretically have a flat spectral density function $\hat{\sigma}_e^2/2\pi$ throughout the frequency range $.0 \le f \le .5$. In practice, of course, this will rarely be the case. Nonetheless, there is an interesting application of prewhitening.

Theoretically, the advantage of prewhitening is that it can produce windowed spectral estimates with smaller bias at no cost in variance. This is true because we would theoretically minimize bias by windowing a reasonably flat spectral density function, which is the contribution of the whitened series

$$e_t = \mathrm{AR}(B)x_t \tag{19.16}$$

It is thus possible to prewhiten the data using an AR model. The residual series, y_t, can then be used as follows. First, $p_y(f)$, the periodogram of y_t, can be computed. Next, a windowed estimate can be computed, for example, using the Daniell window:

$$\hat{p}_{\mathrm{Daniell},y}(f) = \frac{1}{2m+1} \sum_{k=-m}^{m} \hat{p}\left(f + \frac{k}{T}\right) \tag{19.17}$$

Finally, the fundamental theorem is employed to compute

$$\hat{p}_x(f) = \frac{\hat{p}_{\mathrm{Daniell},y}(f)}{|\mathrm{AR}(e^{i2\pi f})|^2} \tag{19.18}$$

Recall that the exact expression for the denominator was given in Chapter 18.

If the AR model is a good fit to the data, this estimate of $p_x(f)$ should resemble the windowed estimate directly computed from the data. The idea can be summarized as follows. We compare the spectral density estimate that would be obtained if the model actually is our estimated AR model with the actual spectral density estimate.

APPENDIX 19A

Linear least-squares autoregressive model fitting

Linear least-squares AR(2) fitting

In the AR(2) model we can write each point as follows:

$$\begin{aligned}
x_3 &= a_1 x_2 + a_2 x_1 + e_3 \\
x_4 &= a_1 x_3 + a_2 x_2 + e_4 \\
x_5 &= a_1 x_4 + a_2 x_3 + e_5 \\
&\vdots \\
x_N &= a_1 x_{N-1} + a_2 x_{N-2} + e_N
\end{aligned}$$

where e_t is noise.

In matrix form this is a linear model

$$\begin{bmatrix} x_3 \\ x_4 \\ x_5 \\ \vdots \\ x_N \end{bmatrix} = \begin{bmatrix} x_2 & x_1 \\ x_3 & x_2 \\ x_4 & x_3 \\ \vdots & \vdots \\ x_{N-1} & x_{N-2} \end{bmatrix} \begin{bmatrix} a_1 \\ a_2 \end{bmatrix} + \begin{bmatrix} e_3 \\ e_4 \\ e_5 \\ \vdots \\ e_N \end{bmatrix}$$

$$\mathbf{Y} = \mathbf{X}\boldsymbol{\beta} + \mathbf{E}$$

with the least-squares solution

$$\begin{bmatrix} \hat{a}_1 \\ \hat{a}_2 \end{bmatrix} = (\mathbf{X}'\mathbf{X})^{-1}\mathbf{X}\mathbf{Y}$$

where

$$\mathbf{X}'\mathbf{X} = \begin{bmatrix} \sum_{t=2}^{N-1} x_t^2 & \sum_{t=2}^{N-1} x_t x_{t-1} \\ \sum_{t=2}^{N-1} x_t x_{t-1} & \sum_{t=1}^{N-2} x_t^2 \end{bmatrix}$$

and the residual variance is

$$\hat{\sigma}_e^2 = \frac{1}{N-2} \sum_{t=3}^{N} (x_t - \hat{a}_1 x_{t-1} - \hat{a}_2 x_{t-2})^2$$

General case of AR(p) fitting

In the general AR(p) case we can write

$$\begin{aligned} x_{p+1} &= a_1 x_p + a_2 x_{p-1} + \cdots + a_p x_1 + e_{p+1} \\ x_{p+2} &= a_1 x_{p+1} + a_2 x_p + \cdots + a_p x_2 + e_{p+2} \\ &\vdots \\ x_N &= a_1 x_{N-1} + a_2 x_{N-2} + \cdots + a_p x_{N-p} + e_N \end{aligned}$$

$$\begin{bmatrix} x_{p+1} \\ x_{p+2} \\ \vdots \\ x_N \end{bmatrix} = \begin{bmatrix} x_p & x_{p-1} & \cdots & x_1 \\ x_{p+1} & x_p & \cdots & x_2 \\ \vdots & \vdots & & \vdots \\ x_{N-1} & x_{N-2} & \cdots & x_{N-p} \end{bmatrix} \begin{bmatrix} a_1 \\ a_2 \\ \vdots \\ a_p \end{bmatrix} + \begin{bmatrix} e_{p+1} \\ e_{p+2} \\ \vdots \\ e_N \end{bmatrix}$$

$$\mathbf{Y} = \mathbf{X}\boldsymbol{\beta} + \mathbf{E}$$

$$\begin{bmatrix} \hat{a}_1 \\ \hat{a}_2 \\ \vdots \\ \hat{a}_N \end{bmatrix} = (\mathbf{X}'\mathbf{X})^{-1}\mathbf{X}'\mathbf{Y}$$

$$\hat{\sigma}_e^2 = \frac{1}{N-p} \sum_{t=p+1}^{N} \left(x_t - \sum_{i=1}^{p} \hat{a}_i x_{t-1} \right)^2$$

APPENDIX 19B

Least-squares autoregressive model fitting: asymptotic normal distribution theory

Anderson's (1971) presentation of least-squares fitting of an AR(p) model is based on assumptions of normality. He assumed a model

$$\varepsilon_t = \sum_{i=0}^{p} \beta_i x_{t-i} + \sum_{j=1}^{q} \gamma_j z_{jt} \tag{19B.1}$$

where the ε_t is normally, independently distributed, normal white noise, the x_t are the observed series, and the z_{jt} are *known* deterministic, time-dependent values, and the β_i and γ_j are the unknown parameters to be estimated. Symbolically, we have $\varepsilon_t = \mathrm{AR}(B)x_t +$ deterministic part. As usual, we can set $\beta_0 = 1$ (see Chapter 14 for this representation of an AR model). The last term in equation (19B.1) could be a fixed mean, or a linear trend, or any polynomial trend. We want the best estimates for the β_i and the γ_j in the least-squares sense, based on our observation of the time series $(x_1, x_2, \cdots, x_T)$. To minimize the sum of squares of the ε_t's, we want to minimize

$$\sum_{t=1}^{T} \left(x_t + \sum_{i=1}^{p} \beta_i x_{t-i} + \sum_{j=1}^{q} \gamma_j z_{jt} \right)^2 \tag{19B.2}$$

If we write this in matrix form using the definitions (see Appendix 14A)

$$\mathbf{x}(t-1) = \begin{bmatrix} x_{t-1} \\ x_{t-2} \\ \vdots \\ x_{t-p} \end{bmatrix}, \qquad \mathbf{z}(t) = \begin{bmatrix} z_{1t} \\ z_{2t} \\ \vdots \\ z_{qt} \end{bmatrix}$$

and $\boldsymbol{\beta}$ and γ are the equivalent parameter column vectors, and denote the row

$$[\mathbf{x}'(t-1), \mathbf{z}'(t)] \quad \text{by} \quad \mathbf{Y}(t)$$

then the least-squares solution is

$$\begin{bmatrix} \hat{\boldsymbol{\beta}} \\ \hat{\gamma} \end{bmatrix} = \left(\sum_{t=1}^{T} \mathbf{Y}_t' \mathbf{Y}_t \right)^{-1} \left(- \sum_{t=1}^{T} \mathbf{x}_t \mathbf{Y}_t' \right) \tag{19B.3}$$

In the text of this chapter I presented a form of this solution for a stepwise procedure. This is an application of a more general solution given by Anderson [1971, p. 187, equation (31)].

Anderson showed that the parameter vector has a limiting normal distribution. Define three matrices:

$$\mathbf{A}_T: \quad a_{ij} = \sum_{t=1}^{T} x_{t-i}x_{t-j} \qquad (i, j = 1, \ldots, p)$$

$$\mathbf{L}_T: \quad l_{ij} = \sum_{t=1}^{T} x_{t-i}z_{jt} \tag{19B.4}$$

$$\mathbf{M}_T: \quad m_{ij} = \sum_{t=1}^{T} z_{it}z_{jt}$$

Then the variance–covariance matrix of the parameter vector is approximately σ^2 times the inverse of the matrix

$$\begin{bmatrix} \mathbf{A}_T & \mathbf{L}_T \\ \mathbf{L}'_T & \mathbf{M}_T \end{bmatrix} \tag{19B.5}$$

More specifically, if we define

$$\mathbf{A}^*: \quad a^*_{ij} = \lim_{T \to \infty} \frac{a_{ij}}{T}$$

$$\mathbf{L}^*: \quad l^*_{ij} = \lim_{T \to \infty} \frac{l_{ij}}{T} \tag{19B.6}$$

$$\mathbf{M}^*: \quad m^*_{ij} = \lim_{T \to \infty} \frac{m_{ij}}{T}$$

The Anderson theorem is that, asymptotically, the vector

$$\sqrt{T}\begin{bmatrix} \hat{\boldsymbol{\beta}} - \boldsymbol{\beta} \\ \hat{\boldsymbol{\gamma}} - \boldsymbol{\gamma} \end{bmatrix} \tag{19B.7}$$

is normally distributed with vector mean zero and variance–covariance matrix

$$\sigma^2 \begin{bmatrix} \mathbf{A}^* & \mathbf{L}^* \\ \mathbf{L}^{*\prime} & \mathbf{M}^* \end{bmatrix}^{-1} \tag{19B.8}$$

This all holds under the following conditions:

1. The mean process is bounded: that is,

   $\mathbf{z}'(t)\mathbf{z}(t) \leq D \qquad$ for any time

   where D is some constant. Anderson noted that this can be generalized for $\mathbf{z}'(t)\mathbf{z}(t)$ that do not increase too rapidly in t.
2. The autoregressive terms are stationary; this is ensured if the characteristic polynomial $\sum_{k=0}^{p} \beta_k w^{p-k}$ has roots greater than 1 in absolute value.
3. The epsilons are identically, independently distributed with mean zero, variance σ^2, and bounded fourth moment.
4. The limits $\mathbf{A}^*$, $\mathbf{L}^*$, and $\mathbf{M}^*$ exist, and the inverse of the matrix in equation (19B.8) exists.

In practice, σ^2 is estimated as

$$s^2 = \frac{1}{T - p - q} \sum_{t=1}^{T} \hat{\varepsilon}_t^2 \tag{19B.9}$$

Edge effects

Given the time series expressed as a vector

$$\mathbf{z}' = (z_1, z_2, \ldots, z_t)' \tag{19B.10}$$

we wish to model an autoregression

$$z_t = a_1 z_{t-1} + a_2 z_{t-2} + \cdots + a_p z_{t-p} + e_t \tag{19B.11}$$

To estimate the parameters a_i for a fixed p, form the vectors

$$\begin{aligned} \mathbf{v} &= (z_k, z_{k+1}, \ldots, z_T) \\ \mathbf{x}_1 &= (z_{k-1}, z_k, \ldots, z_{T-1}) \\ &\vdots \\ \mathbf{x}_p &= (z_1, z_2, \ldots, z_{T-p}) \end{aligned} \tag{19B.12}$$

and then perform the linear multiple regression indicated by the model

$$\mathbf{Y} = \mathbf{XA} + \mathbf{E} \tag{19B.13}$$

where $\mathbf{Y}$ is the original time-series realization $(Y_k, Y_{k+1}, \ldots, Y_T)$ and $\mathbf{X}$ is the matrix of lagged regressors,

$$\mathbf{X} = (\mathbf{x}_1, \mathbf{x}_2, \ldots, \mathbf{x}_p) \tag{19B.14}$$

and $\mathbf{A}$ is vector of unknown parameters with least-square estimates

$$\begin{aligned} \hat{\mathbf{A}} &= (\mathbf{X}'\mathbf{X})^{-1}\mathbf{X}'\mathbf{Y} \\ &= (\hat{a}_1, \hat{a}_2, \ldots, \hat{a}_p) \end{aligned} \tag{19B.15}$$

$\hat{\mathbf{A}}$ is the least-squares estimator for the a_i in the reduced $(T - p)$-dimensional space (only $T - p$ observations are used).

This estimation of the autoregression is equivalent, except for "edge effects," to solving the Yule–Walker equations. Note in

$$\hat{\mathbf{A}} = (\mathbf{X}'\mathbf{X})^{-1}\mathbf{X}\mathbf{Y} \tag{19B.16}$$

that the elements of $\mathbf{X}'\mathbf{X}$ are the cross-products of the lagged variables. Specifically,

$$\hat{\mathbf{A}} = \begin{bmatrix} c_{11} & c_{12} & \cdots & c_{1p} \\ c_{21} & c_{22} & \cdots & c_{2p} \\ \vdots & \vdots & & \vdots \\ c_{p1} & c_{p2} & \cdots & c_{pp} \end{bmatrix}^{-1} \begin{bmatrix} c_{1y} \\ c_{2y} \\ \vdots \\ c_{py} \end{bmatrix} \tag{19B.17}$$

where the c_{ij}'s and c_{iy}'s are the obvious cross-products. This is the same equation as the covariance form (as contrasted with the correlation form) of the Yule–Walker equations *except* that because of the way the vectors in X are lagged, the covariances are computed slightly differently. Indeed, $c_{11} \neq c_{22} \neq \cdots \neq c_{pp}$, $c_{12} \neq c_{23} \neq c_{p-1,p}$, and so on. For example,

$$c_{11} = \sum_{t=p-1}^{T-1} z_t^2 \qquad \text{but} \qquad c_{22} = \sum_{t=p-2}^{T-2} z_t^2$$

However, for any realistic values of p and T, these edge effects will be negligible.

20 Box–Jenkins model fitting: the ARIMA models

By combining differencing, autoregressive, and moving-average representations in one model, Box and Jenkins (1970) were able to propose a parsimonious class of models called the autoregressive, integrated, moving-average (ARIMA) models. This chapter is an introduction to their methods. The Box–Jenkins procedure has been found useful by many researchers, but extremely complex, requiring quite a bit of experience and judgement to use effectively. The Wu–Pandit method is an attempt to simplify the procedures of Box and Jenkins for model building. Pandit and Wu found a subset of the ARIMA models particularly useful. The Wu–Pandit method is briefly reviewed in this chapter.

This chapter introduces the Box–Jenkins ARIMA models. The reader is referred to the book by Box and Jenkins (1970) for a complete discussion.

The ARMA models

Box and Jenkin's (1970) suggestion was precisely that great parsimony can be obtained by combining the autoregressive and moving-average models. Let z_t denote the series, and e_t denote white noise. For example,

$$\text{ARMA}(1,1)\text{: } z_t = az_{t-1} + e_t - be_{t-1} \tag{20.1}$$

combines the first-order autoregressive and the first-order moving-average model, and it is therefore called the first-order autoregressive moving-average model.

Similarly, the ARMA(1,2) model has a first-order autoregressive and a second-order moving-average component:

$$\text{ARMA}(1,2)\text{: } z_t = az_{t-1} + e_t - b_1e_{t-1} - b_2e_{t-2} \tag{20.2}$$

These models have few parameters, but they are capable of very complex behavior. The representation of these models can be simplified by employing operator notation.

Operator notation

Using operators, the ARMA(1, 1) process can be written

$$(1 - a_1 B)z_t = (1 - b_1 B)e_t$$

or

$$z_t = \left(\frac{1 - b_1 B}{1 - a_1 B}\right)e_t \tag{20.3}$$

This representation assumes that the autoregressive component is stationary. In general, the ARMA(p, q) process can be written

$$(1 - a_1 B - a_2 B^2 - \cdots - a_p B^p)z_t = (1 - b_1 B - b_2 B^2 - \cdots - b_q B^q)e_t \tag{20.4}$$

Assuming that the MA component is invertible, this can be written compactly as

$$\left[\frac{\text{AR}(B)}{\text{MA}(B)}\right]z_t = e_t \tag{20.5}$$

where AR(B) and MA(B) are the autoregressive and moving-average polynomials on the left-hand and right-hand sides of equation (20.4), respectively. Once again, this operator notation makes sense only when AR(B) is stationary or MA(B) is invertible, which is the case when the roots of AR(B) = 0 and MA(B) = 0 lie outside the unit circle (see Chapter 13).

Parsimony of the model

The fewer the number of parameters we have to estimate, the simpler and more parsimonious the time-series model will be. Glass et al. (1975) examined 116 time series from education and the social sciences. The series varied in length from 20 to more than 200 points. They wrote:

> The series reflect a variety of things observed (e.g., a person, a city, a nation) and a range of applications: alpha brain waves, crime rates, examination scores, stock prices, word association test scores, student's time spent studying, learning curves, etc. Most of the data were taken from published reports. The two judges (Glass and one of his students) independently identified each series as one of the ARIMA(p, d, q) class. In cases of disagreement, the series was double checked by both judges working together and a consensus was reached. [P. 116]

Of the 116 series, 21 were identified as seasonal, and they were not analyzed. This was unfortunate, because it is possible that they may have been capable of representation with higher-order AR or MA models. The remaining 95 series were identified and approximately 82% of these series had p, d, and $q \leq 1$ (d is defined below). Hence, for a wide range of education and social science applications, it may be possible to find low-order ARIMA models for a parsimonious representation of the time-series process. Fewer parameter models might also be useful when a relatively few number of observations are available. It is on this assumption that this chapter is based. However, *when parsimony is not a concern*, there are procedures for computing the order and estimating the coefficients of an AR(p) process of sufficiently high order.

ACF and PACF of the ARMA model

Recall that for an AR(p) model the PACF truncates after p lags, whereas the ACF is a mixture of decaying exponentials or exponentially damped sine waves. Recall also that for an MA(q) process the reverse is true, that is, the ACF truncates after q lags, whereas the PACF is a mixture of decaying exponentials or exponentially damped sine waves. In this section the ACF and the PACF will again be employed in identifying p and q in an ARMA(p, q) model. Assume that the degree of differencing, d, has been properly identified, a task that will be discussed later in this chapter.

Once again, by definition, the ARMA(p, q) process can be written

$$z_t = a_1 z_{t-1} + \cdots + a_p z_{t-p} + e_t - b_1 e_{t-1} - \cdots - b_q e_{t-q} \tag{20.6}$$

where the mean is assumed to be subtracted so that $E(z_t) = 0$ and e_t is usually (but not necessarily) assumed to be independently distributed $N(0, \sigma_e^2)$. We can, as usual, multiply by z_{t+k} and take expected values to obtain

$$\gamma_k = a_1 \gamma_{k-1} + \cdots + a_p \gamma_{k-p} + \gamma_{ze}(k) - b_1 \gamma_{ze}(k-1) - \cdots - b_q \gamma_{ze}(k-q) \tag{20.7}$$

In equation (20.7), γ_k is the autocovariance at lag k and $\gamma_{ze}(k)$ is the cross-covariance between z_t and e_t at lag k. However, z_{t-k} depends only on the random shocks that have occurred up to time $(t - k)$, so that

$$\gamma_{ze}(k) \begin{cases} = 0, & k > 0 \\ \neq 0, & k < 0 \end{cases} \tag{20.8}$$

Hence, for $k > q$,

$$\gamma_k = a_1 \gamma_{k-1} + a_2 \gamma_{k-2} + \cdots + a_p \gamma_{k-p} \tag{20.9}$$

and, dividing by γ_0,

$$\rho_k = a_1 \rho_{k-1} + a_2 \rho_{k-2} + \cdots + a_p \rho_{k-p} \tag{20.10}$$

which is simply a set of Yule–Walker equations, which hold only for $k > q$.

This means that *for $k > q$, the ACF behaves exactly like that of an AR(p) process.* For earlier lags, this will not be the case. Hence, there will be *q anomalous values,* $(\rho_1, \rho_2, \ldots, \rho_q)$, before the ACF resembles an AR(p) process.

The PACF will look like that for an MA process after p lags, which once again preserves the duality between MA and AR processes. This implies that, after p lags, the PACF will decay (i.e., consist of damped exponentials and/or damped sine waves). The p values $\rho_q, \rho_{q-1}, \ldots, \rho_{q-p+1}$ provide the necessary start-up values for equation (20.10), where $k > q$, which determines all the autocorrelations at later lags.

If $q < p$, the entire ACF will be a mixture decaying exponentials or damped sine waves, or both. If $q \geq p$, on the other hand, there will be $q - p$ initial values $\rho_1, \rho_2, \ldots, \rho_{q-p}$ that are "anomalous" (i.e., do not follow this pattern). On the other hand, the number of anomalous terms in the PACF is $p - q$.

These facts are useful in identifying p and q, and they are summarized as follows. The number of anomalous terms in the ACF equals $(p - q)$, and the number of anomalous terms in the PACF equals $(q - p)$. The ACF for an ARMA(p, q) process is a mixture of decaying exponentials and damped sine waves after the first $q - p$ lags, and the PACF is a mixture of decaying exponentials and damped sine waves after the first $p - q$ lags.

The reader should be cautioned that all these steps are by no means automatic. It is not always so easy to decide when terms are anomalous, or if the ACF or PACF truncates or decays exponentially. Most cases you will encounter will not be very clear cut. Unfortunately, nature is not so obliging. For that reason, considerable artfulness and experience is required to employ these facts, and it is in part for this reason that autoregressive models have great appeal.

As an example of ARMA identification, consider the computer-simulated realization of an ARMA(2, 4) process displayed as Figure 20.1. The process chosen for simulation was

$$x_t = .75x_{t-1} - .50x_{t-2} + e_t + 1.50e_{t-1} + 1.30e_{t-2} + 1.10e_{t-3} + 1.40e_{t-4}$$

The autoregressive part of the process is an AR(2) with $a_1^2 + 4a_2 = -1.44$, which is negative, so this is a pseudoperiodic process, whose peak frequency is $\cos 2\pi f = |a_1|/2(-a_2)^{1/2} = .53$, so that $f = .16$. The ACF of the AR(2) process should show a damped sine wave characteristic of a region 4 AR(2) process.

Since $q - p = 4 - 2$, we would expect two anomalous values for the ACF before the characteristic AR(2) behavior begins. Figure 20.2 illustrates the expected pattern. There are two anomalous values of the ACF, $r_1 = .78$ and $r_2 = .32$, before the damped oscillations characteristic of the AR(2) component begin.

Figure 20.1. Realization of an ARMA (2, 4) process.

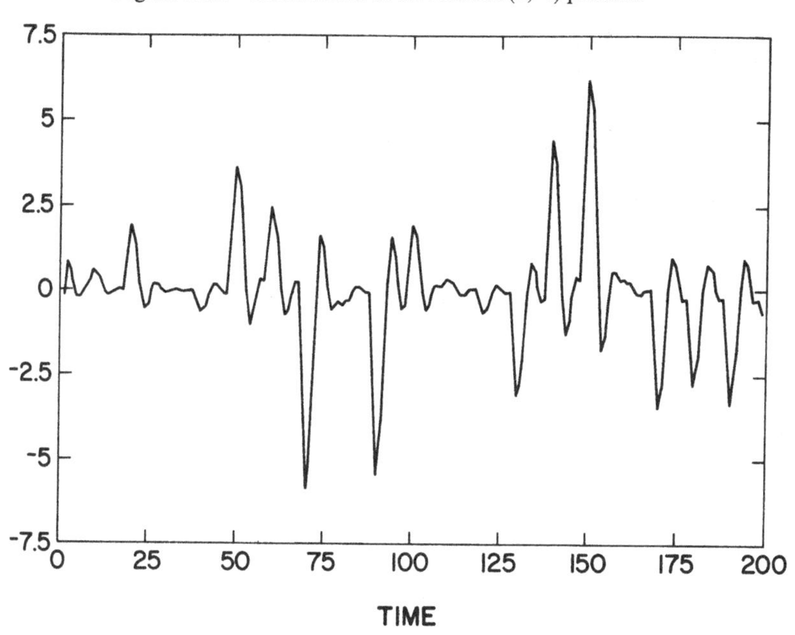

Figure 20.2. ACF of the ARMA (2, 4) process.

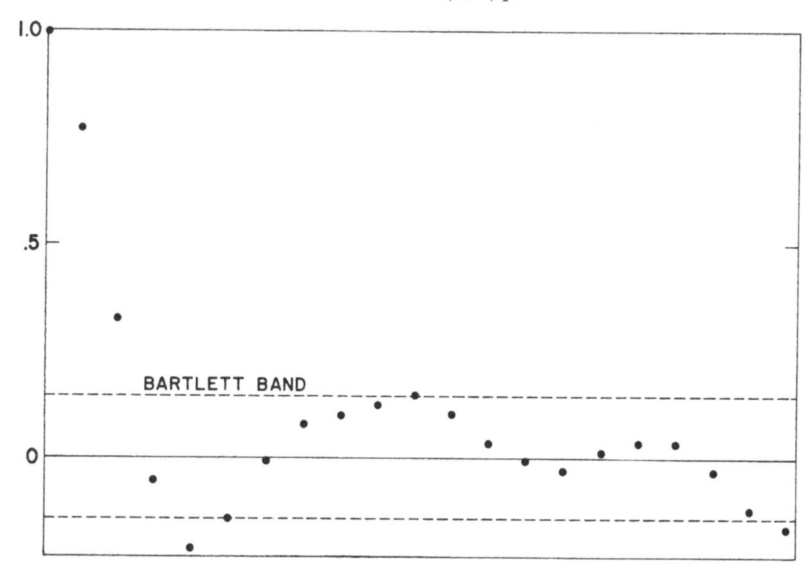

Table 20.1. *AR fitting of a realization of an ARMA(2,4) process*

	Order of AR fit													
Coefficients	1	2	3	4	5	6	7	8	9	10	11	12	13	14
a_1	.774	1.313	1.491	1.467	1.468	1.467	1.491	1.476	1.476	1.476	1.476	1.476	1.480	1.480
a_2		−.696	−1.032	−.938	−.938	−.923	−.957	−.923	−.923	−.919	−.920	−.912	−.917	−.916
a_3			.256	.121	.112	.134	.141	.109	.117	.120	.120	.106	.107	.106
a_4				.091	.096	−.044	−.065	−.059	−.068	−.092	−.092	−.090	−.085	−.085
a_5					−.004	.215	.360	.347	.349	.377	.377	.371	.371	.371
a_6						−.149	−.380	−.293	−.296	−.302	−.302	−.273	−.272	−.272
a_7							.157	.022	.048	.057	.057	.020	.010	.010
a_8								.091	.050	−.023	−.023	−.015	−.001	−.002
a_9									.027	.145	.145	.134	.130	.132
a_{10}										−.008	−.081	.008	.012	.012
a_{11}											.001	−.142	−.176	−.176
a_{12}												.100	.152	.148
a_{13}													−.037	−.032
a_{14}														−.003
Residual variance	.894	.461	.431	.427	.427	.418	.407	.404	.404	.401	.401	.397	.397	.397

Note: Initial variance = 2.229; $2/\sqrt{N} = .141$.

The reader may find it of interest to examine what order of AR process will fit these data. Table 20.1 shows the autoregressive coefficients of the PACF (recall that this is the last coefficient at each stage of fitting) and the residual variance at each stage. An examination of this table would lead one to select an AR(7) model because the PACF is less than $2/\sqrt{N} = .141$ beyond order 7. This model accounts for 81.74% of the initial variance. Note that it might be erroneous to stop at an AR(3); even though the next two PACF coefficients are not significant, they regain strength for lags 6 and 7. Nonetheless, an AR(3) fit may be adequate for many purposes. In this case the AR(7) fit has only one more parameter than the ARMA(2, 4) model.

How differencing incorporates trend

If the time series is nonstationary in the mean, it may be the case that first differences of the series can be represented by an ARMA(p, q) process. In that case

$$\nabla z_t = z_t - z_{t-1} \tag{20.11}$$

is a realization of an ARMA(p, q) process, and hence the model can be written

$$\mathrm{AR}(B)\nabla z_t = \mathrm{MA}(B)e_t \tag{20.12}$$

This is called an autoregressive, *integrated*, moving-average process, ARIMA(p, 1, q). The 1 is the order of differencing necessary for stationarity. If second differencing is necessary, the model is ARIMA(p, 2, q).

The ARIMA models thus make it possible for a series to have two parts, a *stochastic* part [the ARMA(p, q) part] and a *deterministic* part (the ∇^d part). The result is an ARIMA(p, d, q) model.

Deterministic trend can be included in the model by making $\nabla^d z_t$ be represented by an ARMA process that contains a nonzero constant. When $d = 1$ and the differenced process is a first-order moving-average process, linear trend (also called "deterministic drift") can be introduced by

$$\nabla z_t = b_0 + e_t - b_1 e_{t-1} \tag{20.13}$$

To see that this is indeed representative of a linear trend, note that if

$$z_t = (L + Kt) + M \sum_{1}^{t-1} e_j + e_t \tag{20.14}$$

then

$$\nabla z_t = z_t - z_{t-1} = K + Me_{t-1} + e_t \tag{20.15}$$

Hence, $K = b_0$ and $M = -b_1$. In the expression for z_t above, the deterministic component is the part $(L + Kt)$. Hence, this model is an MA(1) model superimposed on a deterministic trend line.

If the model is

$$\nabla^2 z_t = b_0 + e_t - b_1 e_{t-1} \tag{20.16}$$

the MA(1) model is superimposed on a quadratic $(L + K_1 t + K_2 t^2)$, and so on.

The degree of differencing

Using the information from previous chapters, it is possible to find the most appropriate ARIMA(p, d, q) process. The first stage in model building is called *identification*, which means determining the orders p, d, and q.

The most important of these three is the degree of differencing, d. The degree of differencing is most crucial because it is possible to *overdifference*, which will introduce spurious and meaningless patterns into the transformed, overdifferenced series (see Padia, 1975).

The question to ask is: Does the sample correlogram decay rapidly, [i.e., follow an exponential decay pattern (not linear decline to zero)]? If the answer is yes for some smallest value of d, the order of differencing is determined. One method for estimating d was suggested by Glass (personal communication). For each degree of differencing, d, we compute the correlogram $r_{k,d}$ of this differenced series and calculate:

$$\sum_{k=0}^{T/6} r_{k,d}^2 \tag{20.17}$$

where T is the sample size. The number $T/6$ is arbitrary, but we should not compute too many autocorrelations, so that these sums will be comparable across degrees of differencing. The statistic in equation (20.17) should begin increasing when we have overdifferenced, and this will make it possible to estimate the proper degree of differencing. Although the expression in (20.17) resembles the Box–Pierce statistic, the reader should resist comparing the expression to a chi-square distribution.

Summary flowchart for identification of *p*, *d*, and *q*

Step I is to identify the degree of differencing, d. The goal is to transform the series to a stationary time series, which, it will be recalled (1) varies around a fixed mean, (2) has variance independent of historical time, (3) has an autocovariance structure that is independent of historical time, and (4) has an

Figure 20.3. Summary flowchart for identifying p, d, and q in the ARIMA (p, d, q) model.

Step I. Choose degree of differencing, d such that the differenced series:

1. Varies about a fixed mean,
2. Has variance independent of displacements in time,
3. Has autocovariance independent of displacements in time, and,
4. Has an ACF that dies out after lags of N/5.

Compute $\sum_{K=0}^{T/6} r^2_{K,d}$ as a function of d, and note where the minimum value of d occurs.

Step II. Is the differential series a pure AR or a pure MA process?

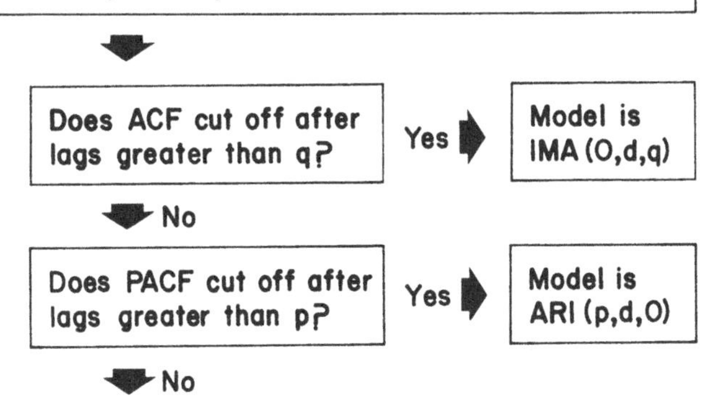

Since both ACF and PACF tail off, Model is ARIMA (p, d, q). To identify p and q, see Step III.

Step III. How many terms of the ACF are anomalous (i.e., not a mixture of exponential decay and damped sine waves)? The number of anomalous terms in the ACF = (q − p). The number of anomalous terms in the PACF = (p − q).

ACF that dies out rapidly. A crude rule of thumb for the last criterion, given in Chapter 8, was that r_k^2 decreases faster than $1/k$. An even cruder rule of thumb is that the ACF dies out after lags greater than $N/5$, where N is the number of observations. This latter criterion is crude indeed, since, if the data are *slowly* cyclic, the ACF will come up again somewhat from its decay after a long lag. These steps are summarized in the first box of the flowchart (Figure 20.3). Recall that differencing is a powerful transformation that does far more than simply eliminate polynomial trend.

Step II of the flowchart is used for the case when the differenced data are not an MA(q), which is the case if the ACF cuts off abruptly after q lags, or an AR(p), which is the case if the PACF cuts off abruptly after p lags.

In step III, identification proceeds to an ARIMA(p, d, q) model. The number of anomalous terms in the ACF is $(q - p)$, and the number of anomalous terms in the PACF is $(p - q)$.

Once again a reminder is in order that this procedure is by no means automatic. Indeed, after a model is identified and its parameters are estimated, Box and Jenkins (1970) pointed out that diagnostic checking of the residual is in order. Recall that the test they recommend for diagnostic checking is the Box–Pierce test (see Chapters 19 and 22).

Parameter estimation

Introduction

Once p, d, and q are determined, the task that remains is to find best estimates of the model parameters. How is this accomplished? The parameters are estimated by methods of nonlinear least squares (which is a maximum likelihood solution if the e_t's are normally distributed). For low-order models, Box and Jenkins recommended a grid-search procedure, which is usually not the most efficient computational procedure. Nonetheless, a grid-search procedure is conceptually clear. In the grid-search procedure, the parameters of the model are successively varied about their stationarity and invertibility regions, the sum of squares of the residuals are computed, and that combination of model parameters is selected that minimizes this sum of squares. The autocorrelation generating function can be used to obtain starting values for the parameters estimates of the autoregressive component.

Any stationary, invertible ARIMA (p, d, q) model can be solved for the random series e_t as follows:

$$\text{ARIMA}(p, d, q): \quad \text{AR}(B)\nabla^d z_t = \text{MA}(B)e_t \tag{20.18}$$

or, alternatively,

$$\text{white noise:} \quad e_t = \text{MA}^{-1}(B)\text{AR}(B)(1 - B)^d z_t \tag{20.19}$$

It is then possible to write

$$S = \sum_{i=1}^{N} e_i^2 \tag{20.20}$$

which is a function of the unknown parameters, and the quantity we seek to minimize.

However, how is this sum to be calculated? There are several methods. For a review of these method, see Aigner (1971). One method suggested by Box Jenkins (1970) is to calculate the e_i recursively from the observed series. For example, if we have identified the model as an ARIMA(1, 1, 1) model

$$(1 - a_1 B)(1 - B)z_t = (1 - b_1 B)e_t \tag{20.21}$$

we can write

$$e_t = b_1 e_{t-1} + z_t + (1 - a_1)z_{t-1} + a_1 z_{t-2} \tag{20.22}$$

Box and Jenkins (1970) pointed out that "special care is needed when the maximum of the likelihood function may be on or near a boundary" (p. 225), because the quadratic approximation may not be adequate to describe the likelihood. However, for an $S(\beta) = \sum e_i^2$ that is reasonably quadratic in the model parameters, Box and Jenkins (1970, p. 229) derived the result that a $(1 - \alpha)$ confidence region is

$$S(\beta) \leq S(\hat{\beta})\left[1 + \frac{\chi_\alpha^2(k)}{N}\right] \tag{20.23}$$

where k is the number of model parameters and N is the number of observations. Knowing the z_t, we can calculate e_t from e_{t-1}. To start the series we can set e_1 equal to its expected value of zero.[1] A $1 - \alpha$ confidence region is given by the set at (a, b) so that

$$S(a, b) \leq (S_{\text{calculated minimum}})\left[1 + \frac{\chi_\alpha^2(p + q)}{N}\right] \tag{20.24}$$

Then calculate

$$S(a, b) = \sum e_i^2 \tag{20.25}$$

for different values of the parameters a and b. Once again, recall that the grid search method works best when S is a quadratic function with one minimum. One test Box and Jenkins suggest is to use the second differences of S, $\nabla^2 S$, as a function of the parameters as an approximation of the second derivative. This difference, $\nabla^2 S$, should be a constant locally around values of the parameters that minimize S.[2]

An example

Identification. The series *B* IBM data (see Figure 20.4) given by Box and Jenkins (1970) can be identified using the autocorrelations and partial autocorrelations (see Table 20.2). First differencing is necessary because the ACF does not die out. The series ∇z_t is an MA(1) because the PACF is effectively zero after the first lag. The differenced series could be white noise, so an "overfit" would be an MA(1), probably with very small b_1. The model is thus

$$\nabla z_t = e_t - b_1 e_{t-1} \tag{20.26}$$

Estimation. For each of a series of values of b_1, $S(b_1) = \sum e_i^2$ is calculated for the observed time series, z_t. If we define λ as $1 - b_1$, then Table 20.3 gives the values of S as a function of λ, as well as ∇S and $\nabla^2 S$. $S(\lambda)$ is a minimum at around $\lambda = 1.1$, and $\nabla^2 S$ does not change very much around this value, so that the confidence interval around S may be a good approximation (see Table 20.3).

Box and Jenkins (1970) find a minimum value[3] of $S(\lambda) = 19{,}216$, and thus the confidence interval is

$$S(\lambda) \le S(\hat{\lambda})\left[1 + \frac{\chi^2_{1-\alpha}(p+q)}{N}\right] \tag{20.27}$$

$$S(\lambda) \le 19{,}230\left(1 + \frac{3.84}{396}\right) = 19{,}416 \tag{20.28}$$

If we refer to Figure 20.5 and draw this line ($S = 19{,}416$), it defines the confidence limits of λ as $.98 < \lambda < 1.19$.

Figure 20.4. IBM data.

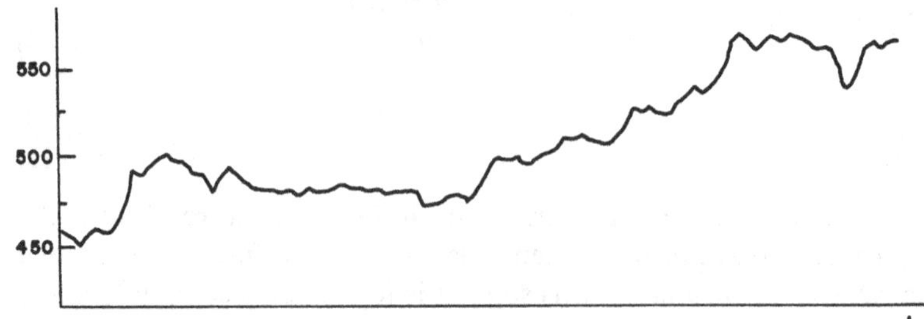

Table 20.2. *ACF and PACF for IBM stock prices series B and its first and second differences*

	Lag																			
Series	1	2	3	4	5	6	7	8	9	10	11	12	13	14	15	16	17	18	19	20
	ACF																			
z	.99	.99	.99	.98	.98	.97	.97	.96	.95	.95	.94	.94	.93	.92	.92	.91	.90	.90	.89	.88
∇z	.08	.00	−.05	−.04	−.02	.13	.07	.03	−.07	.02	.08	.06	−.05	.07	−.07	.12	.13	.05	.05	.07
∇z^2	−.45	−.02	−.04	.00	−.07	.11	−.01	.04	−.11	.02	.04	.04	−.12	.14	−.18	.10	.05	−.04	−.01	.10
	PACF																			
z	.996	−.09	.01	.05	.02	.02	−.12	−.05	−.02	.06	−.05	−.09	−.03	.07	−.08	.06	−.14	−.10	−.01	−.08
∇z	.09	−.01	−.05	−.03	−.02	.13	.05	.02	−.06	.05	.09	.03	−.08	.08	−.06	.14	.10	.00	.17	.08
∇z^2	.45	−.28	−.24	−.20	−.29	−.17	−.13	−.03	−.14	−.16	−.09	.02	−.13	.01	−.19	−.13	−.03	−.10	−.10	−.06

Table 20.3. $S(\lambda)$ *and estimates of its second derivative*

$\lambda = 1 - b_1$	$S(\lambda)$	∇S	$\nabla^2 S$
1.5	23,929	2,334	961
1.4	21,595	1,373	634
1.3	20,222	739	476
1.2	19,483	263	406
1.1	19,220	−143	390
1.0	19,363	−533	422
0.9	19,896	−955	509
0.8	20,851	−1,464	692
0.7	22,315	−2,156	1067
0.6	24,471	−3,223	
0.5	27,694		

Figure 20.5. Sum of squares of the residual for the IBM data with 95% confidence interval.

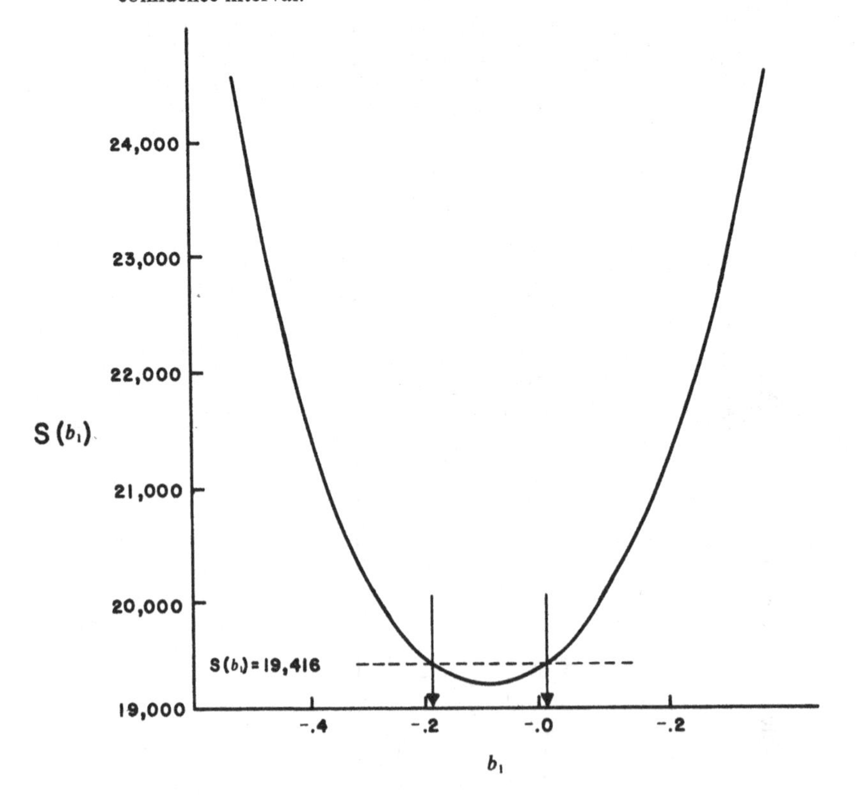

Wu–Pandit method

An elegant modification of the Box–Jenkins method of model fitting was proposed by Pandit and Wu (Wu and Pandit, in preparation), called the dynamic design systems (DDS) method, which is based on sampling of a continuous process and a relationship between differential and difference equations. The DDS method uses ARMA$(2n, 2n-1)$ models to fit the stochastic component of a time series. In a stepwise procedure ARMA(2, 1), ARMA(4, 3), ARMA(6, 5), and so on, models are compared. Parameters are estimated as before using a nonlinear least-squares procedure. If A_1 is the residual sum of squares for the larger model and A_0 is the residual sum of squares for the smaller model, then Pandit and Wu suggest that

$$\frac{(A_1 - A_0)/S}{A_0/(N-r)} \sim F_{S,N-r}$$

when N is the number of observations in the residual series in the larger model that has r parameters, and we are testing to see if S of these parameters are zero. In the stepwise procedure, $r = 4n + 3$ and $S = 4$, when we test an ARMA $(2n, 2n-1)$ against a larger model that is ARMA$(2n + 2, 2n + 1)$.

This ratio is only approximately distributed as F and then only if the smaller model is assumed to be correct. Devor (personal communication) suggests that the stepwise procedure be used with caution because it may lead one to stop when a higher-order model may be appropriate (just as was the case in stepwise fitting of AR models). Devor also recommends always examining the ACF and PACF of the residuals as well as the F ratio. To these recommendations we add an examination of the spectrum of the original series to get a rough estimate of the order of the model and examination of the spectrum of the residuals.

A set of computer programs for the DDS method is available at the University of Illinois, developed by William Kline and Richard Devor.

21 Forecasting

No brief chapter can hope to do justice to the subject of forecasting. It is a literature replete with wisdom and valuable experience. This chapter serves only as an introduction to this interesting field and covers those topics bearing on time-series analysis. Linear least-squares forecasts are discussed for both autoregressive and moving-average models.

Introduction

One of the important goals in time-series analysis is successful prediction estimating x_{t+s} within prescribed limits given the past values $x_t, x_{t-1}, \ldots, x_{t-n}$. Examples of important areas of prediction abound: inventory control, population projections, economic forecasts, quality control, meteorological forecasts, automatic feedback control, and on and on. Chatfield (1975) grouped forecasting techniques into three broad categories: (1) subjective, (2) multivariate, and (3) univariate. Chatfield's classification is useful for our discussion. The first category, *subjective forecasting*, involves, as the name suggests, using an "educated guess." Even though this is a decidedly nonrigorous approach, it may be optimal in terms of cost or it may be the only reasonable method available. In the hands of someone experienced with the data at hand and familiar with how the quantity to be predicted tends to move under certain circumstances, quite adequate forecasts can often be made. This will be the case to the extent that deterministic trends and cycles dominate the stochastic components. For example, in the case of meteorological predictions, all scientific observations may point to the passage of a cold front with concomitant precipitation, but at the same time give no clear clue at all as to whether the precipitation will be a relatively benign rain or destructive sleet and icing. The meteorologist is left to make a potentially very expensive forecast.

The multivariate, or regression, technique may be familiar to most people with statistical backgrounds. In this method the sample realization vector $X(t)$ is fit by least squares to a set of independent variables evaluated at the appropriate times t; that is, we assume

$$X(t) = \sum_{j=1}^{k} B_j Y_j(t) + e_t \tag{21.1}$$

where $Y_j(t)$ is the jth independent variable evaluated at time t, B_j is the jth regression coefficient, and e_t is the random-error term. In other words, our observed time series, $X(t)$, is fit by least squares to a set of appropriately weighted time series $Y_j(t)$.

Regression models are frequently poor predictors, and often to make a model perform even moderately well it may be necessary to incorporate a large number of *independent* regressors. Having a large number of regressors increases the cost of computing the B_j's and aggravates mathematical problems such as linear dependence among the regressors. A further problem that complicates the application of a regression model is that the e_t are often correlated, even for a large number of regressors, thus biasing the variance of the estimates of the B_j's.

The univariate technique is an approach that has often been found to be useful in forecasting. In this technique the fundamental forecasting model is taken to be of the autoregressive form

$$x_t = \sum_{j=0}^{k} a_j x_{t-j} + e_t \tag{21.2}$$

In some cases good success has been obtained with values of k as small as 1 or 2. Methods of arriving at satisfactory coefficients a_j range from simple trial and error to the rather elaborate computations involved in Box–Jenkins models. Not only do univariate models frequently yield sufficiently accurate predictions, but some of the computational schemes for the a_j's are short and inexpensive and lend themselves to easily updating the model by utilizing newly observed x_t's.

More complex "lagged models," combining these last two methods, are possible if the past of $X(t)$ or the pasts of $Y_j(t)$ are included in the model. If there is no clearly specified dependent measure $X(t)$, the model in equation (21.1) represents a *system* of equations in which each time series plays the role of dependent variable. In this case an equation just like equation (21.1) can be written for each of the variables being considered [i.e., $X(t)$, $Y_1(t)$, $Y_2(t), \ldots, Y_k(t)$]. In this situation a system of equations must be solved. This complex situation is discussed in Chapter 27.

Forecasting moving-average processes

A measure of the goodness of our prediction of x_{t+s} (called the s-step prediction) is the prediction variance. If $\dot{x}_{t+s}$ denotes our estimate, this variance is

$$\sigma_s^2 = E[(\hat{x}_{t+s} - x_{t+s})^2] \tag{21.3}$$

A good prediction usually means a small prediction variance. If the series x_t is white noise, $N(0, \sigma_e^2)$, we cannot do any better than predicting the expected value of the process, which is zero, and the prediction variance, which is σ_e^2. In other words, you cannot expect to learn anything from the past if it is white noise.

But what if x_t has some covariance structure? Then the answer is that you can do better than guessing the mean. If we know that the process is an infinite MA,

$$x_t = \sum_{j=0}^{\infty} b_j e_{t-j}, \qquad b_0 = 1 \tag{21.4}$$

then it can be shown that the best linear least-squares predictor of x_{t+s} is

$$\hat{x}_{t+s} = \sum_{j=0}^{\infty} b_{j+s} e_{t-j} = b_s e_t + b_{s+1} e_{t-1} + \cdots \tag{21.5}$$

The prediction variance is

$$\sigma_s^2 = (1 + b_1^2 + b_2^2 + \cdots + b_{s-1}^2)\sigma_e^2 \tag{21.6}$$

In practice, b_{j+s} would be estimated by the methods of Chapter 10 and the e_{t-j} would be estimated by the residuals from fit at time $t - j$. Moreover, the prediction variance can be estimated by replacing theoretical quantities in equation (21.6) by their estimates. Note that for a stationary process, the variance of the forecast error increases rapidly as s increases, and then converges; nonetheless, our confidence in a forecast becomes worse as we step into the future, away from our data. For this reason it is best to forecast only one or two steps ahead and to update the forecasts as data become available.[1]

Forecasting autoregressive processes

If the process is an AR(1) process,

$$x_t = a x_{t-1} + e_t \tag{21.7}$$

then it is easy to show, simply by representing x_t as an infinite moving-average process, that the best one-step forecast is

$$\hat{x}_{t+1} = a x_t \tag{21.8}$$

Note that since $|a| < 1$ for stationarity, this means that the forecasts drift toward the mean.

In general, if x_t is an AR(p) process, the best one-step prediction is

$$\boxed{\hat{x}_{t+1} = \sum_{s=1}^{p} a_s x_{t-s+1}} \tag{21.9}$$

For example, if we have estimated an AR(2) model and are standing at time $t = 48$, the forecast for x_{49} is

$$\hat{x}_{49} = \hat{a}_1 x_{48} + \hat{a}_2 x_{47}$$

Forecasting more than one step at a time requires a recursive procedure. If we wish to predict s steps ahead, the estimate $\hat{x}_{t+s}$ is given by

$$\hat{x}_{t+s} = \sum_{j=1}^{s-1} a_j \hat{x}_{t+s-j} + \sum_{j=s}^{p} a_j x_{t+s-j} \tag{21.10}$$

This is a formidable equation. It has two parts; the first part is recursive, goes back to time $t + 1$, and uses other forecasts, and the second uses real data. It may be helpful to write out each part:

$$\boxed{\begin{aligned}\hat{x}_{t+s} &= (a_1\hat{x}_{t+s-1} + a_2\hat{x}_{t+s-2} + \cdots + a_{s-1}\hat{x}_{t+1}) \\ &\quad + (a_s x_t + a_{s+1}x_{t-1} + \cdots + a_p x_{t+s-p})\end{aligned}} \tag{21.11}$$

For example, if the model is an AR(4), the $s = 3$ step-ahead forecast is

$$\hat{x}_{t+3} = (a_1\hat{x}_{t+2} + a_2\hat{x}_{t+1}) + (a_3x_t + a_4x_{t-1}) \tag{21.12}$$

Suppose that an AR(2) model has been fit to the data. The s-step-ahead forecast is

$$\hat{x}_{t+s} = a_1\hat{x}_{t+s-1} + a_2\hat{x}_{t+s-2} \tag{21.13}$$

For $s = 2$, this yields

$$\hat{x}_{t+2} = a_1\hat{x}_{t+1} + a_2x_t$$

for $s = 3$, this yields

$$\hat{x}_{t+3} = a_1\hat{x}_{t+2} + a_2\hat{x}_{t+1}$$

for $s = 4$, this yields

$$\hat{x}_{t+4} = a_1\hat{x}_{t+3} + a_2\hat{x}_{t+2}$$

and so on.

Forecasting headache pain

The headache data will be employed in this section to compare the one-step-ahead forecast with the k-step-ahead forecast. A second-order autoregressive model was fit to these data, where based on the first 48 observations, the values of a_1 and a_2 were .5378 and $-.0614$, respectively, and hence

$$\hat{x}_{48+1} = .5378(x_{48}) + (-.0614)(x_{47}) = .1531$$

Once the true value of x_{49} is known to be .8039, *the model is refit*, and a new set of estimates for $\hat{a}_1$ and $\hat{a}_2$ are obtained, .5308, and $-.0518$, respectively. Then a new one-step-ahead forecast is computed:

$$\hat{x}_{49+1} = .5308(x_{49}) + (-.0518)(x_{48}) = .4088$$

This procedure is continued throughout. The result is plotted in Figure 21.1c. It can be seen that the one-step-ahead forecast follows the general

Figure 21.1. Forecasting headache pain using autoregressive models.

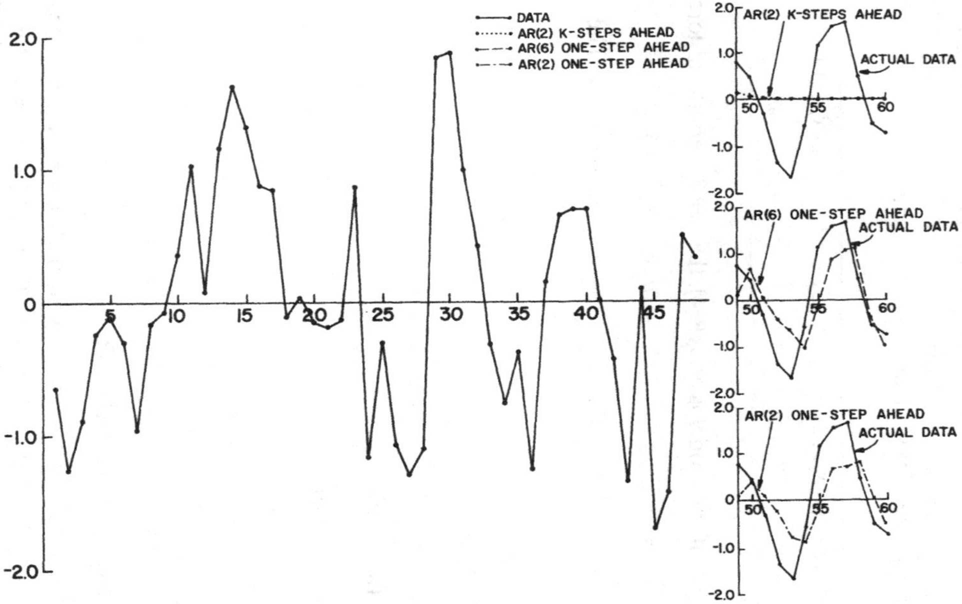

shape of the data. The sample error variance, $\sum_{t=49}^{60}(\hat{x}_t - x_t)^2/11$, was $6.8854/11 = .6259$.

The k-step-ahead forecast is computed quite differently: the same AR(2) model parameters are used throughout the forecast as follows. The first step is identical to the one-step-ahead case. The second step is computed as follows:

$$\hat{x}_{48+2} = \hat{a}_1\hat{x}_{48+1} + \hat{a}_2\hat{x}_{48} = .0605$$
$$\hat{x}_{48+3} = \hat{a}_1\hat{x}_{48+2} + \hat{a}_2\hat{x}_{48+1} = .0229$$
$$\hat{x}_{48+4} = \hat{a}_1\hat{x}_{48+3} + \hat{a}_2\hat{x}_{48+2} = .0085$$
$$\hat{x}_{48+5} = \hat{a}_1\hat{x}_{48+4} + \hat{a}_2\hat{x}_{48+3} = .0031$$

and so on. Notice that the estimates decrease to the mean of zero. The forecast error was $13.41/11 = 1.2191$, and the shape does not parallel the data. With more data points, the updated model parameter estimates change considerably, so that at the 12th step they are .6526 and $-.1992$ for a_1 and a_2, respectively, which represents a model more like a pseudoperiodic AR(2) than an AR(1).

Note that it may be possible to improve on the forecast itself using a higher-order AR model. For example, using all the data, an AR(2) accounts for 32.96% of the total variance, and an AR(6) accounts for 40.44%. Figure 21.1a contains a plot of the one-step-ahead forecast using AR(6) model. The error variance in this forecast was .4272, which is an improvement over the AR(2) one-step forecast. However, notice that the improvement is not enormous. It should also be noted that this example employed a purely stochastic model; forecasting is most impressive when the deterministic components are relatively large and well fit, that is, when this method can be combined with the regression method described earlier.

Confidence intervals

The prediction variance of the s-step-ahead predictor can be evaluated by writing the AR model as an MA model, whose prediction variance is

$$\sum_{k=0}^{s-1} b_k^2\sigma_e^2 \tag{21.14}$$

where σ_e^2 is the variance of the noise process that generates the MA process, and where the coefficients, b_k, can be evaluated using operator notation as

$$\mathrm{AR}(B)\mathrm{MA}(B) = 1 \tag{21.15}$$

Equation (21.15) can be solved to yield the expressions

$$\sum_{j=0}^{k} b_j c_{k-j} = 0, \qquad k = 1, 2, \ldots, s-1 \tag{21.16}$$

where $c_0 = 1$ and $c_j = -a_j$, for $j > 0$, are the autoregressive coefficients.

For the headache data, after an AR(2) fit to the first 48 points, $\hat{a}_1 = .5378$, $\hat{a}_2 = -.0632$, and the residual variance $\hat{\sigma}_e^2$ was .5882, which can also be computed from the Yule–Walker equations using

$$\hat{\sigma}_e^2 = \hat{\sigma}_x^2(1 - \hat{a}_1 r_1 - \hat{a}_2 r_2)$$

where $\hat{\sigma}_x^2$ is the variance of the data up to the point of forecasting. Now employ equation (21.15) for $s =$ two steps ahead ($k = 1$) as follows:

$$b_0 c_1 + b_1 c_0 = 0$$

where $c_0 = 1$, $b_0 = 1$, so

$$c_1 + b_1 = 0$$

so

$$b_1 = -c_1 = a_1$$

and hence the step 2 prediction variance is estimated by

$$\hat{b}_0^2 \hat{\sigma}_e^2 + \hat{b}_1^2 \hat{\sigma}_e^2 = \hat{\sigma}_e^2(1 + \hat{a}_1^2) = .7583$$

For $s =$ three steps ahead ($k = 2$), equation (21.15) is

$$b_0 c_2 + b_1 c_1 + b_2 c_0 = 0$$

or

$$-a_2 + a_1(-a_1) + b_2(1) = 0$$

or

$$b_2 = a_2 + a_1^2$$

The prediction variance at the third step is then estimated by

$$\hat{\sigma}_e^2(1 + \hat{b}_1^2 + \hat{b}_2^2) = .7884$$

At $s =$ four steps ahead ($k = 3$), we obtain

$$b_0 c_3 + b_1 c_2 + b_2 c_1 + b_3 c_0 = 0$$

which can be solved for b_3 since $c_3 = -a_3 = 0$, as follows:

$$b_3 = a_1 a_2 + a_1(a_2 + a_1^2) = a_1^3 + 2a_1 a_2$$

The prediction variance at the fourth step is estimated as

$$\hat{\sigma}_e^2(1 + \hat{b}_1^2 + \hat{b}_2^2 + \hat{b}_3^2) = .7929$$

Proceeding in this manner, the prediction variance at each subsequent step can be computed as .7931, .7937, and so on. Note that this series of numbers converges.

If the time-series data are normally distributed, and the estimated prediction variance at the sth step is denoted $\hat{\sigma}_s^2$, a confidence interval for x_{t+s} with probability $(1 - \alpha)$ of including x_{t+s} is approximately

$$\hat{x}_{t+s} - (Z_{\alpha/2})\hat{\sigma}_s \leq x_{t+s} \leq \hat{x}_{t+s} + (Z_{\alpha/2})\hat{\sigma}_s \qquad (21.17)$$

where $Z_{\alpha/2}$ is the upper $\alpha/2$ critical point of a normal distribution.

22
Model fitting: worked example

This chapter contains practical advice in applying many of the techniques of the first 21 chapters. It includes worked examples in which the computations were performed either by computer or by hand.

In this chapter we discuss the process of fitting a model to the first 60 points of the New England traffic fatalities data (plotted in Figure 2.5). We will follow the flowchart of Figure 22.1 in these analyses.

Detect and remove trend

The first available procedure is to plot the data and visually examine the plot (see Figure 2.5). The total variance of the raw data is .675. There is a downward linear trend in these data. The periodogram for these data (Figure 2.6) showed some, although not a great deal of, elevation at zero. This means that the trend line accounts for some of the variance in the series, but not a great deal.

Next, we can remove the trend. In this case we performed a least-squares fit to the data and obtained equation (9.7). The residuals were then plotted

Figure 22.1. Flowchart for model fitting.

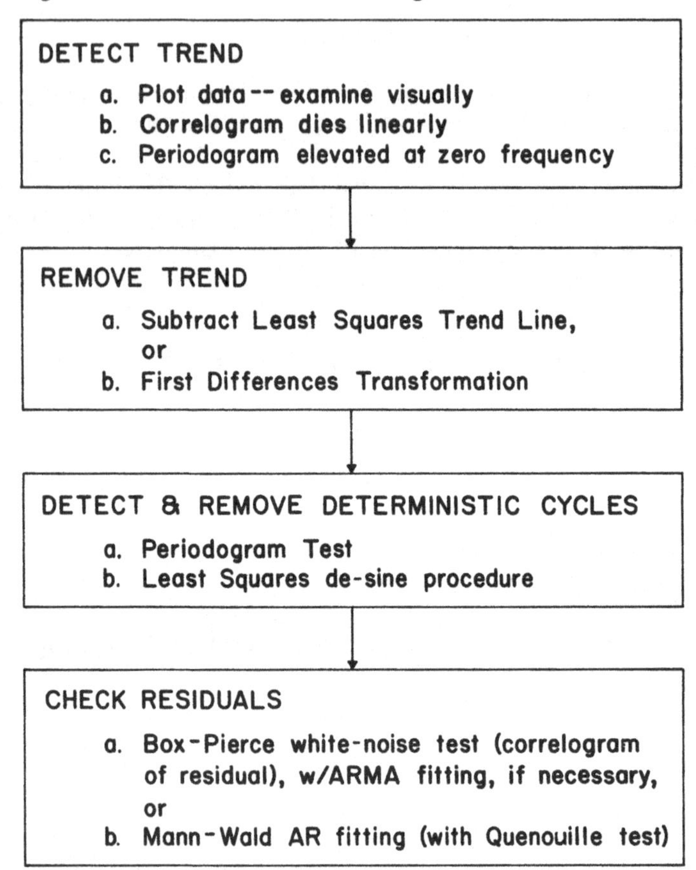

(Figure 9.8). The residual variance is .617. The next step is to search for and remove deterministic cycles.

Detect and remove deterministic cycles

Figure 22.2 is a graph of the periodograms of all 60 points and of the first 40 points of the detrended data. The ratio of these sample sizes is 60/40 = 1.5. The variance accounted for by the peak at .075 is 1.56 and .88, respectively, for 60 and 40 points. This ratio is 1.77, which is close to the ratio of the sample sizes. We thus have evidence of a deterministic cycle with a period of 12

Figure 22.2. Periodogram of the detrended traffic fatalities data for (a) all 60 observations, and (b) the first 40 observations.

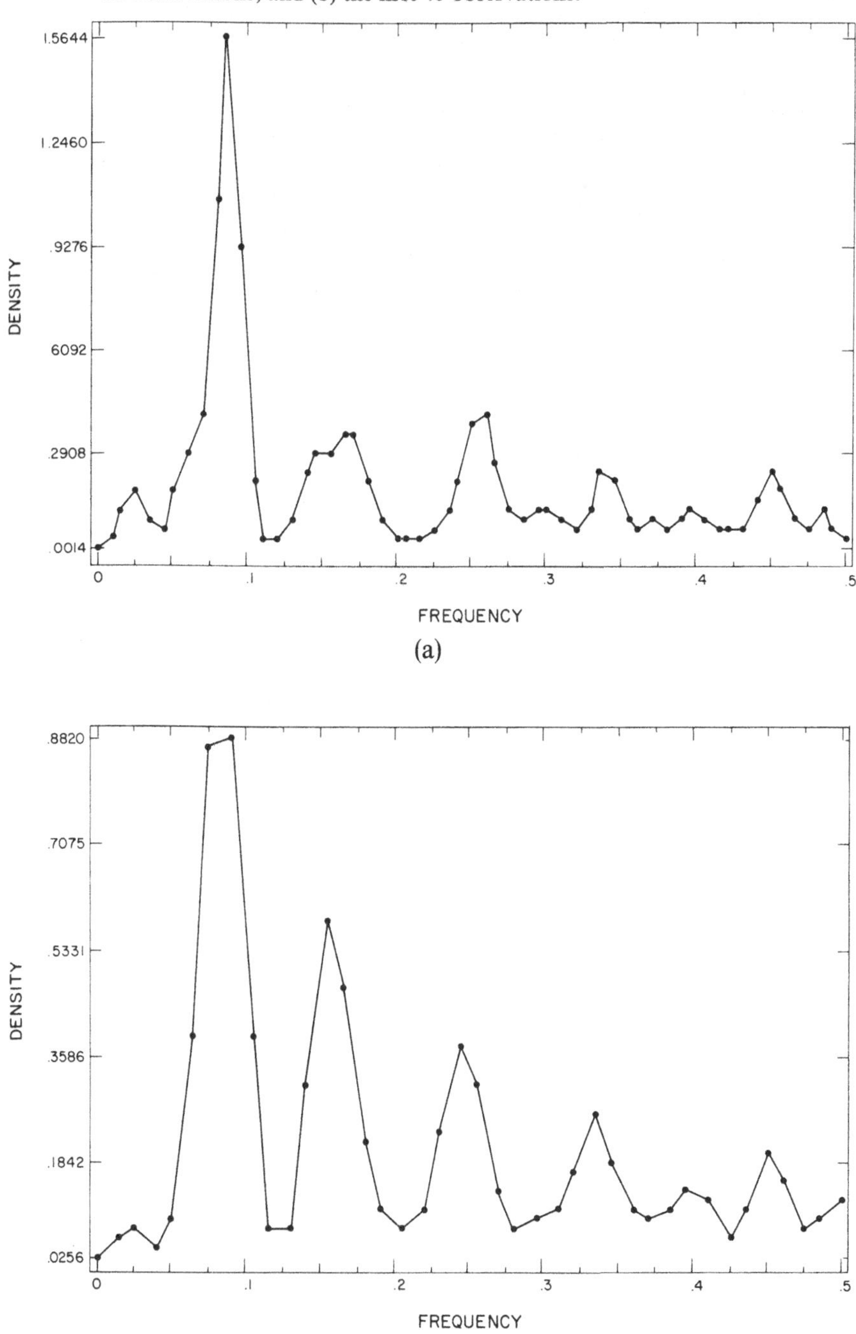

Figure 22.3. ACF of the de-sined and detrended traffic fatalities data.

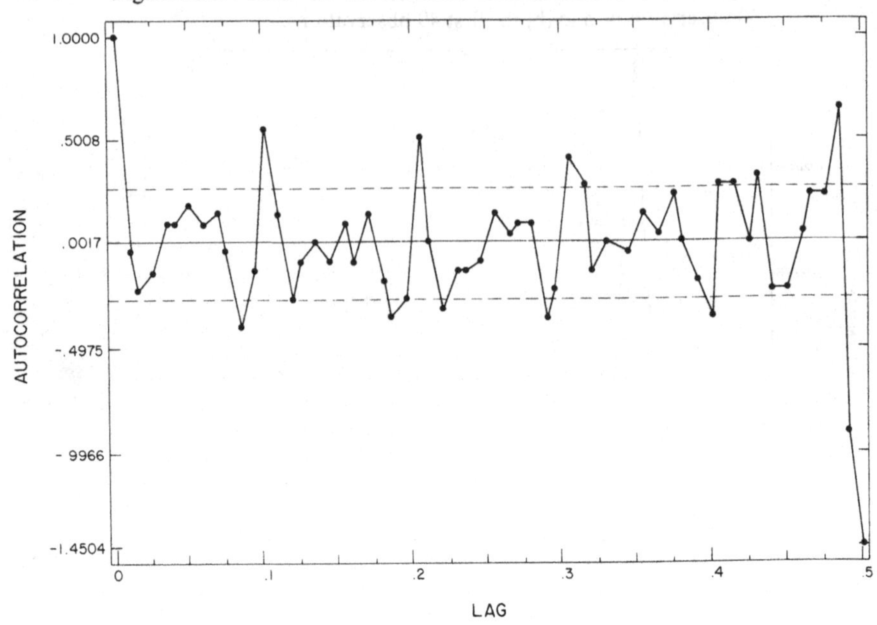

Figure 22.4. Spectral density function estimate of the detrended, de-sined traffic fatalities data.

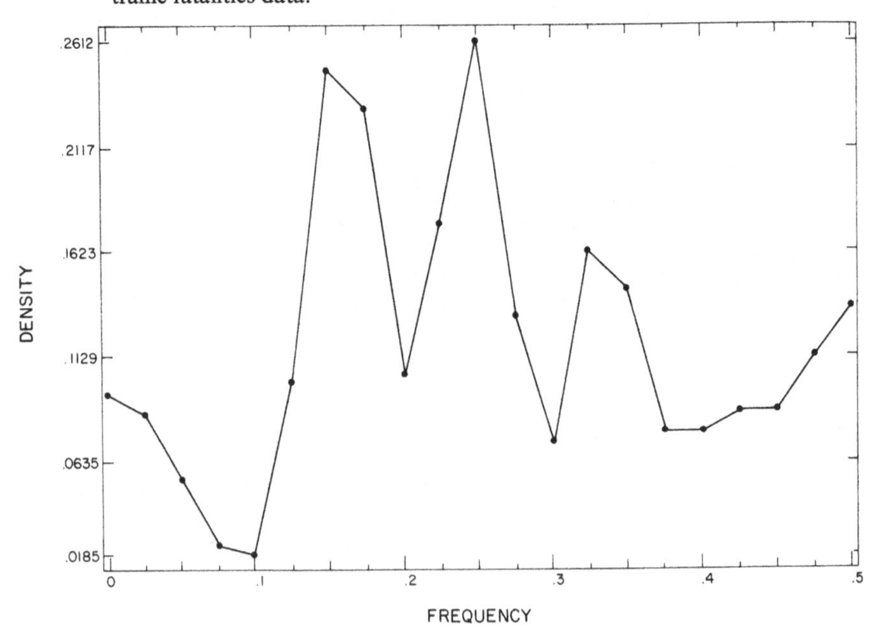

months. We next removed this cycle using a least-squares fit to A and B in

$$A \sin \frac{2\pi}{12} t + B \cos \frac{2\pi}{12} t$$

and obtained $\hat{A} = -.276$ and $\hat{B} = .674$. The residual variance is .371, so the annual cycle accounted for almost half of the variance of the data.

Figure 22.3 is a graph of the autocorrelations of the de-sined and detrended data. We can see that the correlations do not die down to within the $2/\sqrt{N}$ band drawn on the figure. This suggests that there is still power at other frequencies. The spectral density (computed using 20 lags, a Tukey–Hanning window, with a bandwidth of .10) of the detrended, de-sined data (Figure 22.4) shows two major remaining peaks; one is at $f = .161$, and the other at $f = .246$ (i.e., periods of 6.2 months and 4.1 months, respectively). If we wish to continue with a frequency-domain model, we could de-sine these two frequencies (4 and 6 months) and examine the residuals. We could continue in this way, although the appropriateness of this may be questioned; these cycles do not appear deterministic, and there is no obvious externally produced 6- or 4-month cyclicity, as there is at 12 months.

Analyze residuals

After removing the periodicities we decide to remove, we perform an autoregressive fit to the residual. Table 22.1 is an illustration of the results of taking these various options. The full model accounts for 69.5% of the total variance of the data.

If we were to consider the parsimony and explanatory power of these alternative models, we might reason as follows. Locate the (*) in Table 22.1. This refers to the model with trend, three sinusoids, and an AR(1) residual stationary stochastic process. The (*) model has nine parameters, two for trend, six for sinusoids, and one for AR(1) residual. The variance unexplained by this model is .227, or 66.4% of the variance explained. The other nine parameter models, (**), (***), and (****), each leave more unexplained residual variance; they explain 56.6%, 58.2%, and 49.3%, respectively. The simplicity of the (*) model is appealing because there is only one autoregressive term. However, parsimony and explanatory power are not the only considerations in selecting a model. We must be able to interpret the components. In the case of the (*) model we would have to explain the declining trend, the seasonal cycles, and the negative first-order autocorrelation. These explanations can lead to the development of theory. It makes a great deal of sense to select the most *interesting* model, even if it explains less of the variance than

Table 22.1. *Results of various options in modeling the traffic fatalities data*

		Autoregressive parameters						Sinusoidal parameters					
								$T = 12$		$T = 6$		$T = 4$	
Components of model	Residual variance	a_1	a_2	a_3	a_4	a_5	a_6	$\hat{A}$	$\hat{B}$	$\hat{A}$	$\hat{B}$	$\hat{A}$	$\hat{B}$
Trend	617												
De-sine $T = 12$	.371							−.276	.674				
AR(1)	.371	−.023											
AR(2)	.350	−.029	−.234										
AR(3)	.342	−.064	−.239	−.150									
AR(4)	.342	−.063	−.237	−.150	.008								
AR(5) (****)	.342	−.063	−.234	−.145	.009	.017							
AR(6)	.330	−.066	−.236	−.118	.053	.029	.187						
Trend	.617												
De-sine $T = 12$, $T = 6$	.305							−.276	.674	−.249	.262		
AR(1)	.298	−.147											
AR(2)	.282	−.181	−.235										
AR(3) (***)	.282	−.187	−.240	−.027									
AR(4)	.271	−.182	−.193	.010	.197								
AR(5)	.267	−.206	−.194	.033	.219	.122							
AR(6)	.263	−.222	−.223	.029	.245	.149	.130						

Trend	.617												
De-sine $T = 12$, $T = 4$	.302							−.276	.674			−.340	.369
AR(1)	.302	−.040											
AR(2)	.299	−.044	−.090										
AR(3) (**)	.293	−.056	−.096	−.145									
AR(4)	.288	−.076	−.109	−.153	−.136								
AR(5)	.287	−.071	−.103	−.148	−.132	.040							
AR(6)	.240	−.087	−.049	−.088	−.091	−.069	.407						
Trend	.617												
De-sine $T = 12$, $T = 6$, $T = 4$	.237							−.276	.674	−.249	.262	−.340	.369
AR(1) (*)	.227	−.200											
AR(2)	.227	−.209	−.041										
AR(3)	.225	−.204	−.020	.100									
AR(4)	.224	−.211	−.019	.113	.064								
AR(5)	.224	−.211	−.019	.113	.065	.006							
AR(6)	.206	−.213	−.038	.081	.070	.066	.283						

Note: Refer to the text for an explanation of the asterisks.

Table 22.2. *Reanalysis of the traffic fatalities data with cycles at 12, 6, and 3 months*

		Autoregressive parameters						Sinusoidal parameters					
								$T = 12$		$T = 6$		$T = 4$	
Components of model	Residual variance	a_1	a_2	a_3	a_4	a_5	a_6	$\hat{A}$	$\hat{B}$	$\hat{A}$	$\hat{B}$	$\hat{A}$	$\hat{B}$
Trend	.617												
De-sine $T = 12, 6, 3$	.250							−.276	.674	−.249	.262	−.151	.294
AR(1) (*****)	.249	−.087											
AR(2)	.243	−.101	−.153										
AR(3)	.237	−.125	−.170	−.160									
AR(4)	.210	−.072	−.112	−.118	.337								
AR(5)	.203	−.131	−.092	−.098	.350	.176							
AR(6)	.201	−.110	−.051	−.109	.339	.161	−.117						

Note: Refer to the text for an explanation of the asterisks.

other models, whose terms we cannot even begin to understand. For example, in Figure 22.4, the third peak has a period of 3 months, which corresponds to the four seasons. Suppose that we propose the following explanation for these components: (1) the annual cycle is explained by the peak summer vacation road traffic in August; (2) the 6-month cycle is explained by the fact that the winter months and the spring rains in New England are responsible for poor road conditions, which produce an independent peak in fatalities; and (3) the 3-month cycle is a function of peaks in travel on (and near) Memorial Day, Labor Day, Thanksgiving, and Easter. If we assume that this model is appealing for theoretical considerations, we can reanalyze the data ignoring the 4-month cycle. Table 22.2 presents the results of this analysis. The model (*****) compares reasonably well with the model (*) in Table 22.1 in terms of variance accounted for. Furthermore, in both models we have a small negative first-order autocorrelation (see the slight rise at the Nyquist frequency in Figure 22.4). This could be due to the fact that a high incidence of traffic fatalities in an area in one month has a temporary effect on drivers' speeds for a brief period (say one week) and that for the rest of the month they return to their normal speeds. This would result in slight oscillations with a two-month Nyquist cycle, of low power.

In this manner models may be selected from the array of reasonable models.

"Quick and dirty" time-series modeling by hand

This part of the chapter will discuss two problems: (1) univariate frequency-domain modeling and (2) univariate time-domain modeling. The discussion will make use of the migraine headache severity data referred to in Chapter 12. An inexpensive, nonprogrammable electronic statistical hand calculator was used for these computations.

Frequency-domain modeling by hand

For this purpose the Buys-Ballot table as used by Whittaker and Robinson (1924) will be employed to compute a function that is proportional to the periodogram. A visual examination of the data shows that there are approximately seven local peaks (around days 14, 23, 30, 39, 44, 49, and 57). For this reason Buys-Ballot tables were computed for periods from 4 through 11. Table 22.3 illustrates how these tables were constructed (see also Chapter 15 for a review of Whittaker and Robinson's data). In each case the table is constructed simply by writing the data out in order by rows that are P entries long. The column means are computed, and the statistic η is the ratio of the variance of these means to the variance of the series. A graph of η versus these

Table 22.3. *Frequency of the headache data using Buys-Ballot tables*

$p = 4$

1	2	3	4
1.61	1.00	1.38	2.05
2.21	2.00	1.35	2.17
2.25	2.71	3.39	2.50
3.56	4.00	3.76	3.29
3.28	2.31	2.50	2.33
2.27	2.35	3.33	1.33
2.19	1.44	1.24	1.44
4.40	4.50	3.56	3.00
2.29	1.83	2.24	1.39
2.82	3.29	3.33	3.33
2.71	2.29	1.38	1.00
1.33	3.28	3.12	3.59
3.24	2.56	1.44	1.17
2.29	4.00	4.47	4.60
3.37	2.42	2.20	
2.65	2.67	2.58	2.37

Var of means = .014, η = .016.

$p = 5$

	2	3	4	5
1.61	1.00	1.38	2.05	2.21
2.00	1.35	2.17	2.25	2.71
3.39	2.50	3.56	4.00	3.76
3.29	3.28	2.31	2.50	2.33
2.27	2.35	3.33	1.33	2.19
1.44	1.24	1.44	4.40	4.50
3.56	3.00	2.29	1.83	2.24
1.39	2.82	3.29	3.33	3.33
2.71	2.20	1.38	2.83	1.00
1.33	3.28	3.12	3.59	3.24
2.56	1.44	1.17	2.29	4.00
4.47	4.60	3.37	2.42	2.20
2.50	2.43	2.40	2.74	2.81

Var of means = .028, η = .028/.89 = .031.

$p = 6$

1	2	3	4	5	6
1.61	1.00	1.38	2.05	2.21	2.00
1.35	2.17	2.25	2.71	3.39	2.50
3.56	4.00	3.76	3.29	3.28	2.31
2.50	2.33	2.27	2.35	3.33	1.33
2.19	1.44	1.24	1.44	4.40	4.50
3.56	3.00	2.29	1.83	2.24	1.39
2.28	3.29	3.33	3.33	2.71	2.29
1.38	2.83	1.00	1.33	3.28	3.12
3.59	3.24	2.56	1.44	1.17	2.29
4.00	4.47	4.60	3.37	2.42	2.20
2.66	2.78	2.47	2.31	2.84	2.39

Var of means = .040, η = .04/.89 = .04.

$p = 7$

1	2	3	4	5	6	7
1.61	1.00	1.38	2.05	2.21	2.00	1.35
2.17	2.25	2.71	3.39	2.50	3.56	4.00
3.76	3.29	3.28	2.31	2.50	2.33	2.27
2.35	3.33	1.33	2.19	1.44	1.24	1.44
4.40	4.50	3.56	3.00	2.29	1.83	2.24
1.39	2.82	3.29	3.33	3.33	2.71	2.29
1.38	2.83	1.00	1.33	3.28	3.12	3.59
3.24	2.56	1.44	1.17	2.29	4.00	4.47
4.60	3.37	2.42	2.20			
2.77	2.88	2.27	2.33	2.48	2.60	2.71

Var of means = .040, η = .04/.89 = .04.

Table 22.3 *(cont.)*

$p = 8$							
1	2	3	4	5	6	7	8
1.61	1.00	1.38	2.05	2.21	2.00	1.35	2.17
2.25	2.71	3.39	2.50	3.56	4.00	3.76	3.29
3.28	2.31	2.50	2.33	2.27	2.35	3.33	1.33
2.19	1.44	1.24	1.44	4.40	4.50	3.56	3.00
2.29	1.83	2.24	1.39	2.82	3.29	3.33	3.33
2.71	2.29	1.38	2.83	1.00	1.33	3.28	3.12
3.59	3.24	2.56	1.44	1.17	2.29	4.00	4.47
4.60	3.37	2.42	2.20				
2.28	2.27	2.14	2.02	2.49	2.82	3.23	2.96

Var of means = .160, η = .18.

$p = 9$								
1	2	3	4	5	6	7	8	9
1.61	1.00	1.38	2.05	2.21	2.00	1.35	2.17	2.25
2.71	3.39	2.50	3.56	4.00	3.76	3.29	3.28	2.31
2.50	2.33	2.27	2.35	3.33	1.33	2.19	1.44	1.24
1.44	4.40	4.50	3.56	3.00	2.29	1.83	2.24	1.39
2.82	3.29	3.33	3.33	2.71	2.29	1.38	2.83	1.00
1.33	3.28	3.12	3.59	3.24	2.56	1.44	1.17	2.29
4.00	4.47	4.60	3.37	2.42	2.20			
2.34	3.17	3.10	3.12	2.99	2.35	1.91	2.19	1.75

Var of means = .274, η = .31.

$p = 10$									
1	2	3	4	5	6	7	8	9	10
1.61	1.00	1.38	2.05	2.21	2.00	1.35	2.17	2.25	2.71
3.39	2.50	3.56	4.00	3.76	3.29	3.28	2.31	2.50	2.38
2.27	2.35	3.33	1.33	2.19	1.44	1.24	1.44	4.40	4.50
3.56	3.00	2.29	1.83	2.24	1.39	2.82	3.29	3.33	3.33
2.71	2.29	1.38	2.83	1.00	1.33	3.28	3.12	3.59	3.24
2.56	1.44	1.17	2.29	4.00	4.47	4.60	3.37	2.42	2.20
2.68	2.10	2.19	2.39	2.57	2.32	2.76	2.62	3.08	3.05

Var of means = .100, η = .11.

Table 22.3 *(cont.)*

$p = 11$										
1	2	3	4	5	6	7	8	9	10	11
1.61	1.00	1.38	2.05	2.21	2.00	1.35	2.17	2.25	2.71	3.39
2.50	3.56	4.00	3.76	3.29	3.28	2.31	2.50	2.33	2.27	2.35
3.33	1.33	2.19	1.44	1.24	1.44	4.40	4.50	3.50	3.00	2.29
1.83	2.24	1.39	2.82	3.29	3.33	3.33	2.71	2.29	1.38	2.83
1.00	1.33	3.28	3.12	3.59	3.24	2.56	1.44	1.17	2.29	4.00
4.47	4.00	3.37	2.42	2.20						
2.45	2.34	2.60	2.60	2.64	2.66	2.79	2.66	2.32	2.33	2.97

Var of means = .038, η = .043.

values of P is given in Figure 22.5. This figure suggests a period distributed in some fashion between 8 and 9 days. This compares well with the spectral density function estimate given in Chapter 12 that used the Tukey–Hanning window with 13 lags, which showed a peak at $f = .125$, or a period of 8 days, so the "quick and dirty" method did well in this case.

Figure 22.5. Segment of the periodogram, computed using Whittaker and Robinson's method.

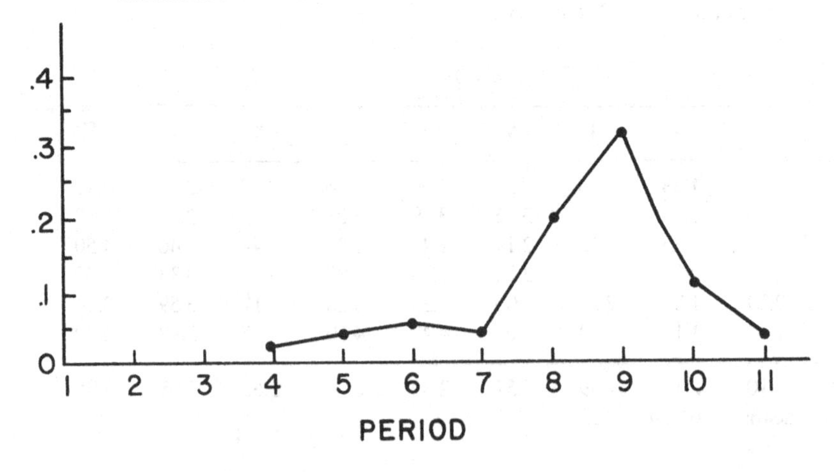

Time-domain modeling by hand

The headache data will now be used to fit an autoregressive model by hand. First, the autocorrelations were computed for 10 lags. They are summarized in Table 22.4. This pattern suggests an AR(2) model with pseudoperiodicity. Next the AR coefficients were computed using the stepwise procedure of Chapter 19. The first step was an AR(2):

$$a_1 = \frac{r_1(1 - r_2)}{1 - r_1^2} = \frac{.55(.85)}{.70} = .663$$

$$a_2 = \frac{r_2 - r_1^2}{1 - r_1^2} = \frac{.15 - .30}{.70} = -.214$$

To compare the period of this representation in the data, one can compute (see Chapter 18)

$$\begin{aligned} f_0 &= \frac{1}{2\pi} \arccos\left[\frac{a_1}{2(-a_2)^{1/2}}\right] \\ &= \frac{1}{2\pi} \arccos\left[\frac{.663}{2(.214)^{1/2}}\right] \\ &= \frac{1}{2\pi} \arccos(.716) = \frac{1}{2\pi}(.887) \\ &= .123 \\ T &= 8.13 \end{aligned}$$

Table 22.4. *Autocovariances and autocorrelations for headache data*

Lag	Autocovariances	Autocorrelations
0	.903	1.000*
1	.517	.573*
2	.165	.182*
3	−.146	−.162
4	−.318	−.352
5	−.289	−.320*
6	−.583	−.065
7	.802	.089
8	.150	.167
9	.126	.140
10	−.034	−.038

Note: Bartlett band $= 2/\sqrt{N} = 2/\sqrt{60} = .258$.

Table 22.5. *Autocovariances and autocorrelations of residuals from an AR(2) fit to the headache data*

Lag	Autocovariances	Autocorrelations
1	−.056	−.096
2	.057	.099
3	−.056	−.097
4	−.121	−.209
5	−.170	−.294
6	.038	.066
7	.012	.021
8	.042	.073
9	.077	.134
10	.032	.055

Figure 22.6. Spectral density function of the residual from an AR(2) fit of the migraine-headache data.

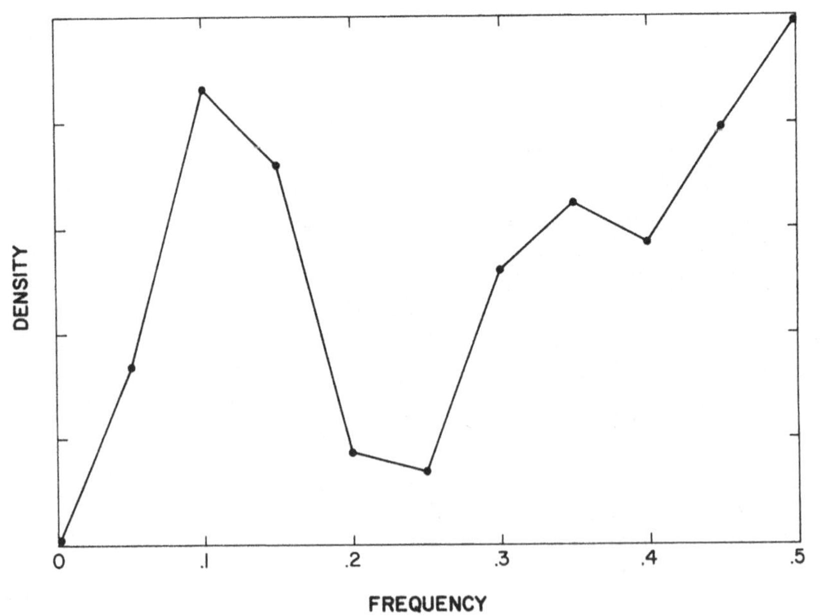

This is reasonably consistent with out other calculations. The model fitting can be continued.

To test the adequacy of the AR(2) fit, the residuals were examined. They were computed by the equation

$$e_t = x_t - (.663)x_{t-1} - (-.214)x_{t-2}, \qquad t = 3, \ldots, 60$$

The autocovariances and autocorrelations of the residuals were then computed. The first 10 are listed in Table 22.5. Note that the AR(2) fit accounted for $[(.89 - .28)/.89] \times 100 = 68.4\%$ of the original variance.

The Box–Pierce test of these residuals was computed by calculating

$$q = n \sum_{j=1}^{20=k} r^2(j) = 58(.046) = 2.67$$

where, if an ARIMA(p, d, q) model was fit, and the number of residual terms was N, $n = N - d$, and Q is computed to a $\chi^2(k - p - q) = \chi^2(20 - 2 - 0) = \chi^2(18)$. The residuals are not significantly different from white noise by this test. The spectral density function was also calculated on these residuals, this time, admittedly *with* the computer's help. The spectral density of the residual is presented as Figure 22.6. This spectral density could very well have been produced by noise: there is no obvious short cycle or band of cycles in this spectral density. These results suggest that an AR(2) model is an adequate fit to these data.

Fitting many deterministic cycles

This chapter used the periodogram, and this practice has become common in my laboratory, due to the work of Duane Steidinger, who worked in my laboratory from 1977 to 1979. This section refers to Steidinger's work and shows that the periodogram can play a useful role in modeling an observed time series. This is true because the spectral density estimates, in weighting periodogram values with various windows, will at times average out sharp peaks that could be detected as deterministic cycles by the periodogram. This may occur even if a fine grid of frequencies is available. The problem is avoided if the periodogram is simply computed as a matter of course. As Jones (1965) pointed out, this is no problem, given the speed of modern computers, if the number of observations is less than 1,000. Many programs compute the periodogram first, using the fast Fourier transform and then smooth, so that the needed computation may be performed anyway.

To be useful, in some cases the periodogram estimates of these deterministic cycles must be taken as initial values in iterative numerical estimation procedures where the actual frequency is fit by least squares. Thus, the removal of deterministic periodicities involves fitting

$$x_t = \sum_{j=1}^{p} (A_j \cos 2\pi f_j t + B_j \sin 2\pi f_j t) + e_t$$

where the A_j's, B_j's *and* f_j's are taken as unknown parameters, to be estimated by *nonlinear* regression methods. Steidinger found that least-squares derivative-free procedures gave the most stable estimates.

The analyses presented in this section also illustrate the need for simulation and checking residuals in approaching a model-fitting task. The data analyzed were the Wolfer sunspot numbers displayed as Figure 15.7. Figure 22.7 shows the periodogram of these data (corrected for the mean). Figure 22.8 shows the result of synthesizing the data with the first two dominant harmonics and then with the first four. These frequencies were computed by Steidinger (1979) by least-squares fitting using Brent's derivative-free method, available in the International Mathematical and Statistical Library (IMSL) with the periodogram values as initial values. Steidinger (1979) noted:

> Even in the case of a fine mesh of harmonic frequencies, the best fit may be "trapped" between adjacent periodogram peaks, and a refinement is necessary. [P. 1]

Figure 22.7. Periodogram of Wolfer sunspot numbers.

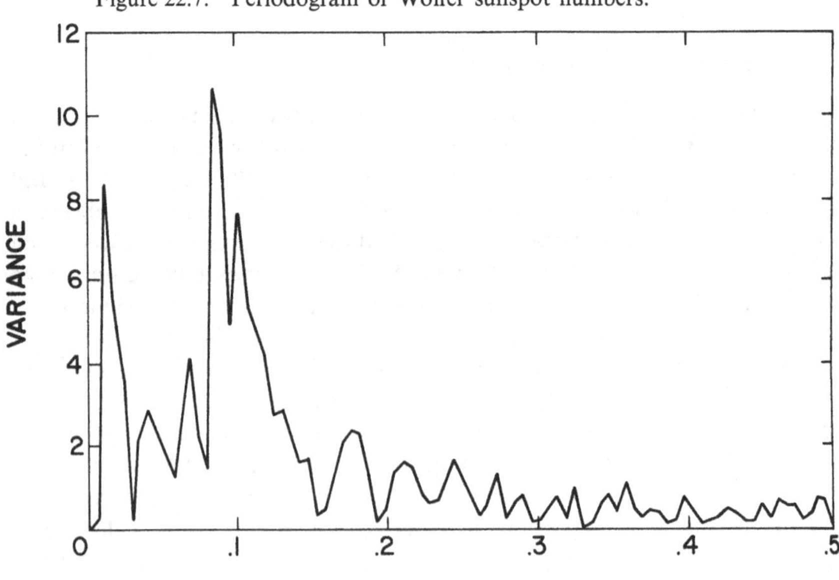

Figure 22.8. Synthesis of the sunspot data with (a) the first two harmonics, and (b) the first four harmonics.

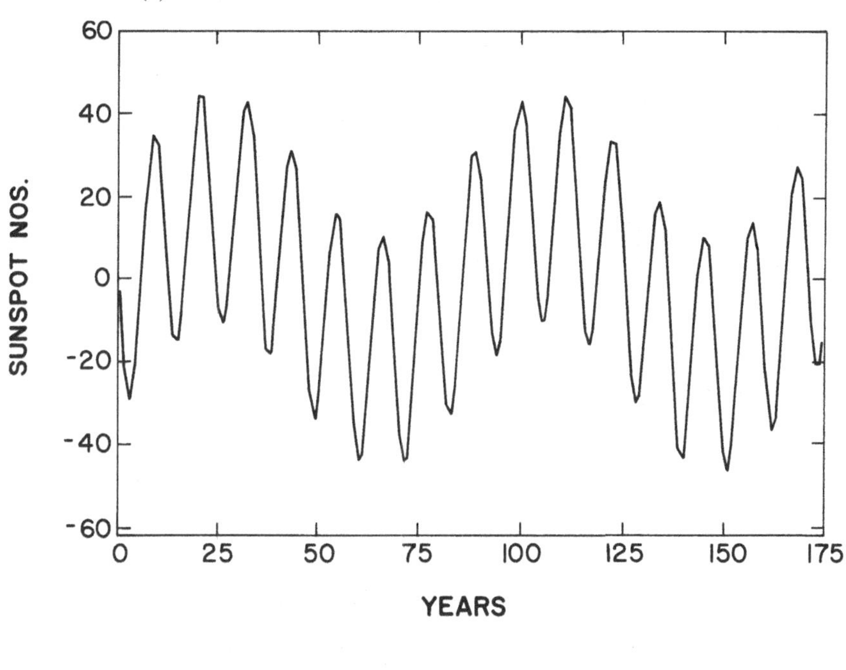

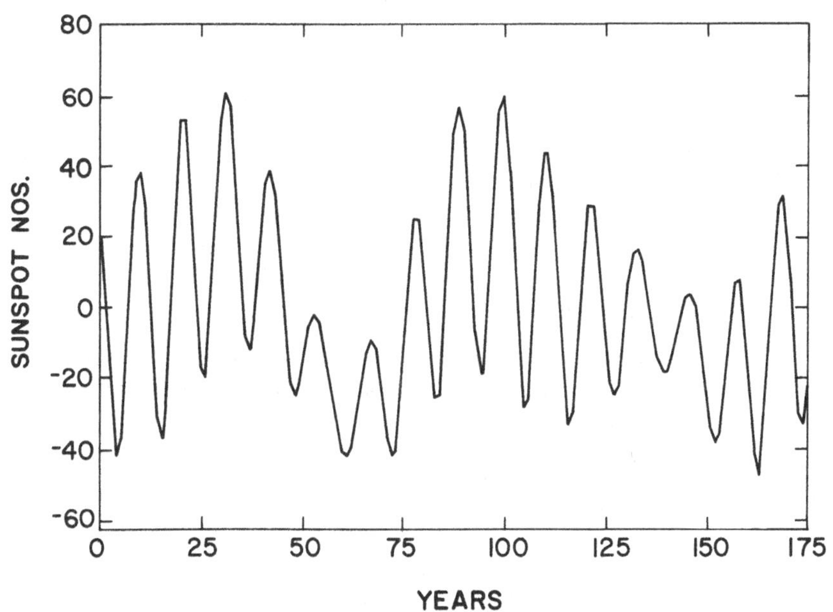

Figure 22.9. Periodogram of the residuals of a four-harmonic fit to the sunspot data.

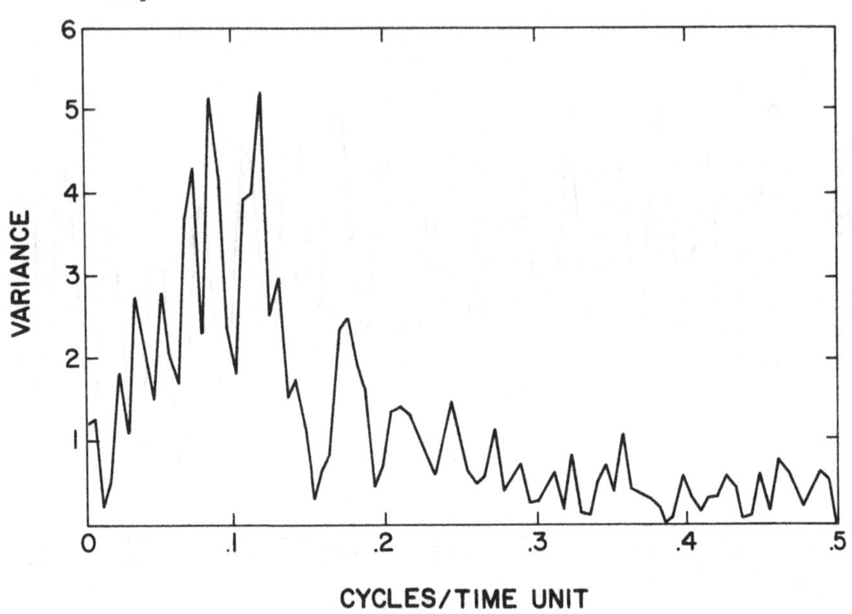

The periodogram of the residuals (Figure 22.9) of the four-harmonic fit to the data does not resemble that of white noise, but a guess at the remaining variation made by Steidinger interpreted the periodogram of the residuals as that of an AR(2) process or an ARMA(2, 1) process. The AR(2) coefficients $a_1 = 1.362$, $a_2 = -.723$, when substituted into equation (17.20) for the peak frequency of a pseudoperiodic AR(2) process, gives a peak of $f = .091$, which is in the correct region. Steidinger used Richard Devor's DDS computer program, which is the Pandit–Wu procedure available at the University of Illinois, to fit an ARMA(2, 1) model to these residuals. The overall residuals and the periodogram of these residuals is given in Figure 22.10. Commenting on this figure, Steidinger (1979) wrote:

> This periodogram is a typical "white noise" periodogram, and also the autocorrelations are white noise correlations. Thus, statistically an adequate fit is given by
>
> $$x_t = \sum_{j=1}^{4} (A_j \cos 2\pi f_j t + B_j \sin 2\pi f_j t) + Z_t$$
>
> where
>
> $$Z_t = e_t + a_1 Z_{t-1} + a_2 Z_{t-2} - b e_{t-1}$$

Figure 22.10. (a) Overall residuals of the full model fit, and (b) periodogram of these residuals.

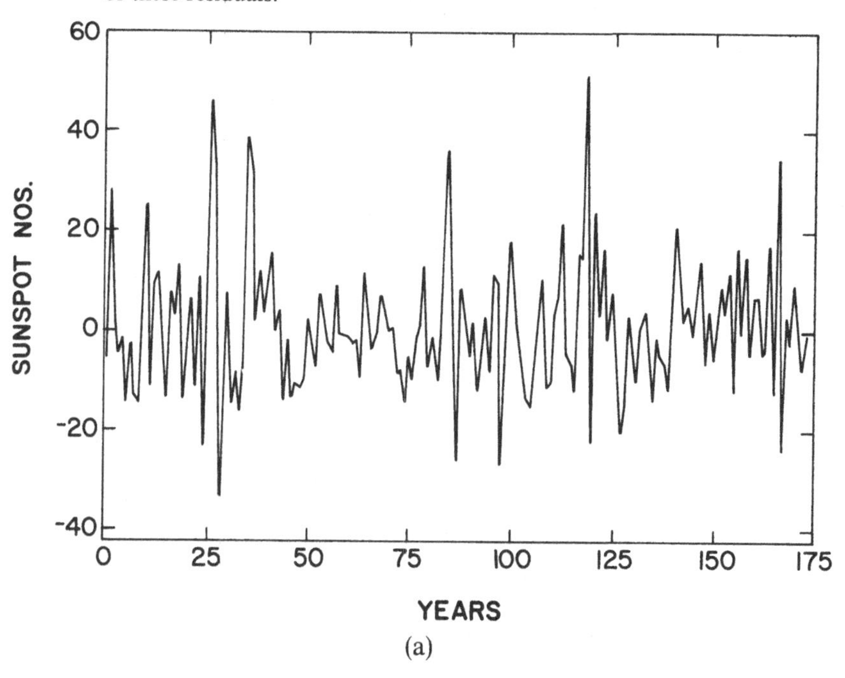

(a)

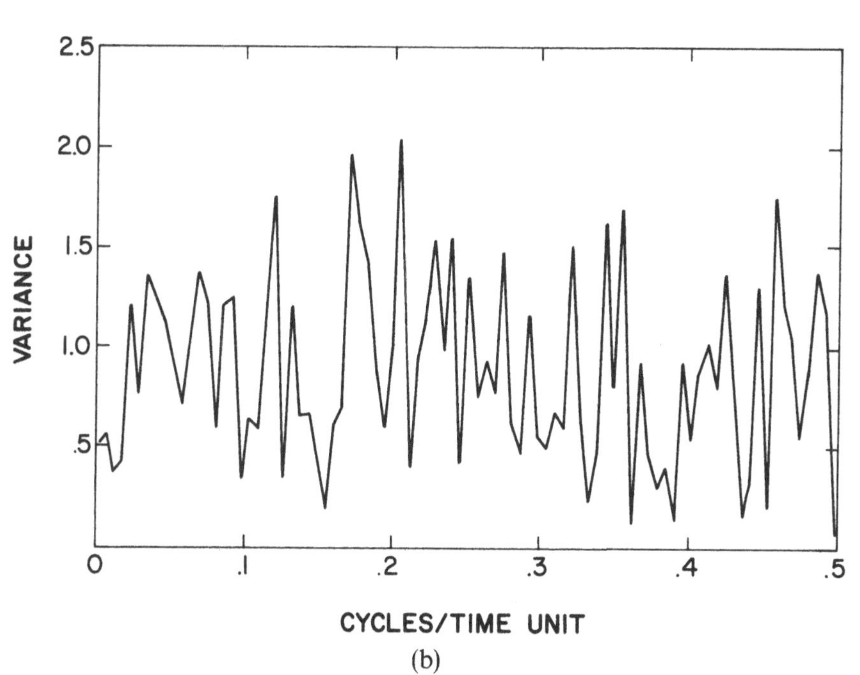

(b)

and

$$
\begin{array}{lll}
f_1 = .01225 & A_1 = -3.824 & B_1 = 18.210 \\
f_2 = .08825 & A_2 = 10.435 & B_2 = -25.480 \\
f_3 = .01975 & A_3 = -1.117 & B_3 = -11.090 \\
f_4 = .10045 & A_4 = 14.220 & B_4 = 2.715
\end{array}
$$

and

$$
\begin{aligned}
a_1 &= 1.362 \\
a_2 &= -.723 \\
b &= .433 \\
\text{SSRES} &= 2.955 \times 10^4 \qquad \text{[Pp. 1–2]}
\end{aligned}
$$

Steidinger also commented:

> This model has considerable merit in that it is parsimonious, and it is also clear how each component contributes to the structure of the data. Moreover, the analysis may have given us new insight into the series. Observe that even with just a 4-harmonic fit the variance in the .01 to .02 frequency range has been largely eliminated, indicating that possibly the long period cycle may be more regular than the more obvious oscillations of about 11 years. It would be interesting to see if the updated series confirms this. [P. 2]

Steidinger's work was based on the mathematical derivations and computer programs developed by Bloomfield (1976). The reader is referred to Bloomfield for a complete discussion of the methods employed.

Part VI

Bivariate time-series analysis

This section of the book reviews methods for studying the relationship between two time series, in both the frequency and the time domains.

23 Bivariate frequency-domain analysis

This chapter reviews the bivariate case in the frequency domain. The generalization is not difficult mathematically, but it is nontrivial conceptually. In ordinary statistics the degree of linear relationship between two variables can be described by one number, the correlation coefficient. However, instead of one statistic, we must use at least two for bivariate time-series analysis. The most common are the coherence and the phase spectra. Both are functions of frequency. Equations for confidence intervals are also quoted in this chapter, and their basis discussed in Appendix 23A.

Introduction: general concepts

Recall that in discussing the spectral density function we needed the concept of spectral decomposition. A time series was represented as a sum of sine waves, each with different amplitudes and phases. Each sine wave was a component of the original time series. Recall that in a Fourier representation of the time with an odd number of observations, the frequencies of the overtone series $[f_k = k/T, k = 1, 2, \ldots, (T-1)/2]$ were used to approximate the frequencies in the data. These frequencies had the useful property of being orthogonal, or statistically independent.

Now imagine how one might discuss the relationship between *two* time series in the frequency domain. Each time series has its own spectral decomposition using the same frequency overtone series. The relationship must be described component by component, since each frequency is independent of the others. In other words, there can only be a relationship between the sinusoid at the same frequency from each series. How can this relationship be described?–by the amplitudes of the two sine waves and by their phase relationship.

A heuristic metaphor may help. Imagine a complex chord represented by the oscillations of a set of pure tones from an overtone series. These oscillations could each occur as vibrations of a set of tuning forks. This is the spectral decomposition of the chord. Now imagine another set of tuning forks resonating with the first set. Suppose that we consider the relationship between the two middle-C tuning forks, for example. To describe this relationship we have to specify the time lag of the second middle-C tuning fork with respect to the first. Perhaps there is a $\frac{1}{2}$-sec delay. This is the phase relationship. We also have to specify the amplitude relationship. Does the second tuning fork vibrate with one-half the amplitude of the first? Or is this cycle somehow amplified?

Across each frequency in our overtone series we have two numbers, one specifying the phase relationship and the other specifying the amplitude relationship. Plotted across the frequency range, this produces *two* spectral density functions, the phase spectrum and the amplitude spectrum. This chapter discusses how these spectral density functions are computed.

The cross-spectrum

Recall from Chapter 15 that we were able to go from the time domain (the autocovariances) to the frequency domain (spectral density) using the Fourier transform, as follows:

$$\text{time domain} \qquad c_k \Leftrightarrow I(f_i) \qquad \text{frequency domain}$$

In a similar way we can create the *cross-spectrum* from the Fourier transform of the cross-covariance.

$$\text{time domain} \qquad c_{yx}(k) \Leftrightarrow p_{yx}(f_i) \qquad \text{frequency domain} \tag{23.1}$$

The analogy is somewhat more complicated in the bivariate case because the cross-covariance is not symmetric (see Table 23.1). An entirely different cross product is formed in predicting y from x at lag k than in predicting x from y at lag k.

This is an important asymmetry we have to notice about cross-correlation between two time series that did not occur when we considered autocorrela-

Table 23.1. *Asymmetry of the cross-covariance*

t	x_t	y_t	$x_t y_{t+1}$	$y_t x_{t+1}$
1	.5	−1	.5	0
2	.0	1	0	−.5
3	−.5	0	0	0
4	.0	0	0	0
5	.5	−1	+.5	0
6	0	+1	0	−.5
7	−.5	0	—	—
Average	0	0	.14	−.14
			$c_{yx}(1)$	$c_{xy}(1)$

tion within one time series. The autocorrelation was the same regardless of the direction of the lag; that is, $r_k = r_{-k}$. This will not be generally true for the cross-correlation because we will obtain different arrays if x_t leads y_t than we will if y_t leads x_t. This is illustrated in Table 23.1. In this table both x_t and y_t are zero mean processes. We computed the sample estimate of the lag 1 cross-covariances, first with x_t leading y_t, and then with y_t leading x_t. Note that they are not equal. This lack of symmetry of the cross-covariance makes bivariate time-series analysis twice as complicated as univariate. We shall see why later.

The same spectral analysis as was carried out for one series applies for several. We can describe both x_t and y_t as the sum of independent frequency components. The additional result here, though, is that the component of x_t of frequency f is independent of all the components of x *and* y at other frequencies. Thus, we need only examine, frequency by frequency, the relationship of corresponding components of x and y. Suppose that we have broken down the first series x_t into its frequency components and have a resultant spectral density function for x_t, $p_x(f_i)$. Similarly, we have a spectral density for a second series $p_y(f_i)$. For each frequency, this relationship can be characterized by two numbers, the cross-amplitude and the relative phase.

Let us consider one such frequency: say that y_t^* and x_t^* are the components at frequency f^*. Then we can represent the components as

$$\begin{aligned} x_t^* &= A_x^* \cos(2\pi f^* t + \theta_x^*) \\ y_t^* &= A_y^* \cos(2\pi f^* t + \theta_y^*) \end{aligned} \tag{23.2}$$

where A_x^* and A_y^* are random, possibly correlated amplitudes and θ_x^* and θ_y^* the random phases. The cross-amplitude measures the covariance of A_x^* and A_y^*. More commonly, the coherence, a normalized version giving the square of the correlation, is discussed. The phase shift is the average lead of x over y, the average of $\theta_y^* - \theta_x^*$.

Figure 23.1 illustrates this concept of the spectral decomposition of two time series, and their relationship, frequency component by frequency component. The input x_t is resolved in this figure using the three frequency components f_1, f_2, and f_3. The same is true for y_t. Each of the double arrows in the lower half of Figure 23.1 illustrates the fact that the relationship between each component must be described by two numbers, the amplitude relationship and the phase relationship.

To be somewhat more precise, it is useful to recall the complex exponential notation of Chapter 16. The equations (23.2) become

$$\begin{aligned} x_t^* &= A_x^* e^{i(2\pi f^* t + \theta_x^*)} \\ y_t^* &= A_y^* e^{i(2\pi f^* t + \theta_y^*)} \end{aligned} \tag{23.3}$$

The cross spectral-density is then the covariance

$$E(x_t^* y_t^*) = E(A_x^* A_y^*) E[e^{i(\theta_x^* - \theta_y^*)}] \tag{23.4}$$

but then if we plot the cross-amplitude relationship as a function of frequency, we have a function that is called the cross-amplitude spectrum, $A_{yx}(f)$; similarly, the plot of the phase relationship as a function of frequency is called the *phase spectrum*, $\phi_{yx}(f)$.

The cross-spectral density can thus be written

$$p_{yx}(f) = A_{yx}(f) e^{i\phi_{yx}(f)} \tag{23.5}$$

A function similar to a regression coefficient can be defined, called the gain spectrum, $G_{yx}(f)$:

$$G_{yx}(f) = \frac{A_{yx}(f)}{p_x(f)} \tag{23.6}$$

An alternative to the gain spectrum is a function that varies between zero and 1 and is similar to the square of the correlation coefficient. It is also a function of frequency and is called the *coherence*, computed as

$$\rho_{yx}^2(f_i) = \frac{|p_{yx}(f_i)|^2}{p_x(f_i) p_y(f_i)} \tag{23.7}$$

Figure 23.1. Concept of the spectral decomposition of two time series, illustrating component-by-component relationships.

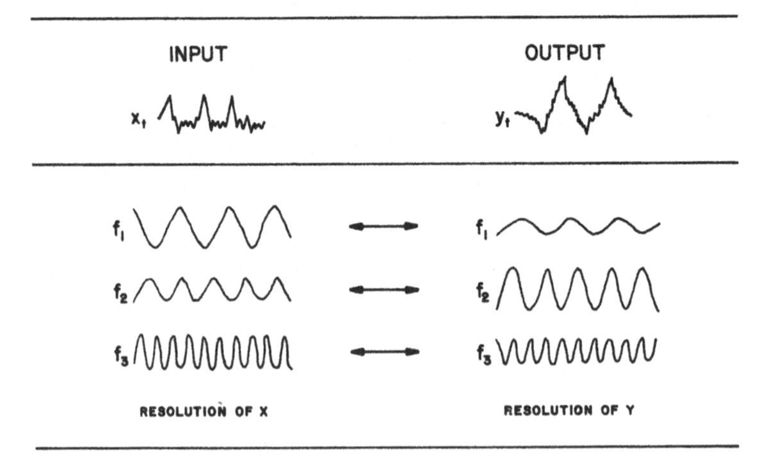

The coherence is a complicated function, but it is frequently used in bivariate frequency-domain analysis. This book follows Koopmans's (1974) notation, using the Greek letter rho to suggest correlation.

In what sense is the coherence spectrum a squared correlation? Recall that the correlation is a measure of the best *linear* relationship between two random variables. In the case of two time series, a linear relationship is known as a *linear filter*. If we want to approximate a zero-mean stationary series $y(t)$ using linear filters of another zero-mean stationary series, $x(t)$, we can form all possible linear filters $\hat{y}(t)$, as follows:

$$\hat{y}(t) = \sum_{i=-\infty}^{\infty} a_i x(t-i)$$

and assess how small we can make the error

$$E\{[y(t) - \hat{y}(t)]^2\} = \sigma^2$$

by a suitable choice of the a_i. Note that σ^2 will not depend on t because $x(t) - y(t)$ is stationary. Suppose that we can find a filter that minimizes σ^2, leaving σ^2_{RES}. Then the amount of linear association between $x(t)$ and $y(t)$ can be assessed by comparing σ^2_{RES} with the amount of variance of $y(t)$ itself, $\sigma^2_y = E[y^2(t)]$. The ratio gives us the proportion of variance unaccounted for, and one minus this ratio, the proportion of variance accounted for. The coefficient of coherence is the latter ratio, the proportion of variance accounted for at each frequency. The coherence is a symmetrical measure of linear association. The coherence is like the square of the correlation coefficient, and it always varies between zero and 1.

The phase spectrum is a measure of the extent to which each frequency component leads the other. In fact, the phase shift can be converted to time units as follows. A component of $x(t)$ *leads* a component of $y(t)$ if the phase is *negative*, and the actual time delay (in whatever units t is measured in) of the component of $y(t)$ at frequency f in cycles/time units is $-\phi(f)/2\pi f$. Consider what this implies if $x(t)$ leads $y(t)$ by a constant time t_0:

$$x(t) = y(t + t_0)$$

which means that whatever happens to $x(t)$ happens to $y(t)$, but t_0 time units later. If $-\phi(f)/2\pi f$ is a constant, namely t_0, then

$$\phi(f) = 2\pi t_0 f$$

which means that across the entire frequency range, the phase spectrum is a straight line with negative slope, and the slope is proportional to the time lag of $y(t)$.

Later in this chapter I quote the result that the variance of the phase is proportional to $(1 - \rho^2)/\rho^2$, where ρ^2 is the coherence. This implies that the confidence interval for the phase will be huge if the coherence is low and small if the coherence is high. This implies that the phase is interpretable only at frequencies of high coherence.

What does high coherence mean? Consider the case of heart rate and respiration. When we inhale, our heart rate increases slightly, and when we exhale, our heart rate decreases slightly. There is thus a cyclic component that the heart-rate spectrum and the respiration spectrum share. However, it is important that the two series influence each other. It is quite possible for two series to both peak at a frequency and yet have no influence on one another. The coherence measures this influence, although the measure is not directional. Also, the coherence spectrum is invariant to any linear transformation performed on either series because it measures the best linear association between the two series.

Note also that more complex models for the phase are possible. For example, Granger and Hatanaka (1964) proposed a two-component model in which there is one lead–lag relationship for slower frequencies and another for faster frequencies. The slope of the phase spectrum would thus be different in the two frequency ranges. Recall that we can also evaluate the time lag

Figure 23.2. Spectral density of both realizations of the x and y processes.

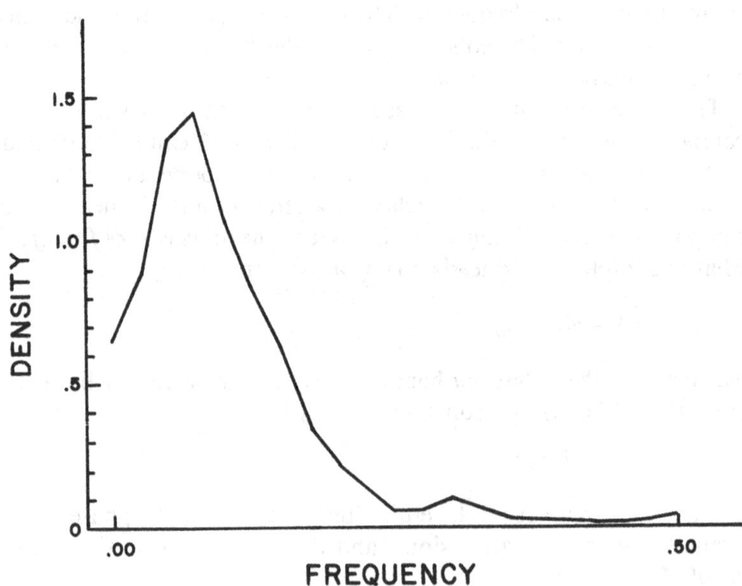

at a frequency and not only across a band of frequencies. Two simple models are most useful.

No correlation. Two uncorrelated white-noise processes will have zero cross-covariance, and hence zero cross-amplitude spectrum. However, the phase spectrum, whose expected value is zero, will be uniformly distributed across the frequency range.

Constant time lag. If the series are white noise processes but lagged by J time units, the cross-amplitude spectrum will be a constant and the phase spectrum will be a straight line whose slope equals the time lag. To illustrate this, I simulated:

$$x_t = 1.1x_{t-1} - .5x_{t-2} + e_t$$
$$y_t = x_{t+1}$$

The series x_t and y_t are both AR(2) processes. Figure 23.2 is a plot of the spectral density of both processes, and Figure 23.3 is a plot of the phase spectrum

Figure 23.3. Phase spectrum between the realizations of the x and y processes, illustrating a constant time lag.

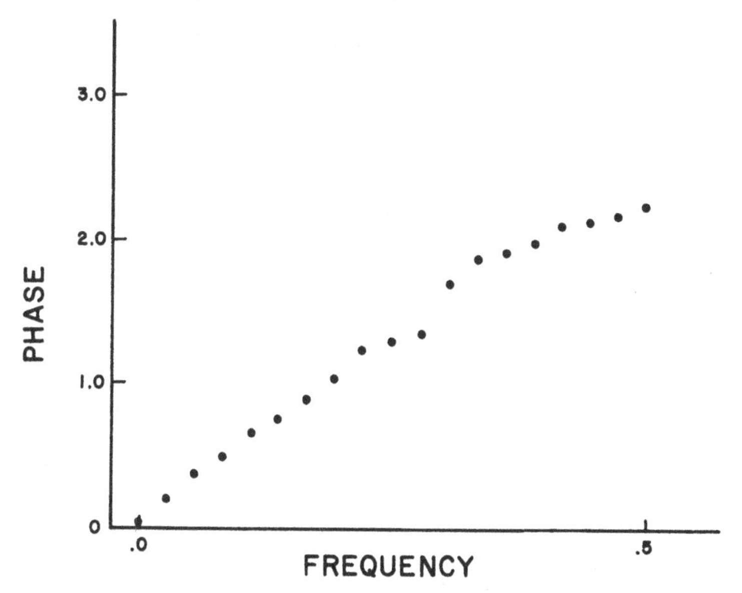

Smoothing and significance testing

For reasons similar to univariate spectral analysis, smoothing of sample spectral estimates has great advantages.[1] If the window in the frequency domain is denoted by W and I is the integral (the area under the curve) of the square of W, the effect of smoothing is to reduce the variances and covariances of sample unsmoothed estimators by a factor of I/T, where T is the sample size (Jenkins and Watts, 1968, p. 376).

See Jenkins and Watts (1968, chaps. 9 and 10) and Koopmans (1974) for derivations of confidence intervals and variances of the cross-spectral function discussed in this chapter. This chapter will quote selected results for the gain, coherence, phase, and amplitude spectra.

Coherence. If r_{yx}^2 is the smoothed sample estimate of the coherence ρ_{yx}^2, then

$$\text{var}(\rho_{yx}^2) \approx \frac{I}{2T}\left[4r_{yx}^2(1 - r_{yx}^2)^2\right] \tag{23.8}$$

Jenkins and Watts (1968, p. 279ff.) suggest that the inverse hyperbolic tangent, arctanh $(|\hat{\rho}_{yx}|)$ has the variance $I/2T$, which is independent of frequency, and recommend asuming that this variable is approximately normal. Note this is just Fisher's z transformation for the correlation coefficient. Koopmans (1974) reviewed this recommendation and concluded that for the coherence, for $n > 20$ and $.4 \leq \rho^2 \leq .95$, $\text{arctanh}(\hat{\rho}_{yx})$ is approximately normally distributed with mean

$$\text{arctanh}(\rho_{yx}) + \frac{1}{2(n-1)}$$

and variance

$$\frac{1}{2(n-1)}$$

where $2n$ is the equivalent degrees of freedom of the estimate of the cross-spectral density (see Koopmans, 1974, p. 283).

It is useful to have a test for whether the coherence is zero. Koopmans (1974, p. 284) points out that

$$\frac{(n-1)\hat{\rho}^2}{1-\hat{\rho}^2} = F_{2,\,2(n-1)}$$

if $\rho = 0$. Thus, we reject $\rho = 0$ if

$$\frac{(n-1)\hat{\rho}^2}{1-\hat{\rho}^2} > F_{2,2(n-1)}(1-\alpha)$$

or equivalently, if

$$\hat{\rho}^2 > \frac{F}{(n-1)(1+F)} \tag{23.9}$$

Phase. The variance of the smoothed phase estimator $\hat{\phi}_{yk}$ is given by

$$\text{var}(\hat{\phi}_{yx}) \approx \frac{I}{2T}\left(\frac{1}{r_{yx}^2} - 1\right) \tag{23.10}$$

Cross amplitude. If α_{yx} is the expected value of the smoothed cross-amplitude estimate A_{yx}, then

$$\text{var}(A_{yx}) \approx \frac{I}{2T}\alpha_{yx}^2\left(1 + \frac{1}{r_{yx}^2}\right) \tag{23.11}$$

The approximate $100(1-\alpha)\%$ confidence intervals for phase are (see also Koopmans, 1974, pp. 285ff.):

$$\hat{\phi}_{yx} \pm \arcsin[t_{\nu-2}^*(1-\alpha)]\left[\frac{1}{\nu-2}\frac{1-r_{yx}^2(f)}{r_{yx}^2(f)}\right]^{1/2} \tag{23.12}$$

where $t_{\nu-2}^*(1-\alpha)$ is the upper $100(1-\alpha)\%$ point of the Student's t distribution, and ν is the equivalent degrees of freedom of the cross-spectral density estimate.

Gain. The approximate $100(1-\alpha)\%$ confidence intervals for gain, $\hat{G}(f)$, are

$$\hat{G}(f) \pm \hat{G}(f)[t_{\nu-2}^*(1-\alpha)]\left[\frac{1}{\nu-2}\frac{1-r_{yx}^2(f)}{r_{yx}^2(f)}\right]^{1/2} \tag{23.13}$$

where $t_{\nu-2}^*(1-\alpha)$ has the same meaning as it previously had.

Cautions. If the coherence is low, the variances of the cross-amplitude and phase spectra will become very large. Thus, Jenkins and Watts (1968) wrote:

> The sampling properties of the phase and cross amplitude estimators may be dominated by the uncontrollable influence of the coherency spectrum rather than by the controllable smoothing factor I/T. [P. 379, italics removed]

They also noted that the estimates of coherence which use windowed estimators will be biased if the phase changes rapidly within that window and recommend "aligning" the two series (Jenkins and Watts, 1968, pp. 396–400, or Koopmans, 1974, p. 309). See also Appendix 23B.

APPENDIX 23A

The basis of the confidence intervals of cross-spectral estimates

In normal theory, in estimating any parameter, A, with a sample estimate, $\hat{A}$, it is preferable if the variance of the error of the estimate, $(A - \hat{A})$, is constant throughout the interval of the parameter. Often, this is not the case, and then it is useful to consider a *variance stabilizing transformation*, f, and use $f(\hat{A})$ to estimate $f(A)$.

To estimate the variance of $f(\hat{A})$ if $\hat{A} \sim N(A, \sigma^2)$, if the variance of the estimate, σ^2, is much less than the true value of the parameter, A, we can use a Taylor series expansion of $f(\hat{A})$ to show that

$$f(\hat{A}) \sim N\{f(A), [f'(A)]^2\sigma^2\} \tag{23A.1}$$

where $f'(A)$ is the derivative of f at A.

Jenkins and Watts (1968) showed that the variance of the sample estimate of the coherence is

$$\text{var}(|\hat{\rho}|) = \frac{1}{\text{EDF}}(1 - \rho^2)^2 \tag{23A.2}$$

where EDF is the equivalent degrees of freedom of the estimate (which varies with the bandwidth and the spectral window). If the coherence is high (close to 1.0), this variance is small. To select a variance stabilizing transformation, we wish to find a transformation f such that

$$\text{var}[f(|\hat{\rho}|)] = \text{constant}$$

That is, we need to find f such that

$$[f'(A)]^2(1 - \hat{\rho}^2)^2 = \text{constant} = 1 \quad \text{(say)}$$

$$f'(A) = \frac{1}{1 - \hat{\rho}^2}$$

$$f(A) = \frac{1}{2}\log\left(\left|\frac{\hat{\rho} + 1}{\hat{\rho} - 1}\right|\right) = \tanh^{-1}(|\hat{\rho}|)$$

Thus, we have shown that

$$\tanh^{-1}(|\hat{\rho}|) \sim N\left[\tanh^{-1}(|\rho|), \frac{1}{\text{EDF}}\right] \tag{23A.3}$$

The 95% confidence interval for the $\tanh^{-1}(|\hat{\rho}|)$ is thus

$$\tanh^{-1}(|\hat{\rho}|) \pm \frac{1.96}{(\text{EDF})^{1/2}}$$

To transform back into ρ space, take the tanh of the end points. Koopmans recommends using the Student's t point rather than the normal z point.

For the phase, ϕ, Jenkins and Watts (1968) derive the fact that

$$\text{var}(\hat{\phi}) = \frac{1}{\text{EDF}}\frac{1 - \rho^2}{\rho^2} \tag{23A.4}$$

is the asymptotic variance when the true coherence is not zero. Note that the smaller ρ^2 is, the less meaningful in the phase value. We do not require a variance-stabilizing transformation because the variance does not involve the phase, ϕ. However, a problem with interpreting the phase is its circularity, that is, $-\pi \leq \phi \leq \pi$, which means that at $-\pi$ the phase will flip to π.

An alternative (Hannan, 1970) is to use the sine. If

$$\phi - \hat{\phi} \sim N\left[0, \frac{1}{\text{EDF}}\frac{1 - \rho^2}{\rho^2}\right]$$

Then

$$\sin(\phi - \hat{\phi}) \sim N\left[0, \frac{1}{\text{EDF}}\frac{1 - \rho^2}{\rho^2}\right] \tag{23A.5}$$

The variance is unchanged because the square of the derivative of the sine at zero is 1.0. Hence,

$$|\phi - \hat{\phi}| \leq \sin^{-1}\left\{(t_{\alpha/2,\,\text{EDF}})\left[\frac{1}{\text{EDF}}\frac{1 - \rho^2}{\rho^2}\right]^{1/2}\right\}$$

As noted in the chapter, Koopmans recommends modifying the EDF by subtracting 2, since two spectral parameters are being estimated.

APPENDIX 23B
Alignment of the series to reduce bias in estimating the coherence

Jenkins and Watts (1968, p. 385) show that if one series leads another, and thus the plot of the cross-covariance is asymmetrical (i.e., does not peak at zero lag), the weighted estimates of the coherence will be severely biased. This is because the windowed cross-spectrum weights the estimates around zero lag the most, and may thus miss the peak in the cross-covariance.

The problem is easily solved by recalling that the coherence is not the square of the correlation between the two series, but of the *best possible* linear filter of the input with the output. Thus, the coherence is invariant to any linear transformation of the input or the output.

One such linear transformation is to shift one of the series so that the two are aligned. The shift parameter can be estimated by the lag at which the cross-covariance peaks, or by the slope of the phase spectrum.

24
Bivariate frequency example: mother–infant play

This chapter presents an example applying bivariate frequency-domain analysis to data obtained from the quantitative scoring of videotapes of mother–infant play.

Introduction

Researchers in adult–infant social interaction have been concerned with whether or not the 3- to 6-month-old infant's attention in face-to-face interaction during social play is cyclic. They have also been concerned with the mother's sensitivity to her infant's cycles of attention and inattention. Spectral time-series analysis would be useful in answering questions of cyclicity and synchronicity. Psychologists interested in social development have also been interested in when, in the infant's development, the social play becomes reciprocal, or bidirectional.

Bidirectionality has been studied in pioneering research by Brazelton and his colleagues (e.g., Brazelton et al., 1974; Tronick et al., 1977). In this research the behavior of mother and infant are recorded by two cameras and merged on a split screen with a time code that makes it possible to code a 5-minute play session in slow motion or frame by frame. This detailed quantitative analysis of nonverbal behavior was a major breakthrough in this area. An additional breakthrough was the univariate scaling of the micro codes on a dimension of affective involvement. The detailed micro codes were categorized into "monadic phases" that were then scaled on a dimension that ranged from maximum negative involvement to maximum positive involvement and excitement. This scaling made it possible to examine overall patterns in interaction that might not have been tapped by one isolated code, and it simultaneously gave specific behavioral referents to the scale (see Brazelton et al., 1974). Tronick et al. (1977) observed three mother–infant dyads and summarized their analyses by the time-series graphs displayed in Figure 8.4.

This chapter will present the spectral analyses of the Tronick et al. (1977) data. The data in Figure 8.4 are not stationary; a first differences transformation was necessary to achieve stationarity. The original discussion of the data by Tronick et al. also suggests that differencing is appropriate. In examining running correlations for synchronicity, Tronick considered agreement not between the affective levels of mother and baby, but rather *agreement in how they were moving*, or, equivalently, how well the *difference* between successive observations agreed. Thus, considering differences is completely consistent with the original Tronick et al. (1977) analysis.

Review of bivariate spectral analysis

When a bivariate time-series analysis is conducted, the individual spectra are examined to see if the two series cycle at the same frequencies. If they do not, there can be little bivariate relationship. However, it is possible for one series to peak within a frequency band and for the other series to have some, but not most, of the power within that band. In this case a bivariate relationship is still possible.

Recall that bivariate spectral time-series analysis provides two additional pieces of information. First, the *coherence spectrum*, $\rho^2(f)$, gives the square of the correlation between the random amplitudes for the two series at each frequency. Like correlation, coherence is not a directed relationship, so we must seek an analog of regression to consider asymmetry.

Second, the *phase spectrum*, $\phi(f)$, describes the lead–lag relationship at each frequency. If two series $x(t)$ and $y(t)$ are considered, with $x(t)$ the input series, a negative phase indicates that $x(t)$ leads $y(t)$. If $x(t)$ denotes the mother's series, this situation could be interpreted to mean that the mother is leading and the baby is responding to the mother when considering cyclicity at a specific frequency. In fact, we can compute, for each frequency component, the baby's response time by dividing the phase $\phi(f)$ by $-2\pi f$. It is assumed here that the phase is given in radians and the frequencies in cycles per unit time (Koopmans, 1974, p. 95). If the ratio is zero, the series are perfectly in phase and synchronous. The phase can be examined at those places in the frequency range where mother and baby are cycling together (i.e., where their individual spectral densities peak at the same frequency). If the phase spectrum is a straight line, this means that *throughout the entire frequency range* the same time lead–lag relationship holds. A positive slope indicates that the baby leads; a negative slope indicates that the mother leads.

Phase information is especially meaningful at those frequencies where the mother and baby are cycling together (i.e., where their individual spectral densities peak together and where the coherence is high). Indeed, the phase spectrum is interpretable only when the coherence is high. Recall that Jenkins and Watts (1968) showed that the variance of the sample estimate of the phase is proportional to $(1 - \rho^2)/\rho^2$. Thus, if ρ^2 is close to 1.0, the variance of the phase estimate is small; as ρ^2 decreases, the variance increases. So low coherence suggests that any lead–lag relationships are accidental and not indicative of cross-correlational patterns.

Of course, we know that in practice it is only possible to estimate these spectral parameters for a small set of frequencies, the overtone series. It should also be noted that these estimation procedures require a rather large amount of data. The Tronick series consisted of roughly 175 observations each. This was by no means excessive.

To summarize, the following use will be made of spectral time-series analysis. The spectral density estimates should be examined to find which cyclicities dominate the series. Regions of high coherence indicate cross-correlation at these frequencies, but do not control directly for autocorrelation. In such regions the phase spectrum can indicate asymmetric time delays. Although these methods do not control specifically for autocorrelation, they will assist in building time-domain models that do (see Chapter 25).

Spectral analysis of the Tronick data

This section will demonstrate the application of spectral time-series analysis, beginning with the examination of the sample estimates of the spectral density functions for mother and baby, followed by the examination of the sample estimates of the coherence spectrum and the phase spectrum. Recall that we are examining the differences between observations, so we are considering cyclicity and the synchronicity of the patterns of change in the data, not in the data themselves.

For dyad 1, note that the spectral densities show the same broad outlines, but that the baby tends to be cycling slightly faster. For example, the mother's spectrum peaks in the $f = .14$ to .18 and .36 to .44 Hz (cycles per second) ranges, and the child's spectrum peaks at the slightly higher .20 to .22 and .42 to .46 ranges (see Figure 24.1). The coherence is low[1] over both these ranges, indicating that these similar cyclicities are more a result of autocorrelation than cross-correlation (see Figure 24.2). Only at very low frequencies, which contribute relatively little to the variances of the two series, is the coherence high. Nonetheless, examining the phase spectrum suggests that there may be a general linear trend throughout the phase spectrum. This is illustrated in Figure 24.2 by the parallel lines. The slope of these lines is 17.45, which, when divided by 2π, gives a time lag of 2.78 sec. Thus, we can conclude that, although most of the variation in these series involves autocorrelation, there is some indication that at slow cyclity the two series are interrelated, with the mother responding to the baby at about a 3-sec lag. In view of the fact that the coherence is generally low and the phase spectrum does not control for autocorrelation, this apparent time delay should be considered as a hypothesis to be tested rather than a proven interpretation of the data.

Figure 24.1. Mother and infant spectral densities, dyad 1.

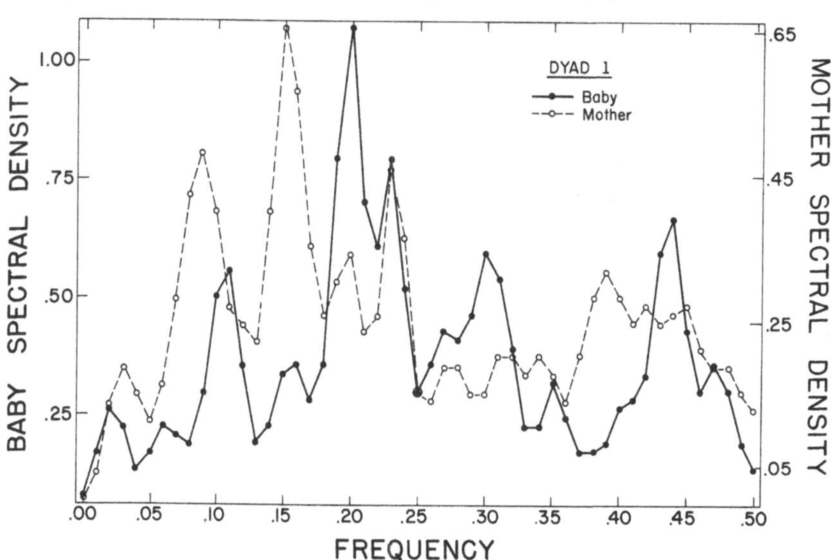

Figure 24.2. Coherence and phase spectra, dyad 1.

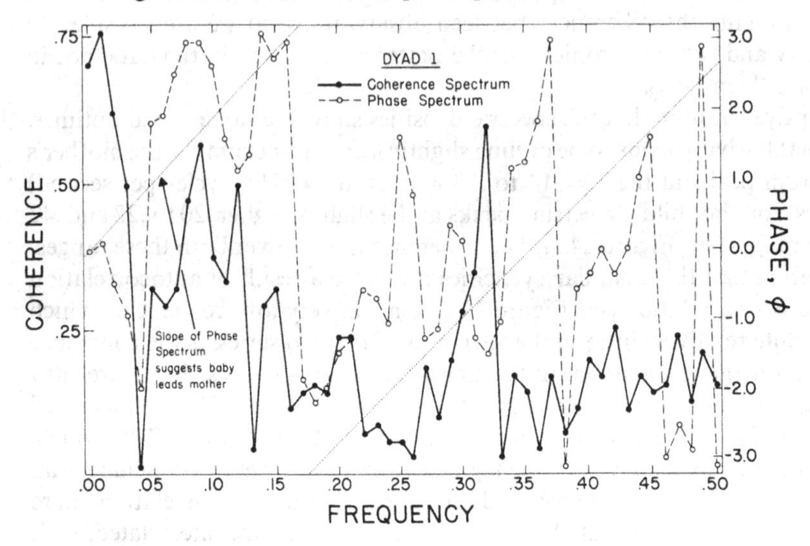

Figure 24.3. Mother and infant spectral densities, dyad 2.

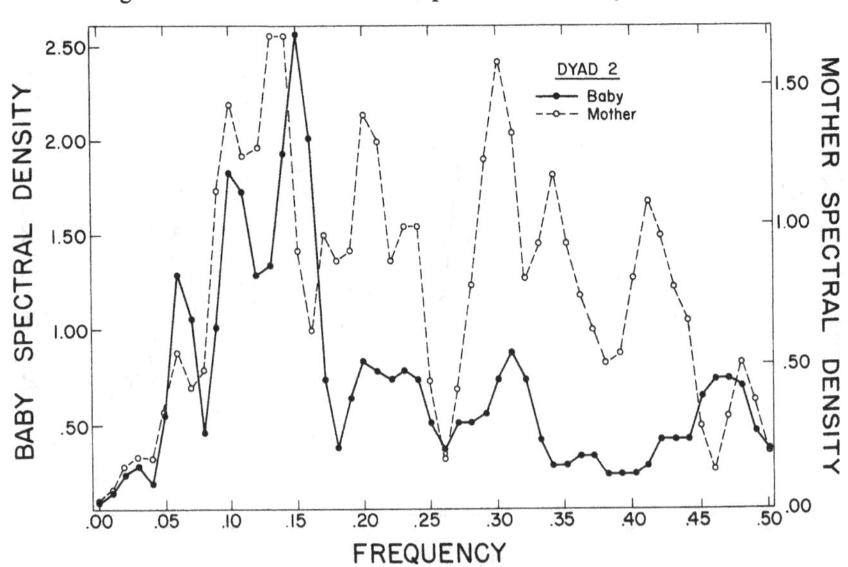

Figures 24.3 and 24.4 illustrate the facts that the spectral densities for the mother and baby for the second dyad both peak in the neighborhood of $.10 \leq f \leq .15$, and this is the frequency range of highest coherence. Other peaks appear, but the alignment between the two spectra is less obvious or not present; only at $f = .30$, where the coherence is low, do the spectra peak together again. If we examine the phase spectrum in the frequency range of highest coherence, we can conclude that the baby leads the mother; by computing the time delay at the center of this frequency range, we find that the mother is responding to the baby with a time delay of less than 1 sec. Thus, for dyad 2 we could hypothesize that there is evidence of strong cross-correlation, but that once again it is the mother who responds to the baby (i.e., the direction of influence is asymmetric). Once again, we must correct for autocorrelation to test this hypothesis (Chapter 25).

The third dyad is much more difficult to analyze spectrally. The spectral densities do not appear to peak together (see Figure 24.5). The coherence is high near $f = .04$, where the baby is cycling strongly but the mother is not, and near $f = .20$, where the opposite is true (see Figure 24.6). At $f = .04$ the time delay is -2.20 sec, and $f = .20$, the time delay is -0.60 sec, which suggests that the baby follows the mother. This is interesting, but we must be cautious in directly interpreting the phase spectrum. The frequency $f = .20$ Hz corresponds to a cycle with a period of 5 sec. Thus, a time difference of $-.6$ sec could also be interpreted as a delay of $-.6 + 5 = 4.4$ sec or $-.6 - 5 = -5.6$ sec. It is not clear which is appropriate.

Figure 24.4. Coherence and phase spectra, dyad 2.

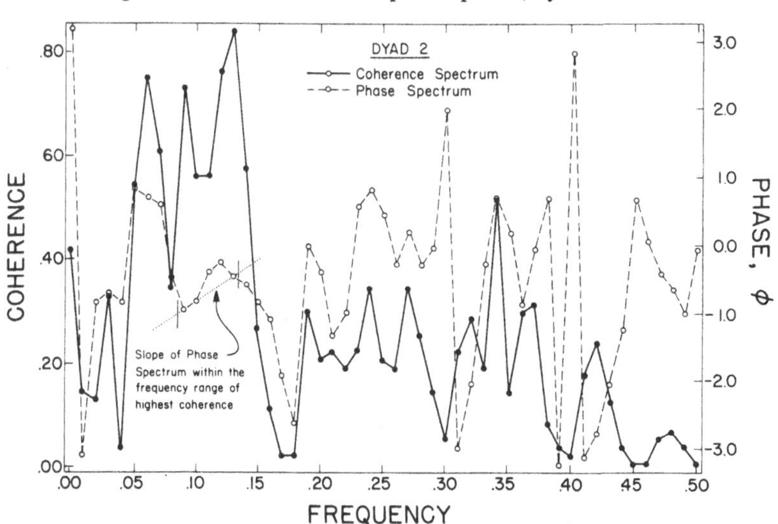

Figure 24.5. Mother and infant spectral densities, dyad 3.

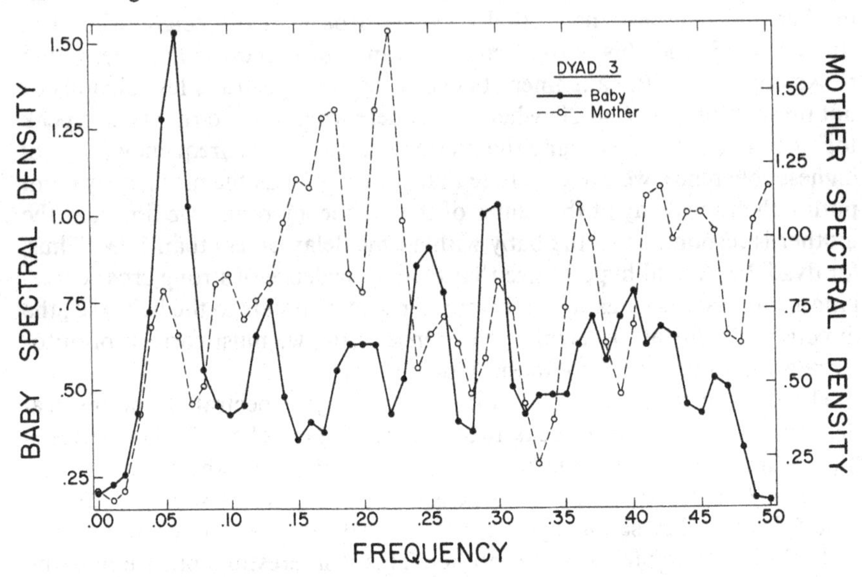

Figure 24.6. Coherence and phase spectra, dyad 3.

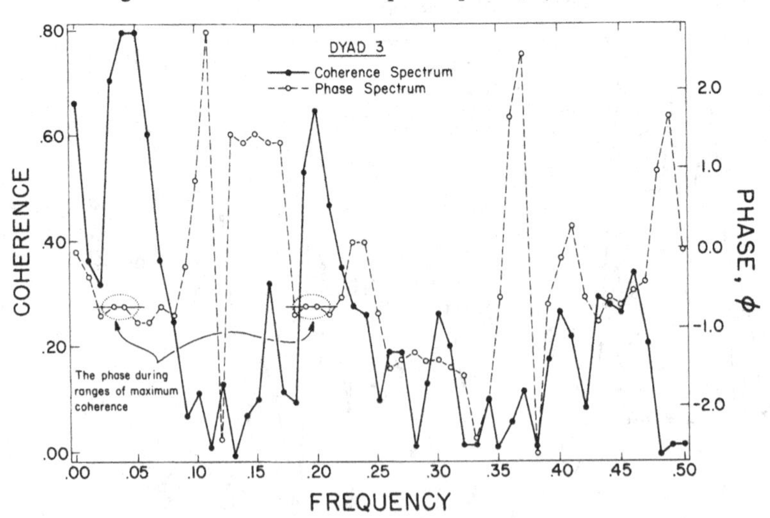

25
Bivariate time-domain analysis

This chapter reviews time-domain procedures for studying the relationship between two series, by introducing transfer functions. The problem of autocorrelation within each series in inferring connection among the series is considered. A procedure using a likelihood-ratio test is introduced and applied to data on mother–infant interaction.

Introduction

This chapter addresses a ubiquitous problem in time-series analysis. In 1927, G. Udny Yule (1971c) presented a graph of two time series, the proportion of Church of England marriages to all marriages for the years 1866–1911, and standardized mortality rates per 1,000 persons for the same years. They were highly correlated ($r = .9512$). Yule wrote:

> Now I suppose it is possible, given a little ingenuity and good will, to rationalize very nearly anything. And I can imagine some enthusiast arguing that the fall in proportion of Church of England marriages is simply due to the Spread of Scientific Thinking since 1866, and the fall in mortality is also clearly to be ascribed to the Progress of Science; hence both variables are largely or mainly influenced by a common factor and consequently ought to be highly correlated. But most people would, I think, agree with me that the correlation is simply sheer nonsense; that it has no meaning whatever. [P. 326]

More recently, in a monograph on time-series regression analysis, Ostrom (1978) discussed the problem of relating the rising pattern of U.S. defense expenditures over the past 25 years to the U.S. reaction to USSR defense expenditures. An additional problem is encountered–that a regression equation based on the hypothesis that the U.S. expenditure is a fixed ratio of the USSR plus uncorrelated error is found to be incorrect because the errors are correlated with one another. In fact, Ostrom noted that "in most political and economic data the serial correlation is likely to be positive, because the same random factors tend to operate on at least two successive period's errors." [P. 29]

The general problem of making inferences from one time series to another is addressed in this chapter. Box and Jenkins's (1970) transfer functions are reviewed. The use of "prewhitening the input" is discussed as a method for obtaining an equation relating two time series. For a special case that could have considerable usefulness, an alternative method employed by Gottman and Ringland (1981) is discussed and applied to data on mother–infant play from the paper by Tronick et al. (1977) previously discussed in this book. Time-series regression analysis is not discussed in this chapter but is reviewed briefly in Chapter 27.

Cross-covariance

Suppose that we have two stationary time series x_t and y_t and we want to assess the extent to which we can use the past of x_t to predict y_t. One way of conceptualizing this assessment is to consider x_t the *input* and to consider how it is transformed or is capable of being represented as a transformation into something that resembles y_t, which we call the *output*. The simplest of these transformations is, of course, linear, although that need not be the best representation. Nonetheless, it is the one that has been studied most, and it is (unfortunately) also called a *transfer function*.

To be consistent, we will always use x_t to predict y_t, so that x_t will always try to behave as a lead indicator of y_t. If the processes are zero mean, then we define the cross-covariance as the expected value of the lagged cross product:

$$\gamma_{yx}(k) = E(x_t y_{t+k}) \tag{25.1}$$

which is read "the cross-covariance of y given x at lag k" and is estimated by the consistent estimate

$$c_{yx}(k) = \frac{1}{T} \sum_{t=1}^{T-k} x_t y_{t+k} \tag{25.2}$$

Cross-correlation

The sample cross-correlation of two time series x_t and y_t (with T observations) is estimated as

$$r_{yx}(k) = \frac{c_{yx}(k)}{s_x s_y} \tag{25.3}$$

where s_x and s_y are the estimated standard deviations of the x_t and y_t series, respectively.

Under the assumption that the two processes have no cross-correlation, it was shown by Bartlett (1955) that

$$\operatorname{var}[r_{yx}(k)] \simeq \frac{1}{T-k} \sum_{k=-\infty}^{\infty} r_x(k) r_y(k) \tag{25.4}$$

If one of the processes, say x_t, has been prewhitened, then

$$\operatorname{var}[r_{yx}(k)] \simeq \frac{1}{T-k} \tag{25.5}$$

since the autocorrelation of white noise is zero at all lags but zero.

The concept of prewhitening

If we find a model for a time series x_t, we have a transformation for creating white noise. For example, if x_t is the AR(1) processes

$$x_t = .6x_{t-1} + e_t$$

where e_t is white noise, the transformation

$$x_t - .6x_{t-1}$$

applied to the series will produce white noise. This process is called *prewhitening*.

Given stationary time series, a model-fitting procedure (as described in Chapters 19 and 20) will produce an approximation to a suitable transformation that will prewhiten the series.

Transfer function

If we want to predict y_t from the past of x_t, we would like to express the relationship between y_t and x_t as a *linear filter*, which is of the form

$$y_t = \sum_{s=0}^{M} h_s x_{t-s} + e_t \tag{25.6}$$

where e_t is uncorrelated with x_t. The h_s plotted against s is called the *impulse response function*. This is so because if x_t is zero everywhere except at $t = 0$, when it is 1, x_t is called an *impulse input* and y_t will be equal to h_t. Thus, h_t is the output for an *impulse input*. It is usually *graphed* once the h_s's have been estimated.

Conceptually, the interpretation of the impulse response function is clear. Suppose that we obtained the impulse response function shown in Figure 25.1. This would imply that there was a delayed response of four time units

(0, 1, 2, and 3), a large response for two time units and then an exponentially decaying influence.

Estimating the impulse response function

If we multiply the transfer function equation by x_{t-k}, we obtain

$$y_t x_{t-k} = \sum_{s=0}^{M} h_s x_{t-s} x_{t-k} + e_t x_{t-k} \tag{25.7}$$

Taking expected values, we obtain

$$\gamma_{yx}(k) = \sum_{s=0}^{M} h_s \gamma_x(s-k), \qquad k = 0, 1, 2, \ldots, M \tag{25.8}$$

This results in $M + 1$ equations in $M + 1$ unknowns (i.e., $h_0, h_1, \ldots, h_M$). The γ_x are the autocovariances of the x_t series. In practice, we estimate the cross-covariances and the autocovariances from the data and estimate the impulse response function.

However, Box and Jenkins (1970, p. 379) caution against this procedure. They comment that the method does not result in efficient estimates, and it also requires prior knowledge of M.

An alternative Box and Jenkins suggest is prewhitening the input, which will also be useful in cross-spectral estimation. If the x_t can be represented

Figure 25.1. Impulse response function.

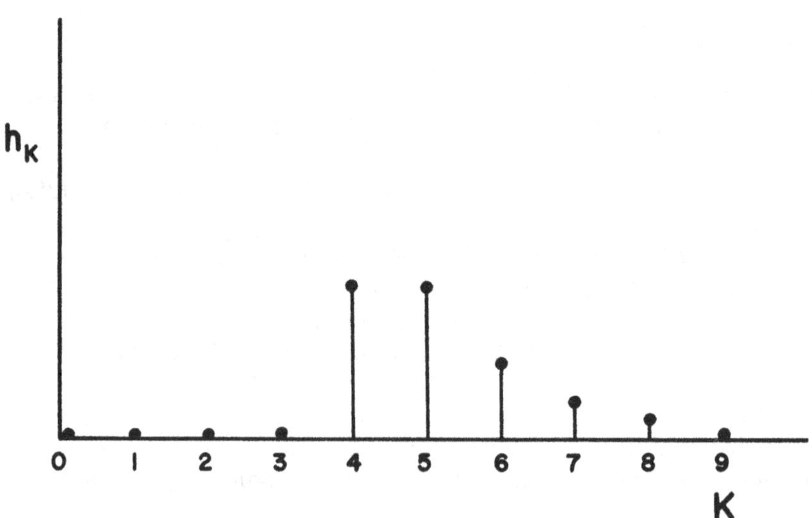

as an AR(p) process, the residual

$$\mathrm{AR}(B)x_t = \alpha_t \tag{25.9}$$

will be white noise. We can then apply the same linear operator to the entire transfer function [equation (25.6)] to obtain

$$\mathrm{AR}(B)y_t = \sum_{s=0}^{M} h_s \mathrm{AR}(B)x_{t-s} + \mathrm{AR}(B)e_t \tag{25.10}$$

or

$$w_t = \sum_{s=0}^{M} h_s \alpha_{t-s} + n_t \tag{25.11}$$

where w_t and n_t are transformations of y_t and e_t with the AR(B) transformation we used to prewhiten the input x_t. If we multiply our last equation by the noise process α_{t-k} and take expectations, we obtain

$$\gamma_{w\alpha}(k) = h_k \sigma_\alpha^2 \tag{25.12}$$

which is easily solved for h_k:

$$h_k = \frac{\gamma_{w\alpha}(k)}{\sigma_\alpha^2} \tag{25.13}$$

In terms of the sample estimates of the cross-correlation, this has the form

$$\hat{h}_k = r_{w\alpha}(k) \frac{s_w}{s_\alpha} \tag{25.14}$$

Cross-correlation and antocorrelation

It is obvious that two time series can be autocorrelated and not cross-correlated. This means we can predict from the past of each series to its present value, but not from one series to the other. It is also the case that two series may have no autocorrelation but may be highly cross-correlated. For example, we may not be able to predict the price of either of two stocks from their pasts, but they may rise and fall as the market does and then their cross-correlations will be high. Similarly, we may not be able to predict the amount of rainfall in either of two locations that are near one another, but because of the proximity of the two locations, the cross-correlation will be high. Although all these examples may exist, most common is the problem of inferring cross-correlation independent of autocorrelation in the series.

Box and Jenkins (1970, p. 377) showed that a white noise series and an autocorrelated series will not be cross-correlated, but the sample cross-correlation function will vary around zero with a standard deviation

$1/(T - k)^{1/2}$ in a *systematic pattern* characteristic of the autocorrelation function of the autocorrelated series (see also Jenkins and Watts, 1968, p. 338). We thus have to be extremely careful in interpreting a sample cross-correlation at a particular lag.

In a recent paper on mother–infant interaction, Gottman and Ringland (1981) discussed the issue of inferring that social influence in the dyad's interaction is bidirectional. They wrote:

> What needs to be demonstrated is more than that predictability (at any lag) exists from mother to infant and from infant to mother. By itself this demonstration is inadequate to infer bidirectionality, because of the problem of autocorrelation. This was recently pointed out by Sackett (1980), who wrote: "The basic issue of autocontingency has not been addressed by students of social interaction. Unfortunately, autocontingency does affect the degree to which crosslag dependencies can occur. In some instances, apparent cross-contingencies may be a total artifact of strong autolag functions" (p. 330). Sackett's point is consistent with Jenkins and Watts (1968) who showed that: "... very large cross covariances, all of them spurious, can be generated *between* two uncorrelated processes as a result of the large autocovariances *within* the two processes" (p. 338). To demonstrate that the behavior of the baby is influenced by the behavior of the mother *we need to show that we can reduce uncertainty in the infant's behavior from our knowledge of the mother's past behavior, over and above our ability to predict simply from the infant's past.* Bidirectionality occurs when we can demonstrate the converse as well, and asymmetry in predictability occurs when we cannot demonstrate this symmetry." [P. 13]

The point has also been made by Pierce (1977) in the econometric literature. Pierce wrote:

> Intuitively, x causes y only if after explaining whatever of y_t that can be explained on the basis of its own past history, y_t, some more remains to be explained by x_s, $s < t$, i.e., by x_t. [P. 14]

The problem is precisely how to accomplish the appropriate analysis. Pierce reviewed the analysis of patterns in cross-correlation when the input is whitened. Gottman and Ringland worked out a procedure that uses likelihood-ratio tests.

The procedure of Gottman and Ringland

Gottman and Ringland (1981) analyzed the data presented by Tronick et al. (1977), displayed as Figure 8.4. They began by using the cross-spectral and autospectral analysis to estimate the number of terms necessary in the transfer function models. The phase spectrum was used to estimate an approximate time delay when a lead–lag relationship was suspected.

To augment these frequency-domain analyses, Box and Jenkins's transfer function models were used. The mother's time series was denoted M_t and the baby's series by B_t; these processes were represented as

$$M_t = \sum_{i=1}^{A} a_i M_{t-i} + \sum_{i=1}^{B} b_i B_{t-i} + e_t \tag{25.15}$$

$$B_t = \sum_{i=1}^{C} c_i B_{t-i} + \sum_{i=1}^{D} d_i M_{t-i} + n_t, \qquad t = 1, 2, \ldots, T \tag{25.16}$$

where we assume that e_t and n_t are independent, normally distributed, with means zero and variances σ_e^2 and σ_n^2, respectively. Gottman and Ringland developed a maximum likelihood procedure for testing the bidirectionality hypothesis.

One way of thinking of the model is to transfer the first summation on the right-hand side of each equation to the left-hand side. These could be translated into English as attempts to *predict the residual from an auto-regression on each series from the past of the other series.* The innovation processes e_t and n_t are the residuals of this prediction. If there is no gain in this prediction, this means that a smaller autoregressive model without the past of the other series is adequate. They thus sought tests for comparing smaller models with bigger models. Because of autocorrelation it is not possible to derive exact F-ratio statistics, but they developed likelihood-ratio tests with asymptotic distribution theory.

There may appear to be an indeterminacy in the model. The mother's behavior depends on her own past as well as on the baby's past, which in turn also depends on the mother's past. However, because of the assumption that the mother and baby are introducing independent innovations e_t and n_t, there are unique estimates for all parameters. Moreover, the estimates of the a_i's and b_i's do not depend on the c_i's or d_i's, so equations (25.15) and (25.16) can be handled separately.

Gottman and Ringland employed Mann and Wald's (1943) least-squares procedure for estimating the a_i and b_i for a given A and B conditional on "start up" observations $M_0, M_{-1}, \ldots, M_{-A+1}$ and $B_0, \ldots, B_{-B+1}$. The first A or B observations (whichever is larger) at the beginning of the play sessions are thus treated differently from those once the session is established. The parameter vector for the a_i's and b_i's is estimated by ordinary least-squares autoregression[1]. Details are given in Appendix 25A. The residual mean square error σ_e^2 was estimated in the usual way. This made it possible to perform individual tests on the a_i's and b_i's. In particular, they fit a model with A and B larger than necessary and used the asymptotic variances of the estimates (see Appendix 25A) to test at the 10% level whether $a_A = 0$ or $b_B = 0$. Note that this was not the usual hypothesis-testing situation: One wishes to include terms if there is *any* evidence at all that they are nonzero. Thus, the analysis is used informally; the α level was chosen to be rather large and they did not worry about multiple comparison problems. If $a_A = 0$ or $b_B = 0$, a smaller model is fit and the testing is repeated, and thus step by step the model is reduced to appropriate size, to find the best full model to describe the series.

Theoretically, the parameters A, B, C, and D may be as large as the number of observations, and it may seem necessary to start the step-down procedure with an excessively large first guess for these values, which could be highly inefficient. Fortunately, the spectral analysis was used to suggest a reasonable number of parameters in the model to start with. Recall that for a univariate autoregression, a model with two parameters can represent one cycle or approximate a series whose spectral density shows one well-defined peak; one of order 4 can represent two cycles, and so on (see Box and Jenkins, 1970). Thus, by examining the spectral densities, one obtains some indication of how many *auto*regressive terms to include. The indication on time delays computed from the phase suggests how many *cross*-regressive terms from the other series to consider in each model. For example, if the time delay is 5 sec, we will need at least five cross-regressive terms.

Likelihood-ratio tests

Employing the least-squares estimation procedure *reduced* "null hypothesis" models for the mother's and the baby's behaviors, which assuming no interaction, can also be fitted:

$$M_t = \sum_{i=1}^{A^*} a_i^* M_{t-i} + e_t^* \tag{25.17}$$

$$B_t = \sum_{i=1}^{C^*} c_i^* B_{t-i} + n_t^* \tag{25.18}$$

A new pair of estimates for the variances of e_t^* and n_t^*, respectively – $(\hat{\sigma}_e^*)^2$ and $(\hat{\sigma}_n^*)^2$– can be found. To test whether the cross-regressive terms in the final models of the form (25.15) and (25.16) significantly help predict the mother's or baby's behavior we will compare model (25.15) with (25.17), using the same value for A and A^* and model (25.16) with (25.18), using the same value for C and C^*, with the likelihood-ratio test procedure described below. This is the more usual testing situation and requires significance at the 5% level before asserting that the cross-terms do contribute, that is, one series is partially predictable from the other. Figure 25.2 gives a summary of the four regions that are possible outcomes of these tests. If $B \rightarrow M$ denotes that the mother's behavior is predictable from her infant's, region 2 represents bidirectionally and regions 1 and 4 represent asymmetry. Gottman and Ringland discussed why, in a context other than infant–mother play, these two regions could represent dominance patterning.

In addition, they compared models of the form (25.15) and (25.16) with different A, B, C, and D to check whether, overall, the initial term-reduction procedure was justified and to compare models of the forms (25.17) or (25.18) with different A^* and C^* to check whether additional terms in a purely autoregressive model give a significantly better fit. These comparisons can also be made using the likelihood-ratio test procedure. This procedure is described in Appendix 25A.

The likelihood-ratio procedure yields a test statistic, Q, which, if the smaller model is true, is distributed as chi-square with degrees of freedom equal to the difference in the number of parameters in two models being

Figure 25.2. The four regions of social influence.

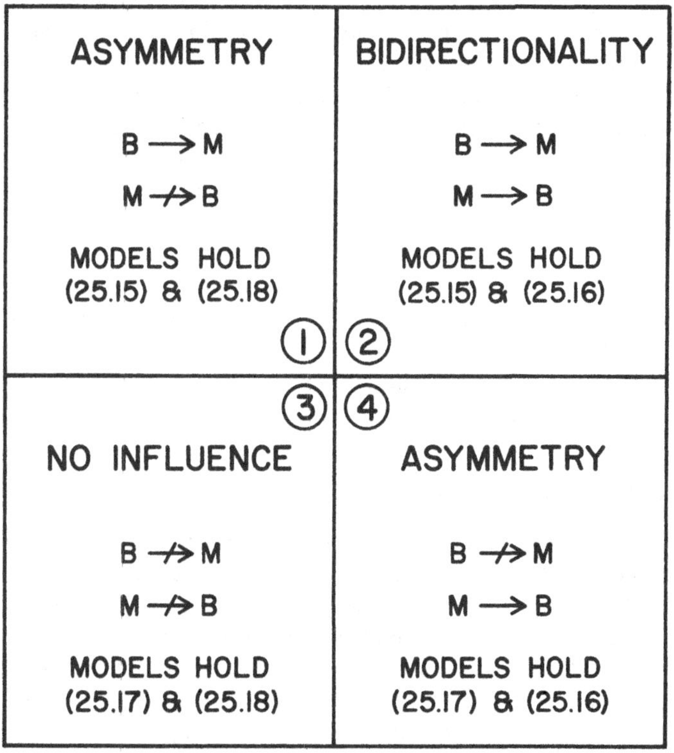

ASYMMETRY	BIDIRECTIONALITY
$B \rightarrow M$ $M \not\rightarrow B$ MODELS HOLD (25.15) & (25.18) ①	$B \rightarrow M$ $M \rightarrow B$ MODELS HOLD (25.15) & (25.16) ②
③ **NO INFLUENCE** $B \not\rightarrow M$ $M \not\rightarrow B$ MODELS HOLD (25.17) & (25.18)	④ **ASYMMETRY** $B \not\rightarrow M$ $M \rightarrow B$ MODELS HOLD (25.17) & (25.16)

compared, the larger and the smaller model. This allows a comparison of the two models, and, if the smaller model is inadequate, provides a measure of its inadequacy. In particular, this gives a test of whether all the b's and d's are zero. The same procedure provides a test of whether the step-by-step reduction used to construct models (25.15) and (25.16) was justified. There should *not* be a significant difference before and after reduction; that is, the Q should be near DF, the difference in the number of parameters, which is the mean of a χ^2_{DF} distribution.

Finally, one can check that no additional terms in equation (25.17) or (25.18) significantly help in predicting the mother's or baby's behavior by comparing $(2T \ln \sigma_e^*)$ before and after adding extra terms. Again, there should not be a significant difference–the value should be near DF.

Note that all models compared are assuming implicitly that T is unchanging. The Gottman–Ringland procedure was to establish the first 10 terms of each series as startup observations (so time = 1 is the 11th observation). The value 10 was chosen as a result of preliminary spectral analyses. This was not mathematically necessary: The same likelihood-ratio tests apply with different T's, and one could take only as many startup observations as needed in any particular model, more for models with more terms, less for shorter models, but one then has some difficulty interpreting results about slightly different stretches of data.[2]

To summarize, they suggested the following time-domain analysis for each of the six series of Figure 8.4 (mother and baby in each dyad):

1. Using spectral estimates, guess appropriate values of A, B, C, and D.
2. Starting with slightly larger A, B, C, and D, step by step remove terms to find an appropriate model of the forms (25.15) or (25.16).
3. Compare the model in step 2 with the one of the forms (25.17) or (25.18), where A^* and C^* are the values in the model computed in step 2.
4. Compare the reduced model in step 3 with one with larger A^* and C^* to check that additional autoregressive terms do not help.

Bidirectionality analysis of the Tronick data

Initial examination of the autocovariance function and the spectral densities suggested nonstationarity; thus, the series were differenced and analysis was carried out on these. Examination of the spectral densities of the differenced series suggested the number of autoregressive terms that would be necessary. For dyad 1, the mother's spectrum showed two peaks, and the baby's spectrum showed four, suggesting, for this dyad, $A = 4$ and $C = 8$. Similarly, for dyad 2, $A = 8$ and $C = 4$, 6, or 8 appear reasonable. Thus, they started the step-down procedure at slightly larger values, $A = C = 10$.

Preliminary estimates of B and D were derived from the phase analysis. In dyad 1 they found the mother responding to the baby with a 3-sec delay. This suggested that B should be at least 3, but it said nothing about D. Indeed, there was no spectral indication about the baby's response to the mother's past behavior. Similarly, for dyad 2, they found B to be at least 1, with no evidence as to what D was. Dyad 3 gave ambiguous results for the time delay, because it was not clear whether the partners were synchronous or whether there was a delay, either way, of about 5 sec. Thus, they took B *and* D to be at least 5. For safety, they started the step-down procedure at the larger values of $B = D = 10$.

The results of (1) the original excessively large model with $A = B = C = D = 10$, (2) the step-by-step reduction, (3) the model fitted by removing the cross-regressive terms from (2), and (4) a model with 10 autoregressive terms and no cross-regressive terms are summarized in Table 25.1. For comparison, the residual variance has also been given for the observations without a fitted model. The relevant likelihood-ratio test results are obtained by comparing (1) the starting model, (2) the best auto- and cross-regressive model, (3) the purely autoregressive model, and (4) the enlarged autoregressive model. Recall that the comparisons of (1) versus (2) and (3) versus (4) should give nonsignificant results, whereas the comparison of (2) versus (3) will indicate the presence or absence of predictability of one series from the other, controlling for autocorrelation.

In each case, note that the step-down procedure behaved as expected, and that adding additional autoregressive terms did not significantly decrease the error variance. For dyad 1, they asserted that $B \rightarrow M$ but did not have sufficient evidence that $M \rightarrow B$. Note, however, that as they set things up, they assumed no cross-relation until it was clearly demonstrated. In dyad 2, $B \rightarrow M$ with really no evidence at all that $M \rightarrow B$ (i.e., no evidence of bidirectionality). Interestingly, in this dyad the spectral analysis showed strong synchrony. This appeared to be the result of the mother very closely following an independent baby. Dyad 3 was both more complicated and more interesting. We clearly have $M \rightarrow B$; here for the first time is a baby responding to the mother. Additionally, we have evidence that $B \rightarrow M$, that is, we have indeed a bidirectional dyad. The baby in dyad 3 was the oldest of the three.

It is interesting to relate their time-domain analyses with their spectral analyses. The number of autoregressive and cross-regressive terms suggested by the spectral considerations were nearly what the fitting procedure gave. In asymmetric dyads 1 and 2, the spectral analysis was fairly clear and agreed with the time-domain models. In the symmetric dyad 3 the spectral analysis was ambiguous and time-domain work was needed to clarify the relationships. Note, however, that the spectral analyses in any case were useful in predicting the sizes of the ultimate time-domain models to avoid overly inefficient model identification.

Table 25.1 *Summary of the time-domain analyses*

Dyad 1	Mother				Baby			
($T = 168$)	A	B	SSE	$T \ln(\mathrm{SSE}/T)$	C	D	SSE	$T \ln(\mathrm{SSE}/T)$
(1)	10	10	212.7	39.62	10	10	345.3	121.02
(2)	3	5	228.4	51.61	8	7	357.3	126.75
(3)	3	0	249.6	66.50	8	0	380.7	137.45
(4)	10	0	240.8	60.48 ·	10	0	372.7	133.85
	0	0	257.0	71.42	0	0	415.0	151.92

	Mother	Baby
(1) vs. (2)	$Q = 11.99$, 12 DF, nog sig.	$Q = 5.73$, 5 DF, not sig.
(2) vs. (3)	$Q = 14.89$, 5 DF, $p < .025$	$Q = 10.70$, 7 DF, not sig.
(3) vs. (4)	$Q = 6.02$, 7 DF, not sig.	$Q = 3.60$, 2 DF, not sig.
Conclusions:	$B \rightarrow M$	$M \nrightarrow B$

Dyad 2	Mother				Baby			
($T = 161$)	A	B	SSE	$T \ln(\mathrm{SSE}/T)$	C	D	SSE	$T \ln(\mathrm{SSE}/T)$
(1)	10	10	625.9	218.61	10	10	511.6	186.13
(2)	8	1	660.2	227.19	10	8	519.4	188.56
(3)	8	0	705.3	237.82	10	0	540.1	194.88
(4)	10	0	703.1	237.32				
	0	0	834.0	264.81	0	0	694.0	235.23

	Mother	Baby
(1) vs. (2)	$Q = 8.58$, 11 DF, not sig.	$Q = 2.43$, 2 DF, not sig.
(2) vs. (3)	$Q = 10.63$, 1 DF, $p < .001$	$Q = 6.31$, 8 DF, not sig.
(3) vs. (2)	$Q = .50$, 2 DF, not sig.	
Conclusions:	$B \rightarrow M$	$M \nrightarrow B$

Dyad 3	Mother				Baby			
($T = 168$)	A	B	SSE	$T \ln(\mathrm{SSE}/T)$	C	D	SSE	$T \ln(\mathrm{SSE}/T)$
(1)	10	10	674.9	233.62	10	10	412.1	150.74
(2)	4	8	688.8	237.05	4	5	447.4	164.55
(3)	4	0	759.3	253.41	4	0	503.4	184.37
(4)	10	0	745.2	250.27	10	0	479.1	176.05
	0	0	810.0	264.28	0	0	515.0	188.19

	Mother	Baby
(1) vs. (2)	$Q = 3.43$, 8 DF, not sig.	$Q = 13.81$, 9 DF, not sig.
(2) vs. (3)	$Q = 16.36$, 8 DF, $p < .05$	$Q = 19.82$, 5 DF, $p < .005$
(3) vs. (4)	$Q = 3.14$, 6 DF, not sig.	$Q = 8.32$, 6 DF, not sig.
Conclusions:	$B \rightarrow M$	$M \rightarrow B$

APPENDIX 25A

Matrix formulation of the procedure of Gottman and Ringland

In this appendix we relate the procedure of Gottman and Ringland, which was described generally in the text to the matrix formula for multiple regression and then develop the likelihood-ratio tests we propose.

Matrix formulation of the models

The basic model

$$M_t = \sum_{i=1}^{A} a_i M_{t-i} + \sum_{j=1}^{B} b_j B_{t-j} + e_t$$

where $t = 1, 2, \ldots, T$, and (allowing for "startup" observations $M_0, M_{-1}, M_{-2}, \ldots$), can be rewritten in matrix form as

$$\mathbf{Y} = \mathbf{X}\boldsymbol{\theta} + \mathbf{E}$$

where $\mathbf{Y}$ is the observation vector, $\boldsymbol{\theta}$ the parameter vector, and $\mathbf{X}$ the design matrix:

$$\mathbf{Y} = \begin{bmatrix} M_1 \\ M_2 \\ \vdots \\ M_T \end{bmatrix}, \qquad \boldsymbol{\theta} = \begin{bmatrix} a_1 \\ a_2 \\ \vdots \\ a_A \\ \hline b_1 \\ \vdots \\ b_B \end{bmatrix},$$

$$\mathbf{X} = \left[\begin{array}{cccc|ccc} M_0 & M_{-1} & \cdots & M_{1-A} & B_0 & \cdots & B_{1-B} \\ M_1 & M_0 & \cdots & M_{2-A} & B_1 & \cdots & B_{2-B} \\ \vdots & \vdots & & \vdots & \vdots & & \vdots \\ M_{T-2} & M_{T-3} & \cdots & M_{T-1-A} & B_{T-2} & \cdots & B_{T-1-B} \\ M_{T-1} & M_{T-2} & \cdots & M_{T-A} & B_{T-1} & \cdots & B_{T-B} \end{array}\right]$$

which has the familiar least-squares parameter estimates

$$\hat{\boldsymbol{\theta}} = (\mathbf{X}'\mathbf{X})^{-1}\mathbf{X}'\mathbf{Y}$$

In large samples, these estimates have approximately a normal distribution with mean $\boldsymbol{\theta}$ and variance–covariance matrix $\sigma_e^2(\mathbf{X}'\mathbf{X})^{-1}$, where σ_e^2 is the variance of the e_t's.

This variance is estimated by the usual mean square error:

$$\hat{s}_e^2 = \frac{1}{T-(A+B)} \sum_{t=1}^{T} \left(M_t - \sum_{i=1}^{A} a_i M_{t-i} - \sum_{j=1}^{B} b_j B_{t-j} \right)^2$$

$$= \frac{1}{T-(A+B)} \sum (\text{residuals})^2$$

$$= \frac{\text{SSE}}{T-(A+B)}$$

It will be convenient, for the discussion of the likelihood-ratio tests, to introduce the symbol $\hat{\sigma}_e^2$ for the estimate of variance with $1/T$ weighting:

$$\hat{\sigma}_e^2 = \frac{\text{SSE}}{T}$$

All of this is nearly the same as the usual multiple regression setting; and, at least in large samples, the same methodology applies.

The estimates of the c's and d's for describing the baby's behavior is handled identically and separately, using

$$\begin{bmatrix} \hat{\mathbf{c}} \\ \hat{\mathbf{d}} \end{bmatrix} = (\mathbf{X}'\mathbf{X})^{-1}\mathbf{X}'\mathbf{Y}$$

where in $\mathbf{X}$ and $\mathbf{Y}$ the roles of M and B are interchanged and C and D replace, respectively, A and B.

Likelihood-ratio tests

The *likelihood function* for a statistical model is the joint probability density of all the random observations, considered as a function of unknown parameters. If we let $L_1(M; a)$ denote the likelihood function for the smaller model (25.17) and $L_2(M, B; a, b)$ be the likelihood function for the more general model (25.15), then, assuming normally distributed errors, we can write

$$L_2(M, B; \hat{a}, \hat{b}) = \frac{1}{(2\pi)^{T/2}} \frac{1}{(\hat{\sigma}_e)^T} \exp\left[-\frac{\sum (\text{residuals})^2}{2\hat{\sigma}_e^2} \right] \tag{25A.1}$$

Noting that $\sum (\text{residuals})^2 = T\hat{\sigma}_e^2$, this reduces to

$$L_2(M, B; \hat{a}, \hat{b}) = (\text{constant}) \frac{1}{(\hat{\sigma}_e)^T} \tag{25A.2}$$

or denoting the natural logarithm of this multiplicative constant by F,

$$\ln L_2(M, B; \hat{a}, \hat{b}) = F - T \ln \hat{\sigma}_e$$

Similarly, with F denoting the same constant,

$$L_1(M;\hat{a}) = F - T\ln\hat{\sigma}_e^* \qquad (25A.3)$$

where $\hat{\sigma}_e^*$ is computed using the residual sum of squares from the model in equation (25.17).

Let R be the likelihood ratio L_1/L_2. If the smaller model is true, then $Q = -2\ln R$, which can be conveniently expressed as the difference $(2T\ln\sigma_e^*) - (2T\ln\hat{\sigma}_e)$, is asymptotically distributed as χ^2_{DF} with DF equal to the difference in the number of parameters in the two models. This allows a comparison of the two models and, if the smaller model is inadequate, gives a measure of its inadequacy. In particular, this gives a test whether $b_1 = b_2 = \cdots = b_B = 0$ and whether $d_1 = d_2 = \cdots = d_D = 0$.

Note also that we assumed normally distributed errors in the construction of this test. Although this provides a justification of the test, the asymptotic χ^2 distribution holds without normality; thus, this procedure will still give valid results.

Part VII

Other techniques

This section contains two chapters, one on the interrupted time-series experiment, and the other on multivariate time-series analyses (including time-series regression analysis). The book has provided the basic techniques necessary for understanding these chapters; however, Chapter 27 will be only an *introduction* to multivariate techniques rather than a thorough exposition.

26
The interrupted time-series experiment

A reasonably general method is presented that makes it possible to employ linear autoregressive models, which greatly simplifies the analysis of interrupted time-series experiments.

Introduction

Throughout this book ideas have been introduced that affect the very nature of scientific inquiry. We have yet to experience the impact that *thinking about processes in time* will have on the way research is designed. As in the bivariate case, this final section of the book is still concerned with *accounting* for variation, and the interrupted time-series experiment (ITSE) is perhaps the most primitive *experimental* design toward this end, in which the researcher exerts one form of control on one process and, then, after a certain time (the interruption), a second form of control, with the goal being to observe its effects on another process. The interrupted time-series experiment consists of a set of observations prior to and after some planned intervention; the observations prior to intervention are usually called the "baseline." When the intervention is not planned as an experiment–for example, when we wish to assess the effects of a political reform–the experiment is usually called a "quasi-experiment" (Campbell and Stanley, 1963). The endeavor is basic to discovering how *systems* operate. A knowledge of time-series analysis can be very helpful toward this end.

We must not assume that every experiment must be analyzed as an interrupted time-series experiment. At times, it is more useful to cast the problem in a bivariate mold. When we exert "control" over some process in an intervention in the social sciences, we hardly ever do so without some stochastic variation. In many experiments the intervention itself is a time series and the fluctuations it creates in a dependent measure can also be studied. An awareness of this point should influence how the data from experiments are analyzed. For example, in a weight-loss program, the intervention may involve a reduction in caloric intake and an increase in exercise. However, the implementation of this intervention will not be an all-or-none phenomenon. One subject may stick rigorously to the program; another may alternate hedonistic with stoic days, and so on. An awareness of this fact must affect the design of the experiment.

Among clinical psychologists who are interested in the modification of behavior, there is a common misconception about analyzing time-series experiments. This misconception is that a prerequisite for interrupted time-series experiments is a *stable baseline* before intervention. The "baseline" is the name for the set of observations prior to intervention. The recommendation is that before intervention, it is necessary for the baseline to be "stable" (i.e., not vary too erratically around the mean, to have low variance, and no trend). In part, this recommendation follows from the "interocular test" that has been advocated among applied behavior analysts (for a discussion, see Gottman and Glass, 1978). Presumably, an intervention effect will be easily detected by visual examination if the baseline is stable. Typically, however, the concept of a stable baseline is not sharply defined, and actual intervention effects are not obvious visually. Judges err as often in the direction of ignoring significant effects as in proclaiming nonsignificant effects as significant. Amazingly, the subject has become the object of controversy. Furthermore, a stable baseline is not even necessary for the analysis of an interrupted time-series experiment. One of the implications of the recommendation that a stable baseline is necessary is the practical problem of how long one should wait until the baseline "settles down" (see Hersen and Barlow, 1976, pp. 744ff.). This is a pseudo issue, because it should be clear to the reader that a stable baseline is *not* a requirement for time-series analysis because deterministic components can be included in the time-series model. For example, one should not inform city officials that the effects of their revision of the juvenile criminal justice system cannot be assessed until their crime rates are "stable."

It is therefore the case that knowledge of the statistics of time-series analysis will dissolve what have been energetic controversies in some literatures. Furthermore, this knowledge will create opportunities by making research design more flexible than is currently specified as acceptable practice. The theme of the research opportunities created by a knowledge of time-series analysis is not new.

Glass et al. (1975) discussed the uses of the ITSE to assess the effects of societal changes (e.g., change in law or policy) when no randomized control group is sensible, and the case when the ITSE is the preferred design. The ITSE has some logical advantages over other designs, among which is the specific consideration of the form of the intervention effect over time. Glass et al. wrote:

> The most important advantage of the time-series design is not that it offers an alternative when a traditional, randomized, comparative experimental design is not feasible, but it offers a unique perspective on the evaluation of intervention (or "treatment") effects. Simultaneous comparative designs in the Fisherian tradition may blind the experimenter to important observations when such designs become

> a thoughtless habit of mind. The Fisherian design which has so captured the attention of social and behavioral scientists was originally developed for use in evaluating agricultural field trials. The methodology was appropriate to comparing two or more agricultural methods with respect to their relative yields. The yields were crops which were harvested when they were ripe; it was irrelevant in this application whether the crops grew slowly or rapidly or whether they rotted six months after harvest. For social systems, there are no planting and harvest times.... The value of an intervention is properly judged not by whether the effect is observable at the fall harvest, but by whether the effect occurs immediately or is delayed, whether it increases or decays, whether it is only temporarily or constantly superior to the effects of alternative interventions. The time-series design provides a methodology appropriate to the complexity of the effects of interventions into social organizations or with human beings. [Pp. 4–5]

Furthermore, variants of the ITSE can be used to generate hypotheses at various levels of causal inferences, from the study of concomitant variation, to post hoc scanning for shifts in a series following specific kinds of critical events, to planned experimentation with appropriate controls. Specific design variations of the ITSE with their logical properties designed to eliminate potential sources of invalidity are discussed by Glass et al. (1975); this work is an extension of the work of Campbell and Stanley (1963).

The problem of autocorrelation

As with most of time-series analysis, the basic problem in assessing the effects of an intervention concerns the existence of autocorrelation in the data. Because of autocorrelation, it is more difficult to determine whether a change was the result of the intervention or simply the normal behavior of the series. If the time series is a realization of a white-noise process, testing for the effects of an intervention would be simple. This fact was discussed in reference to Shewart charts. However, if the series is represented by a stationary stochastic model, *we can use the model to transform the time series to a residual that is a realization of a white-noise process. That is the basic idea of the mathematics of this chapter.* For example, let us suppose that the series is an AR(1) process. Then

$$z_t = a_1 z_{t-1} + e_t \tag{26.1}$$

and if we define

$$y_t = z_t - a_1 z_{t-1} \tag{26.2}$$

then it is obvious that

$$y_t = e_t \tag{26.3}$$

a white-noise process.

In other words, the transformation of z to y created the simplified Shewart situation for the transformed series. The transformation will be called "prewhitening" because it creates a white-noise process.

The logic of this chapter is basically to *transform* the time-series parameter estimation problem (which includes parameters for the intervention effect) into the case of classical linear regression, which has a simple solution. This solution is particularly elegant in matrix notation, and I refer the reader who is unfamiliar with the solution of the regression problem using least squares to the appendices of Chapter 14.

The chapter will illustrate how the transformation works in the simplest case, when autoregressive (AR) models are used. Other approaches to the interrupted time-series experiment (e.g., Box and Tiao, 1965, 1975; Glass et al., 1975; Hibbs, 1974) employ nonlinear models and two-stage least squares. The reader is referred to these references for other approaches to the problem. Two-stage least squares is described in Chapter 27. Autoregressive models are not usually as parsimonious as ARIMA models, and for this reason the nonlinear ARIMA procedures were applied by Glass et al. (1975).

To make the discussion clear, one problem will be analyzed by hand. Then the discussion will be generalized for least-squares AR fitting by computer using equations based on the Mann–Wald procedure.

Analysis of interrupted time-series experiments by hand

This section is intended only as an illustration of the concepts of how fitting models to the data is used in the analysis of the ITSE. I simply hope that a worked numerical example may demystify some of the procedures recommended in this chapter. The procedure suggested in this section is to (1) perform an autoregressive fit to the preintervention data; (2) "prewhiten" the entire data set, pre- and postintervention, with this model; and (3) use a Shewart band on these residuals. Of course, this procedure makes the potentially untenable assumption that the intervention effect does not alter the model, except to change its mean level or slope. Methods for testing this assumption will be discussed. The data set will be called the Ireland data. The data are the percentage of students by year in Ireland who passed intermediate- and senior-level examinations between 1879 and 1971. From 1879 until 1924, funds paid to the Intermediate Board of Education for Ireland were dependent on the number of students taking and passing these examinations, which Glass et al. called essentially a performance contract program. The program was abandoned after 1925. The data are plotted as Figure 26.1, with the break between 1924 and 1925 serving as the point of intervention ($n_1 = 45$, $n_2 = 47$). Glass et al. (1975, pp. 193ff.) identified an ARIMA (0, 1, 1) model; this model has no autoregressive terms, and one moving-average

Figure 26.1. Ireland data.

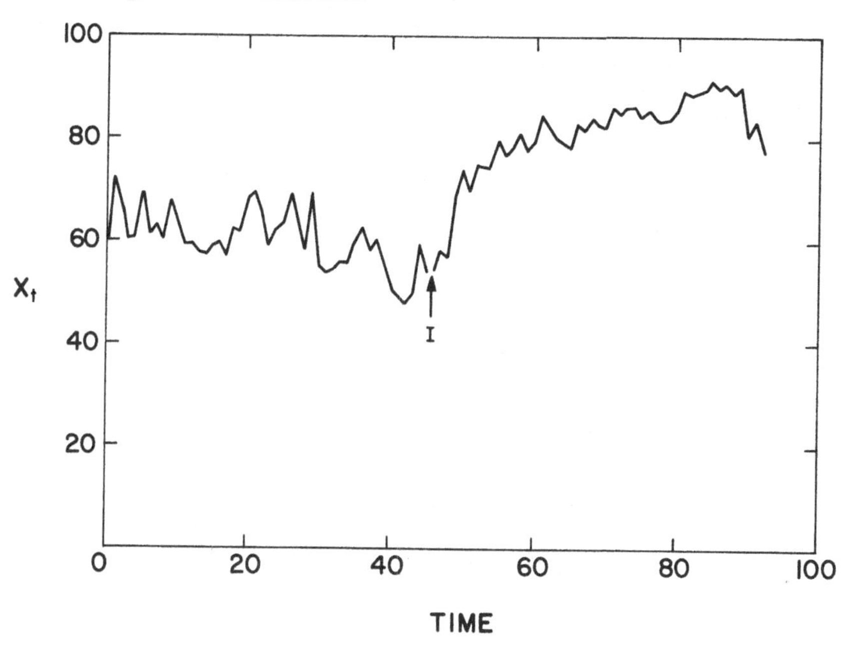

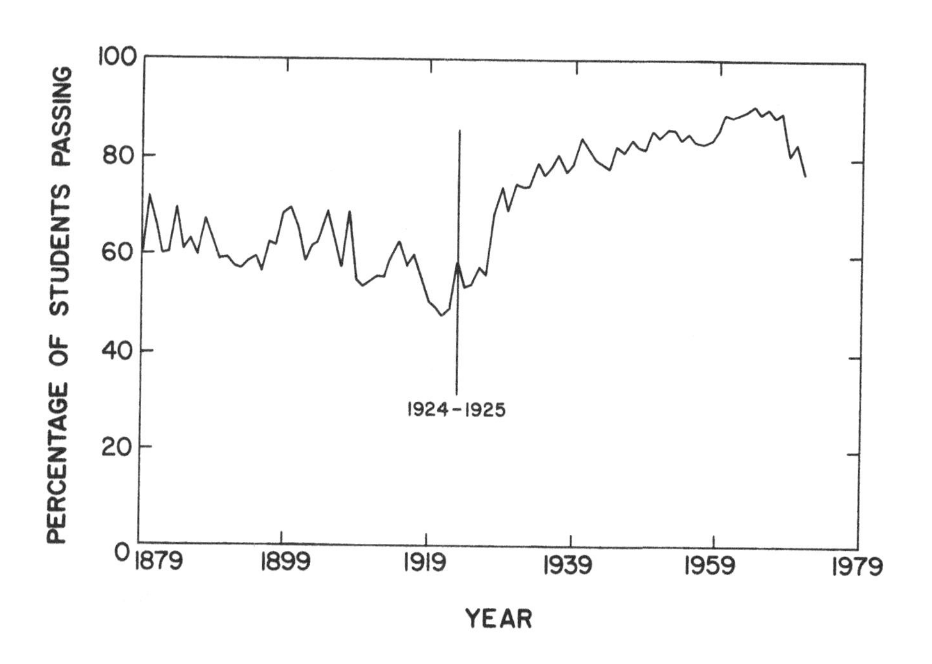

Table 26.1. *ACF and PACF for the Ireland data*

Lag	Autocorrelations (ACF)	Partial autocorrelations (PACF)
1	.536	.536
2	.390	.143
3	.316	.059
4	.230	.092
5	.086	.088
6	.052	.042

Source: Based on Glass et al., 1975, p. 105.

Table 26.2. *Autocorrelations of residuals of Ireland data after an AR(1) fit*

Lag	Autocorrelation
1	−.045
2	.050
3	.124
4	.153
5	−.031
6	.002

Figure 26.2. Spectral density function of the residuals from an AR(1) fit.

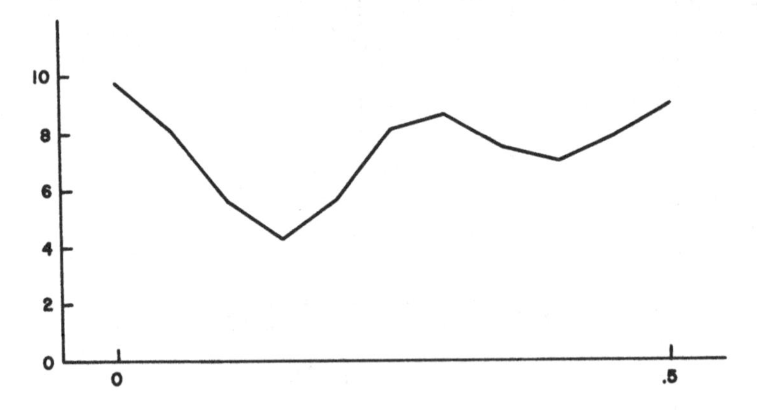

term represents the first differences of the series. A moving-average parameter of $b_1 = .40$ was identified from a maximum-likelihood grid search. The changes in level and slope with this model were not significant.

I will reanalyze this problem using an autoregressive model. Glass et al. (p. 105) presented the autocorrelations and partial autocorrelations for these data (see Table 26.1). The PACF truncates after the first lag, which is evidence for an AR(1) model, but as Glass et al. (1975) pointed out, "the picture here is not crystal clear, and rarely is in any analysis" (p. 106), because the ACF seems to be decaying slower than exponentially. As $r_1 = .536$, this is the estimate one may select for a_1. If L is the mean level to be subtracted, this suggests that the model

$$Y_t - L = .536(Y_{t-1} - L) + e_t \tag{26.4}$$

is worth considering. The residuals from this AR(1) model for the preintervention data (L is 59.97) had the ACF given in Table 26.2. The Box–Pierce test of these residuals is

$$Q = n \sum_{j=1}^{6=k} r^2(j) = 45(.044) = 1.99$$

Since $n = N - d = 45$, and Q is compared to a $\chi^2(k - p - q) = \chi^2(6 - 1 - 0) = \chi^2(5)$, these residuals are not significant. Usually, k should be at least 20, but for illustrative purposes, the discussion will proceed. The spectral density function of the residuals is plotted as Figure 26.2 and appears undistinguished, so we can take a chance and conclude that an AR(1) fit is adequate.

Now the same transformation would be applied to the postintervention data, and for a crude but useful preliminary analysis a Shewart band is plotted around these residuals, as shown in Figure 26.3. This figure suggests that the intervention effect was significant.

Figure 26.3. Shewart chart applied to the residuals of an autoregressive fit.

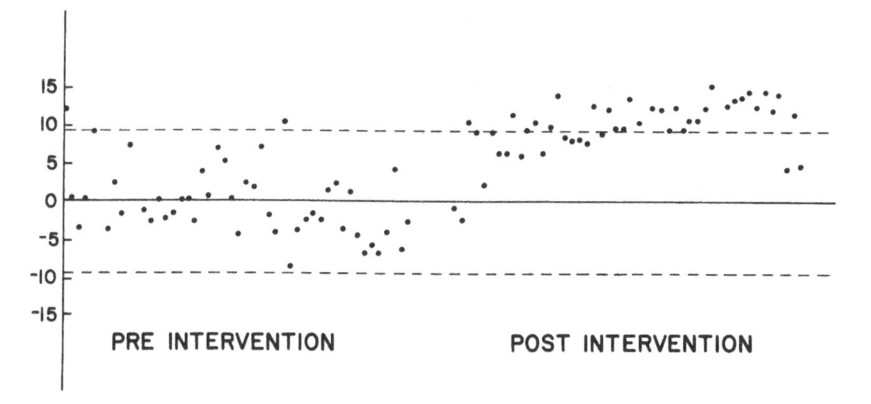

This conclusion would not be warranted unless the postintervention residuals were essentially noise with some deterministic change, such as a level shift or slope change, presumably introduced by the intervention. To test this assumption, the autocorrelations were computed for the linearly detrended residuals (see Table 26.3). The mean of the trend line was 9.966 and the slope was .149. The Box–Pierce test of these residuals is

$$Q = \sum_{j=1}^{6=k} r^2(j) = 46(.041) = 1.89$$

As $n = N - d = 46$, and Q is compared to a $\chi^2(k - p - q) = \chi^2(6 - 1 - 0) = \chi^2(5)$, these residuals are not significant. Thus, it is reasonably safe to conclude that the effect is significant.

This discussion illustrates the simplicity of the recommended procedure. The next section discusses a more general procedure for modeling the intervention and develops the appropriate statistical tests.

Extension of the Mann–Wald procedure

This chapter proposes the use of linear autoregressive (AR) models rather than the nonlinear autoregressive, moving-average (ARMA) models discussed by Glass et al. (1975). Any ARMA model can be approximated arbitrarily closely by a suitable AR model, so the two kinds of models are mathematically equivalent (see Anderson, 1971). In some cases, ARMA models may be more parsimonious (i.e., require fewer parameters) than AR models. However, AR models have the advantage that their parameters can be estimated by simple least-squares linear regression techniques, as de-

Table 26.3. *Autocorrelations of detrended postintervention residuals (from an AR(1) fit) of the Ireland data*

Lag	Autocorrelation
1	.034
2	.092
3	.150
4	−.074
5	.034
6	.050

scribed in Chapter 19. Furthermore, whereas a great deal of skill, artfulness, experience with Box–Jenkins (1970) methodology, and complex computer programs are necessary for fitting the ARMA models, no artfulness and no computer programs are necessary beyond matrix algebra for AR model fitting.

Assume that the data are zero-mean and stationary (i.e., constant mean, variance, and autocovariance); that is, assume that the data have been suitably differenced or detrended. In the stationary case, the time series Y_t is written as an AR model by predicting it linearly from its past Y_{t-1}, Y_{t-2}, $\ldots$, Y_{t-p} as follows:

$$Y_t = a_1 Y_{t-1} + a_2 Y_{t-2} + \cdots + a_p Y_{t-p} + e_t$$

where the residuals of this fit, e_t, are white noise (i.e., uncorrelated) and assumed to be distributed normally with zero mean and constant variance σ_e^2.

In this case it is easy to estimate the order of the model, p, to derive least-squares estimates of the autoregressive parameters, $a_1, a_2, \ldots, a_p$, and to derive asymptotic tests of whether these parameters are significantly different from zero at some α level (see Appendix 19A).

A simple modification will also make it possible to obtain least-squares estimates of the hypothesized intervention effects. No statistical techniques more complicated than matrix algebra are required for this estimation. Furthermore, it is possible to devise asymptotic tests for the significance of the intervention effects, under the null hypothesis of no effects.

Simplest case

Suppose that we have a first-order stationary autoregressive process AR(1), Y_t, which has a zero mean:

$$Y_t = aY_{t-1} + e_t$$

where e_t is normally, independently distributed with zero mean and constant variance, σ_e^2, and e_t is independent of the past of Y_t, that is, the

$$\text{cov}(e_t, Y_{t-k}) = 0 \qquad \text{if } k > 0$$

Then N observations can be written out as follows:

$$\begin{aligned} Y_2 &= aY_1 \quad + e_2 \\ Y_3 &= aY_2 \quad + e_3 \\ &\vdots \\ Y_N &= aY_{N-1} + e_N \end{aligned}$$

or, in matrix form, they can be written

$$\begin{bmatrix} Y_2 \\ Y_3 \\ \vdots \\ Y_N \end{bmatrix} = \begin{bmatrix} Y_1 \\ Y_2 \\ \vdots \\ Y_{N-1} \end{bmatrix} \{a\} + \begin{bmatrix} e_2 \\ e_3 \\ \vdots \\ e_N \end{bmatrix}$$

or

$$\mathbf{Y} = \mathbf{X}\boldsymbol{\beta} + \mathbf{E}$$

which has the linear least-squares solution

$$\hat{a} = (\mathbf{X}'\mathbf{X})^{-1}\mathbf{X}'\mathbf{Y} = \frac{\sum Y_t Y_{t-1}}{\sum Y_t^2} = r_1 \tag{26.5}$$

which is the sample estimate of the first-order autocorrelation coefficient. This result is well known and has been derived in other ways (see Box and Jenkins, 1970). The particular derivation presented here, however, can be generalized to include simultaneous assessment of the linear least-squares estimates of intervention effects.

Suppose now that we have N data points, with n_1 points prior to intervention and $n_2 = N - n_1$ points after intervention. Suppose that it makes sense to assume the simplest possible model holds [i.e., an AR(1) model]. In addition, suppose that we use the simplest model of the intervention effect, namely that the effect of the intervention is to add a value δ to point after the intervention, which accumulates. We can express the data as follows:

$$\begin{aligned}
Y_2 &= aY_1 + e_1 \\
Y_3 &= aY_2 + e_2 \\
&\vdots \\
Y_{n_1} &= aY_{n_1-1} + e_{n_1} \\
\hline
Y_{n_1+1} &= aY_{n_1} + e_{n_1+1} + \delta \\
Y_{n_1+2} &= aY_{n_1+1} + e_{n_1+2} + \delta \\
&\vdots \\
Y_N &= aY_{N-1} + e_N + \delta
\end{aligned} \tag{26.6}$$

Note that the effect of the intervention in this model is not sudden, but accumulates gradually to an asymptotic level. The mean of the first post-intervention point is δ, of the second is $\delta + a\delta$, of the third is $\delta + a\delta + a^2\delta$, and so on, so that the new mean level of the series is asymptotic to the value $(\sum a^k)\delta = \delta/(1 - a)$. Recall that for stationary series, we must have $|a| < 1$. The model would have to be rewritten for a sudden effect, for which it would be nonlinear. The limiting mean $E(Y_\infty)$ must also satisfy

$$E(Y_\infty) = a(E_\infty) + \delta$$

which leads to an alternative derivation of the result:

$$E(Y_\infty) = \frac{\delta}{1 - a}$$

This model can be written in matrix form as

$$\mathbf{Y} = \mathbf{X}\boldsymbol{\beta} + \mathbf{E} \tag{26.7}$$

where

$$\mathbf{Y} = \begin{bmatrix} Y_2 \\ Y_3 \\ \vdots \\ Y_N \end{bmatrix}, \qquad \mathbf{X} = \begin{bmatrix} 0 & Y_1 \\ \vdots & \vdots \\ 0 & Y_{n_1 - 1} \\ \hline 1 & Y_{n_1} \\ \vdots & \vdots \\ 1 & Y_{N-1} \end{bmatrix}, \qquad \boldsymbol{\beta} = \begin{bmatrix} \delta \\ a \end{bmatrix}$$

and

$$\mathbf{E} = \begin{bmatrix} e_2 \\ e_3 \\ \vdots \\ e_N \end{bmatrix}$$

which has the least-squares solution

$$\hat{\boldsymbol{\beta}} = \begin{bmatrix} \hat{\delta} \\ \hat{a} \end{bmatrix} = (\mathbf{X}'\mathbf{X})^{-1}\mathbf{X}'\mathbf{Y} \tag{26.8}$$

where $\mathbf{X}'$ is the transpose of $\mathbf{X}$.

In this case, it is easy to show that

$$\mathbf{X}'\mathbf{X} = \begin{bmatrix} n_2 & \sum_{n_1}^{N-1} Y_t \\ \sum_{n_1}^{N-1} Y_t & \sum_{1}^{N-1} Y_t^2 \end{bmatrix} \tag{26.9}$$

$$\mathbf{X}'\mathbf{Y} = \begin{bmatrix} \sum_{n_1+1}^{N} Y_t \\ \sum_{2}^{N} Y_t Y_{t-1} \end{bmatrix} \tag{26.10}$$

It is easy to invert a 2 × 2 matrix[1] and thus these computations can be easily performed with a hand calculator.

If the first element (first row, first column) of $(\mathbf{X}'\mathbf{X})^{-1}$ is denoted c, it can also be shown (Mann and Wald, 1943) that an asymptotic $N(0,1)$ test can be derived for δ, under the null hypothesis that $\delta = 0$. For small samples the following is a statistic with an approximate t distribution:

$$t(N-3) = \frac{\hat{\delta}}{(S_e)\sqrt{c}} \tag{26.11}$$

where S_e is the square root of the residual variance, computed as

$$S_e^2 = \frac{1}{N-3}(\mathbf{Y} - \mathbf{X}\boldsymbol{\beta})'(\mathbf{Y} - \mathbf{X}\boldsymbol{\beta}) \tag{26.12}$$

The denominator $N - 3$ is used because we have $N - 1$ observations in the $\mathbf{Y}$ vector and are estimating two parameters.

Hand-worked example

This example is presented to demonstrate the fact that the computations presented are not complicated, even if they are unfamiliar. The data in Table 26.4 were generated using an $N(0,1)$ random number table for the e_t, and equation (26.6) with $a = .6$, $\delta = 2.0$, initial startup value $Y_1 = 0.0$, set at its expected value, and with $n_1 = n_2 = 5$. Following equation (26.9), and using an ordinary hand calculator, $\mathbf{X}'\mathbf{X}$ was computed as

$$\mathbf{X}'\mathbf{X} = \begin{bmatrix} 5.00 & 19.01 \\ 19.01 & 93.91 \end{bmatrix}$$

Table 26.4. *Small-sample simulated data set*

t	e_t	What X_t would be if $\delta = 0$	X_t
1	.46	.00	.00
2	.06	.06	.06
3	1.49	1.53	1.53
4	1.39	2.31	2.31
$5 = n_1$	−.53	.85	.85
$6 = n_1 + 1$	.97	1.48	3.48
7	.45	1.34	4.54
8	−.48	.32	4.24
9	1.36	1.55	5.90
$10 = N$	.14	1.07	5.68

Using note 1, its inverse was computed as

$$(\mathbf{X}'\mathbf{X})^{-1} = \frac{1}{108.17}\begin{bmatrix} 93.91 & -19.01 \\ -19.91 & 5.00 \end{bmatrix} = \begin{bmatrix} .87 & -.18 \\ -.18 & .05 \end{bmatrix}$$

It is noted that the c in equation (26.11) is 0.87. Using equation (26.10), $\mathbf{X}'\mathbf{Y}$ was computed:

$$\mathbf{X}'\mathbf{Y} = \begin{bmatrix} 23.84 \\ 102.12 \end{bmatrix}$$

These two matrices were then multiplied:

$$\hat{\boldsymbol{\beta}} = \begin{bmatrix} \hat{\delta} \\ \hat{a} \end{bmatrix} = \begin{bmatrix} .87 & -.18 \\ -.18 & .05 \end{bmatrix}\begin{bmatrix} 23.84 \\ 102.12 \end{bmatrix} = \begin{bmatrix} 2.36 \\ .81 \end{bmatrix}$$

To compute S_e^2:

$$\mathbf{Y} = \begin{bmatrix} .06 \\ 1.33 \\ 2.31 \\ .85 \\ 3.48 \\ 4.54 \\ 4.24 \\ 5.90 \\ 5.68 \end{bmatrix}, \qquad \mathbf{X}\boldsymbol{\beta} = \begin{bmatrix} 0 & .00 \\ 0 & .06 \\ 0 & 1.53 \\ 0 & 2.31 \\ 1 & .85 \\ 1 & 3.48 \\ 1 & 4.54 \\ 1 & 4.24 \\ 1 & 5.90 \end{bmatrix}\begin{bmatrix} 2.36 \\ .81 \end{bmatrix} = \begin{bmatrix} .00 \\ .05 \\ 1.24 \\ 1.87 \\ 3.04 \\ 5.18 \\ 6.04 \\ 5.79 \\ 7.14 \end{bmatrix}$$

$$\mathbf{Y} - \mathbf{X}\boldsymbol{\beta} = \begin{bmatrix} .06 \\ 1.48 \\ 1.07 \\ -1.02 \\ -.44 \\ 1.80 \\ .11 \\ -1.46 \end{bmatrix}$$

$$S_e^2 = \frac{1}{N-3}(\mathbf{Y} - \mathbf{X}\boldsymbol{\beta})'(\mathbf{Y} - \mathbf{X}\boldsymbol{\beta}) = \tfrac{1}{7}(10.37) = 1.48$$

$$S_e = 1.22$$

The t ratio for δ was then calculated using equation (26.11) as

$$t(7) = \frac{2.36}{1.22\sqrt{(.87)}} = 2.08, \qquad p < .05 \quad \text{(one-tailed test)}$$

This demonstrates that the recommended computations are straightforward. However, this example should not be interpreted as a recommendation that time-series analysis can be performed with as few as 10 points. This number was selected to simplify the necessary arithmetic and to illustrate that the calculations do not *require* a computer for simple AR models. The statistics presented here are, moreover, extensions of the asymptotic tests, derived by Mann and Wald (1943), and it is currently not known how many data points are necessary for the statistic in equation (26.11) to be adequately approximated by a Student's t distribution. The answer to this question must await a series of Monte Carlo studies with AR models. This chapter presents the results of one Monte Carlo study with AR models.

A confidence interval for a can be computed using the fact that

$$t_a(N-3) = \frac{\hat{a} - a}{S_e\sqrt{d}}$$

(where d is the last element (second row, second column) in $(\mathbf{X}'\mathbf{X})^{-1}$) has approximately a Student's t distribution. Thus, a 90% confidence interval would be formally

$$\hat{a} - t_{.05;N-3}S_e\sqrt{d} < a < \hat{a} + t_{.05;N-3}S_e\sqrt{d}$$

In our example this becomes

$$.81 - (1.90)(1.22)\sqrt{.05} \le a \le .81 + (1.90)(1.22)\sqrt{.05}$$

or

$$.29 \le a \le 1.33$$

In a practical sense, because we know that $|a| < 1$, this means that our confidence interval is $.29 \le a < 1.00$, which is a large confidence interval; this is to be expected for a small sample. The fact that the interval includes impossible values also demonstrates the weakness of the t approximation in very small samples.

It is unclear how large N should be for asymptotic conditions to be reached, nor is it clear how large N should be for the recommended tests to be powerful. However, the reader should not assume that the situation is necessarily grim. The question has been faced in other areas of statistics. For example, for a 2×2 contingency table to be compared to a $\chi^2(1)$ distribution, a good approximation is obtained if the mean cell size is at least 5; thus, $N = 20$ is equivalent to asymptotic conditions. Also, the t distribution with more than 30 degrees of freedom is almost the same as the $N(0, 1)$ distribution. For time-series analysis, the fact that reasonably adjacent observations are correlated, where each observation provides less information than in the independently distributed case, would seem to make it necessary to have more observations to approach asymptotic conditions. However, as shall be shown, the results of a Monte Carlo study demonstrate this is not always the case. For moderate levels of autocorrelation as few as 10 observations before and 10 observations after intervention are adequate.

Extension of the procedure to AR(p) models

Extension of the Mann–Wald procedure is straight forward for each successive order of the AR(p) model. The order of the model as well as the parameter estimates are derived from the t tests. The model is of order p when the autoregressive parameter estimates of higher-order models beyond p (i.e., $a_{p+1}, a_{p+2}, \ldots$) are not significant. Some new notation will need to be introduced to extend the procedure. The work presented in the remainder of this chapter and in the appendices to this chapter was completed in collaboration with James Ringland.

Notation

We now assume that n_1 data points before intervention consist of a stationary stochastic component, which we will fit with an autoregressive model, *plus a linear trend.* Thus, before intervention ($t = p + 1, p + 2, \ldots, n_1$) the data can be represented as

$$Y_t = m_1 t + b_1 + \sum_{i=1}^{p} a_i Y_{t-i} + e_t \tag{26.13}$$

After intervention ($t = n_1 + 1, \ldots, N$) we will assume that remaining $n_2 = N - n_1$ data points move to a *new asymptote trend line.* We further assume that the autoregressive parameters have not changed as a result of intervention. Thus, after intervention the data can be represented as

$$Y_t = m_2 t + b_2 + \sum_{i=1}^{p} a_i Y_{t-i} + e_t \tag{26.14}$$

The data themselves do not lie on the lines $m_1 t + b_2$ before and $m_2 t + b_2$ after intervention. Rather, before intervention the data follow a "steady-state" trend line of the form $B_1 + M_1 t$ and approach $B_2 + M_2 t$ after intervention, where B_i can be written as a function of b_i and the a_i's, and M_i can be written as a function of b_i, m_i, and a_i (see Appendix 26B).

Furthermore, note that we will never need to require that the e_t are normally distributed. However, if the e_t *are* normally distributed, two facts follow: (1) convergence to the large sample normal distribution of the parameter estimates will be faster than in other cases, and (2) the least-squares estimates are the best (most efficient) available; without normality the least-squares estimates are still reasonable, but more efficient (and more complicated) estimates may be available. Denote the variance of the Y_t by σ_Y^2 and the variance of the residual series e_t by σ_e^2; note that these are necessarily different numbers, as long as autocorrelation exists in the data.

Estimation

The model in equations (26.13) and (26.14) can be written in matrix form as

$$\mathbf{Y} = \mathbf{XB} + \mathbf{E}, \tag{26.15}$$

where

$$\mathbf{Y} = \left[\begin{array}{c} Y_{p+1} \\ Y_{p+2} \\ \vdots \\ Y_{n_1} \\ \hline Y_{n_1+1} \\ Y_{n_1+2} \\ \vdots \\ Y_N \end{array}\right], \quad \mathbf{X} = \left[\begin{array}{cccccccc} 1 & 1 & 0 & 0 & Y_p & Y_{p-1} & \cdots & Y_1 \\ 1 & 2 & 0 & 0 & Y_{p+1} & Y_p & \cdots & Y_2 \\ \vdots & \vdots & \vdots & \vdots & \vdots & \vdots & & \vdots \\ 1 & n_1 & 0 & 0 & Y_{n_1-1} & Y_{n_1-2} & \cdots & Y_{n_1-p} \\ \hline 0 & 0 & 1 & 1 & Y_{n_1} & Y_{n_1-1} & \cdots & Y_{n_1-p+1} \\ 0 & 0 & 1 & 2 & Y_{n_1+1} & Y_{n_1} & \cdots & Y_{n_1-p+2} \\ \vdots & \vdots & \vdots & \vdots & \vdots & \vdots & & \vdots \\ 0 & 0 & 1 & n_2 & Y_{N-1} & Y_{N-2} & \cdots & Y_{N-p} \end{array}\right],$$

$$\mathbf{B} = \begin{bmatrix} b_1 \\ m_1 \\ b_2 \\ m_2 \\ a_1 \\ a_2 \\ \vdots \\ a_p \end{bmatrix} = \begin{bmatrix} \beta_1 \\ \beta_2 \\ \beta_3 \\ \beta_4 \\ \beta_5 \\ \beta_6 \\ \vdots \\ \beta_{p+4} \end{bmatrix}, \quad \mathbf{E} = \begin{bmatrix} e_{p+1} \\ e_{p+2} \\ \vdots \\ \\ \\ \\ \\ e_N \end{bmatrix}$$

As usual, the least-squares estimates are

$$\hat{\mathbf{B}} = (\mathbf{X}'\mathbf{X})^{-1}\,\mathbf{X}'\mathbf{Y} \tag{26.16}$$

If we denote the estimate of σ_e^2 by S_e^2, which is, as usual, $(1/\nu)(\mathbf{Y}-\mathbf{XB})'(\mathbf{Y}-\mathbf{XB})$, where ν, the degrees of freedom for error, is $N - 2p - 4$, and we denote the vector $\mathbf{C}$ as the diagonal of $(\mathbf{X}'\mathbf{X})^{-1}$, then each of the parameters in $\{\beta_i\} = \{b_1, m_1, b_2, m_2, a_1, \ldots, a_p\}$ can be referred to a t distribution with ν degrees of freedom, where

$$t = \frac{\hat{\beta}_i}{S_e\sqrt{c_i}} \tag{26.17}$$

Also, in a manner analogous to linear regression, we can test the hypothesis $m_1 = m_2$ and $b_1 = b_2$ using the F test with $(2, v)$ degrees of freedom.

Note that the parameter estimates cannot take on any value. For the series to be stationary (constant mean, variance, and autocovariance function), the parameters must lie within specific regions. For example, for the AR(2) model the parameters are constrained so that $-1 < a_2 < 1$, $a_1 + a_2 < 1$, and $a_2 - a_1 < 1$. See Chapter 14 for a general discussion of stationarity conditions for the AR(p) model.

Approach to a new steady-state level

Following intervention, the rate of approach to a new steady-state line will be a function of the autoregressive model parameters. Figure 26.4 is a plot of the percent change in level as a function of the time after interruption for three second-order autoregressive models. The model before intervention was $Y_t = a_1 Y_{t-1} + a_2 Y_{t-2} + 1 + e_t$, whereas after intervention the model was $Y_t = a_1 Y_{t-1} + a_2 Y_{t-2} + e_t$. Figure 26.4 thus illustrates the speed of convergence from a mean level of $1/(1 - a_1 - a_2)$ to a mean level of zero, as a function of the model parameters. Convergence is slowest when the sum of a_1 and a_2 is high. This fact points out one of the limitations of employing AR models: They will smooth the transition from one trend line to the next. They thus obscure the form of the intervention effect near the point of intervention. Therefore AR models are useful when the long-range effects and not the immediate effects of intervention are of interest.

Figure 26.4. Asymptotic approach to the postintervention slope line.

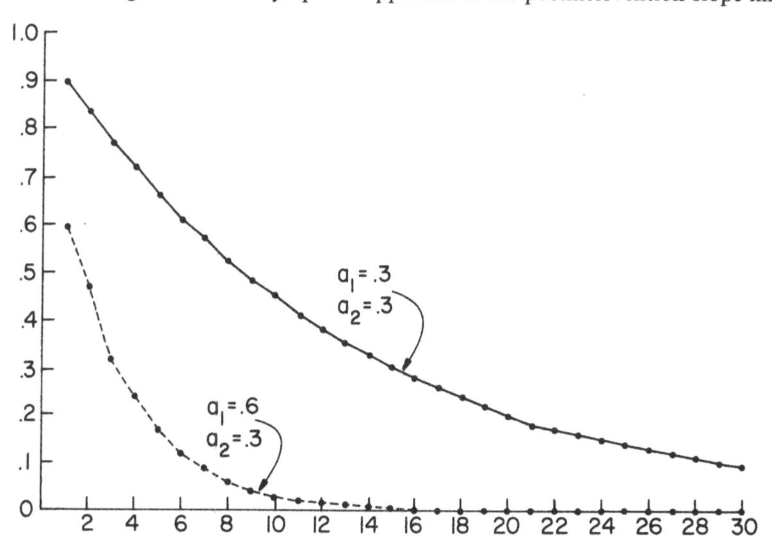

Figure 26.4. *(continued)*

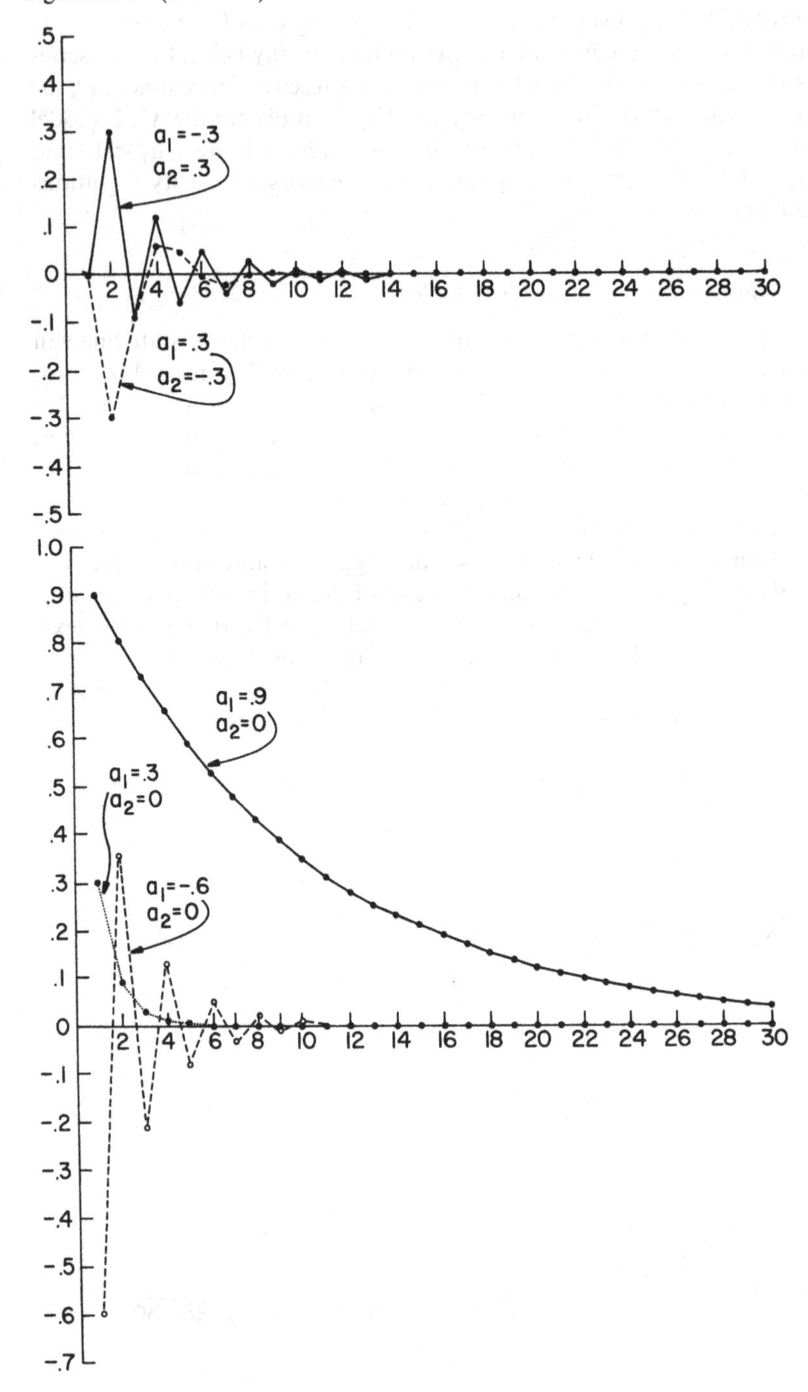

A two-step procedure is now proposed for examining intervention effects. First we must *build the general model* by identifying the number of autoregressive terms. A stepwise procedure is recommended: Fit models with different slopes and intercepts before and after intervention with one, two, or more autoregressive terms. We stop including further autoregressive terms when clearly they do not improve the model fit, that is, when they are not even suggestively significant. Next we *test intervention effects* by testing whether $m_1 = m_2$ and $b_1 = b_2$ in the model built above. The usual $F(2, \nu)$ test is used.

The next section of this chapter addresses the question of how large a sample is needed for testing intervention effects in the interrupted time-series experiment.

Sample-size requirements

The number of observations needed to perform a useful interrupted time-series analysis depends on the circumstances. Broadly stated, there are two questions: *First, how many observations are needed so that the limiting normal theory or the naive use of t or F tests provides valid procedures, that is, so that the actual level of the tests is near the nominal value? Second, how many observations are needed so that the tests are powerful against, that is, sensitive to, intervention effects of practical interest?*

As will be shown, the first question produces no lower limit on the sample size provided that the series exhibits predominantly negative correlation or moderate positive correlation and that the model (i.e., the number of autoregressive terms) is appropriately identified. In short series this means there must be some a priori assumption that the model will require no more than (say) one or two autoregressive terms, just as fitting a polynomial regression to a short stretch of data usually requires the assumption that a linear or quadratic curve will do. Thus, sample size requirements lead to power computations, as is the case for, say, the two-sample t test.

To examine sample-size requirements for adequate level control, a simulation study was carried out. The results of this study are presented in the following section. Approximate sample-size requirements for sufficient power against various-sized intervention effects can be approximated analytically; this is discussed in Appendix 26A.

Validity of the proposed ITSE analysis: a simulation study

Since the interrupted time-series model considers autocorrelated data, the usual regression assumptions do not hold, and as a consequence exact t (or

F) distributions are not obtained for the usual test statistics. All that is available at present are the large-sample normal results obtained by Anderson (1971). Thus, the questions that naturally arise–(1) how large a sample is large enough to allow the use of limiting normal theory? and (2) can the t and F distributions be used as reasonable approximations?–cannot be answered easily by analytic computation. To shed some light on these issues, the following Monte Carlo study was conducted.

Consider models that include only a change in the mean:

$$Y_t = b_1 + \sum_{i=1}^{p} a_i Y_{t-i} + e_t, \qquad t = p+1, p+2, \ldots, n_1$$

$$Y_t = b_2 + \sum_{i=1}^{p} a_i Y_{t-i} + e_t, \qquad t = n_1 + 1, \ldots, n_1 + n_2 = N \quad (26.18)$$

For each of a variety of AR(1), AR(2), and AR(3) models of this form and for each of a variety of values of n_1 and n_2, 5000 series were generated and analyzed. To illustrate the procedure, suppose that we wish to examine the behavior of $(\hat{b}_1 - \hat{b}_2)$ in an AR(2) model under the null hypothesis that $b_1 = b_2$, that is, when in reality there is no change in the mean. We generate a series of length N satisfying, for example, the model

$$Y_t = .6Y_{t-1} + .3Y_{t-2} + e_t, \qquad t = 1, 2, \ldots, N$$

by generating N pseudo-random-normal deviates $e_1^*, \ldots, e_N^*$ and then constructing

$$Y_1 = e_1^*$$
$$Y_2 = .6Y_1 + e_2^*$$
$$Y_t = .6Y_{t-1} + .3Y_{t-2} + e_t^* \qquad \text{for } t \geq 3$$

Such a series is analyzed by estimating as unknowns b_1, b_2, a_1, a_2 and the standard deviation σ_e in the model

$$Y_t = b_1 + a_1 Y_{t-1} + a_2 Y_{t-2} + e_t, \qquad t = p+1, \ldots, n_1$$
$$Y_t = b_2 + a_1 Y_{t-1} + a_2 Y_{t-2} + e_t, \qquad t = n_1 + 1, \ldots, N$$

We then evaluate the t statistic

$$\frac{(\hat{b}_1 - \hat{b}_2)}{\text{estimated standard error of } (\hat{b}_1 - \hat{b}_2)} \qquad (26.19)$$

and find the proportion of the 5000 series for which this value does not fall between the appropriate t distribution .025 (and .005) upper and lower critical points, as would be used in a two-sided level $\alpha = .05$ (and .01) test. We would hope that these proportions would be near the nomial $\alpha = .05$ (or, .01) value. In any case we obtain an approximation to the actual significance level of the t test in the particular situation.

The accuracy of such a simulation technique can be assessed by noting that whether or not the t statistic exceeds the upper or lower critical points can be viewed as a single binomial trial. The total number of series out of the 5000 that (accidentally) give significant results thus has a binomial (5000, p) distribution, where p is the actual significance level of the test. This count is thus a random variable with standard deviation $\sqrt{5000p(1-p)}$. The significance level is estimated by this count divided by 5000, which has standard deviation $\sqrt{p(1-p)/5000}$. If $p = .05$, this value is .0032; if $p = .01$, the value is .0015. Thus, if a proportion outside the range $.05 \pm (2.57)(.0032) = (.042, .058)$ is observed, we can be 99% confident that the value being approximated in the simulation is not .05. Similarly, if a value outside the range (.0064, .0136) is observed, we can be 99% confident that the actual probability is not .01.

Tables 26.5 to 26.10 present the numerical results of the simulation. Five conclusions can be drawn from these results. *The first and most important result from these tables is that sample size makes relatively little difference.* The actual significance levels tend to be a bit below the nominal levels ($\alpha = .05$ or .01) for very small samples such as those having five observations before and after intervention (denoted 5–5). The actual significance levels increase as we

Table 26.5. *AR(1) simulation: nominal $\alpha = .05$*

	$n_1 - n_2$				
a_1	5–5	10–10	20–20	40–40	60–60
−.9		.028	.040	.042	
−.6	.012	.030	.044	.042	
−.3		.033	.041	.048	
0		.041	.048	.045	
.3	.025	.052	.056	.053	
.45		.059	.062	.055	
.6	.036	.079	.072	.070	.064[a]
.7	.042	.093	.087	.073	.061
.8	.047	.116	.100	.088	.082
.9	.039	.130	.158	.137	.109[b]

Note: $\text{prob}[|(\hat{\mu}_2 - \hat{\mu}_1)/\hat{\sigma}_{\hat{\mu}_2-\hat{\mu}_1}| > t_{.025;n_1+n_2-4}]$. Accuracy: ±.008 (with 99% confidence).

[a] For $n_1 = n_2 = 80$ a value of .061 was obtained.

[b] For $n_1 = n_2 = 100$ a value of .087 was obtained.

increase to sample sizes to 10–10 or 20–20, but as the sample sizes increases beyond 20–20, the significance level slowly converges to the nominal value. However, if the significance levels are considerably larger than the nominal values for 10–10 or 20–20, sample sizes of 60–60 or even 100–100 may not yield satisfactory results.

The results show that *the determining factor is the amount of positive autocorrelation present.* For AR(1) models, the significance test becomes questionable when a_1 is too large, with the dividing line somewhere between

Table 26.6. *AR(1) simulation: nominal $\alpha = .01$*

	$n_1 - n_2$				
a_1	5–5	10–10	20–20	40–40	60–60
−.9		.002	.003	.009	
−.6	.001	.002	.007	.007	
−.3		.002	.005	.008	
.0		.005	.009	.008	
.3	<.0005	.005	.008	.011	
.45		.007	.012	.009	
.6	.001	.009	.014	.014	.014[a]
.7	.001	.012	.019	.018	.014
.8	.001	.019	.020	.021	.021
.9	.001	.021	.044	.036	.029[b]

Note: prob$[|(\hat{\mu}_2 - \hat{\mu}_1)/\hat{\sigma}_{\hat{\mu}_2-\hat{\mu}_1}| > t_{.005;n_1+n_2-4}]$. Accuracy: $\pm.004$ (with 99% confidence).
[a] For $n_1 = n_2 = 80$ a value of .013 was obtained.
[b] For $n_1 = n_2 = 100$ a value of .022 was obtained.

Table 26.7. *AR(1) simulation: unbalanced designs*

	$n_1 - n_2$ for nominal $\alpha = .05$			$n_1 - n_2$ for nominal $\alpha = .10$		
a_1	5–15	10–20	20–60	5–15	10–30	20–60
−.6	.036	.043	.043	.004	.007	.008
.3	.051	.054	.055	.006	.008	.008
.9	.085	.128	.112	.013	.030	.029

Note: prob$[|(\hat{\mu}_2 - \hat{\mu}_1)/\hat{\sigma}_{\hat{\mu}_2-\hat{\mu}_1}| > t_{\alpha/2;n_1+n_2-4}]$.

Table 26.8. *AR(2) simulation: nominal* $\alpha = .05$

a_1	a_2	$n_1 - n_2$ 10–10	20–20	40–40	a_1	a_2	$n_1 - n_2$ 10–10	20–20	40–40
.0	−.9	.033	.040	.043	.6	−.6	.043	.059	
	−.6	.033	.047			−.3	.059	.057	
	−.3	.040	.041			.0	.084	.078	.068
	.0	.046	.047			.3	.120	.177	.149[a]
	.3	.066	.069	.069	−.6	−.6	.036	.041	
	.6	.098	.107	.077		−.3	.038	.044	
	.9	.109	.177	.187		.0	.042	.054	
.3	−.3	.044	.048			.3	.047	.058	
	.0	.063	.068	.058	1.2	−.6	.059	.053	
	.3	.089	.098	.080[b]	−1.2	−.6	.029	.040	
−.3	−.3	.041	.047						
	.0	.045	.049						
	.3	.057	.064	.062[c]					

Note: prob$[|(\hat{\mu}_1 - \hat{\mu}_2)/\hat{\sigma}_{\hat{\mu}_2 - \hat{\mu}_1}| > t_{.025; n_1 + n_2 - 6}]$. Accuracy: $\pm .008$ (with 99% confidence).

[a] At $n_1 = n_2 = 100$ a value of .101 was obtained.
[b] At $n_1 = n_2 = 60$ a value of .072 was obtained.
[c] At $n_1 = n_2 = 60$ a value of .057 was obtained.

Table 26.9. *AR(2) simulation: nominal* $\alpha = .01$

a_1	a_2	$n_1 - n_2$ 10–10	20–20	40–40	a_1	a_2	$n_1 - n_2$ 10–10	20–20	40–40
.0	−.9	.002	.007	.006	.6	−.6	.003	.006	
	−.6	.002	.007			−.3	.006	.011	
	−.3	.004	.006			.0	.009	.013	.014
	.0	.004	.009			.3	.018	.044	.040[a]
	.3	.006	.013	.012	−.6	−.6	.005	.007	
	.6	.012	.021	.020		−.3	.003	.007	
	.9	.014	.049	.055		.0	.004	.009	
.3	−.3	.004	.008			.3	.004	.011	
	.0	.008	.013	.011	1.2	−.6	.088	.006	
	.3	.010	.020	.018[b]	−1.2	−.6	.003	.012	
−.3	−.3	.005	.008						
	.0	.004	.006						
	.3	.006	.013	.009[c]					

Note: prob$[|(\hat{\mu}_2 - \hat{\mu}_1)/\hat{\sigma}_{\hat{\mu}_2 - \hat{\mu}_1}| > t_{.005; n_1 + n_2 - 6}]$. Accuracy: $\pm .004$ (with 99% confidence).

[a] At $n_1 = n_2 = 100$ a value of .025 was obtained.
[b] At $n \ = n_2 = 60$ a value of .017 was obtained.
[c] At $n_1 = n_2 = 60$ a value of .011 was obtained.

$a_1 = .5$ and $a_1 = .75$. For the AR(2) models, it is instructive to note which are the worst cases. Near the edge of the stationarity region defined by $a_1 + a_2 = 1$ the behavior is very bad. The worst cases are those for which $a_1 + a_2 = .9$, with those having $a_1 + a_2 = .6$ following; if we compare the four cases for which $a_1 + a_2 = .6$, we find that the worst is $a_1 = 0, a_2 = .6$, followed in order of increasing accuracy by $a_1 = .3$, $a_2 = .3$; $a_1 = .6$, $a_2 = .0$; and $a_1 = 1.2$, $a_2 = -0.6$. Thus, the further back in the past the autocorrolation extends, the worse. Note that the $a_1 = 1.2, a_2 = -.6$ case gives very reasonable results. The negative correlation two steps in the past can thus counterbalance a large positive correlation one step back. The AR(3) models yield the same general results. The $a_1 = .3$, $a_2 = .2$, and $a_3 = .2$ case $(a_1 + a_2 + a_3 = .7)$ is the worst, as would be expected with three positive autocorrelations. The $a_1 = -.3$, $a_2 = .2$, $a_3 = .2$ case is worse than the $a_1 = .3$, $a_2 = .2$, $a_3 = -.2$ case, because in the former the positive autocorrelation extends further into the past.

It should be pointed out that all these results apply to the change in mean of the series. The estimates of the autoregressive coefficients themselves are known to be biased (Johnston, 1972) and it is reasonable to expect that when this bias is especially severe, as is the case when there is a large amount of positive autocorrelation, other more stable estimators will be affected.

Third, *similar results are obtained for either the nominal level of .05 or the nominal level of .01*. If we consider relative errors, for example in the AR(2) case for which $a_1 = 0$, $a_2 = .3$, $n_1 = n_2 = 20$, and $\alpha = .05$, the actual probability is 1.38 times too large, whereas in the corresponding $\alpha = .01$ case, the factor is

Table 26.10. *AR(3) simulation*

a_1	a_2	a_3	Nominal $\alpha = .05$ $(n_1 = n_2 = 20)$	Nominal $\alpha = .01$ $(n_1 = n_2 = 20)$
.0	.0	.0	.055	.008
.3	.0	.0	.070	.014
.3	.3	.0	.095	.021
.3	.3	.2	.130[a]	.029[b]
.3	−.3	.0	.059	.010
.4	−.2	.2	.077	.015
.3	.2	−.2	.061	.010
−.3	.3	.2	.076	.015
.4	−.3	−.2	.053	.011

Note: $\text{prob}[|(\hat{\mu}_2 - \hat{\mu}_1)/\hat{\sigma}_{\hat{\mu}_2 - \hat{\mu}_1}| > t_{\alpha/2; n_1 + n_2 - 8}]$.
[a] For $n_1 = n$ = 40 a value of .093 was obtained.
[b] For $n_1 = n_2 = 40$ a value of .025 was obtained.

1.30. In the most extreme cases, however, the relative error is a bit worse at $\alpha = .01$, as is witnessed in the $a_1 = .6$, $a_2 = .3$, $n_1 = n_2 = 20$ case. Here, in the nominal .05 case, the actual value is three times too large, whereas in the .01 case, the actual value is four times too large.

Next, we note from these tables that *overfitting a model* [e.g., fitting an AR(2) scheme when $a_2 = 0$, so that an AR(1) model would be appropriate] *does not have a serious influence on significance levels.* With a nominal $\alpha = .05$, a difference of only about .01 is seen when comparing the fitting of an appropriate AR(1) model and a too-large AR(2) model.

Finally, *imbalance between the number of observations before and after intervention makes very little difference.* Only in the case of large positive autocorrelation is any effect visible, and there the significance levels in the unbalanced case are *closer* to the nominal values than in the balanced case.

To summarize, we have found that the total positive autocorrelation is the factor that controls the accuracy of the proposed ITSE analysis. Sample size, balance or imbalance in the design, overfitting, and the desired α level have a relatively small influence. Provided that we view with caution (see the example at the end of this chapter) those cases for which the sum of the autoregressive coefficients exceeds 0.6, ITSE analysis with autoregressive models can be used with confidence even in small samples.

The question remains: If ITSE analysis with autoregressive models is used in a small sample, can we hope to detect any but the largest effects? This question of sensitivity is considered next.

Power of the ITSE

Consider, as before, the model corresponding to a simple change in mean:

$$Y_t = b_1 + \sum_{i=1}^{p} a_i Y_{t-i} + e_t, \qquad t = 1, \ldots, n_1$$

$$Y_t = b_2 + \sum_{i=1}^{p} a_1 Y_{t-i} + e_t, \qquad t = n_1 + 1, \ldots, n_1 + n_2 = N_1$$

These equations differ very slightly from those introduced earlier in that here $t = 1, 2, \ldots, n_1$ in the first equation instead of $t = p + 1, \ldots, n_1$ as before. Thus, in this formulation the first two observations in an AR(2) model would be labeled Y_{-1} and Y_0. This was done to make these sample-size calculations resemble the calculations for, say, the two-sample t test. In the new formulation n_1, n_2, and N are the number of observations once the pattern is established. Once we have the variance of $(\hat{b}_1 - \hat{b}_2)$, which we denote by v/N, because it decreases as N increases, we can estimate the number of observations, N, needed to be reasonably sure of detecting an intervention effect of a given size.

Suppose that we want to determine how many observations are needed to have a probability of $\beta = .90$ of detecting a change in the mean as large as the standard deviation of the series (either half, before or after intervention), when using a two-sided test. That is, in the notation introduced earlier, we have $|B_1 - B_2| \geq \sigma_Y$. Because, as derived in Appendix 26B, $b_1 = B_1/(1 - \sum a_i)$, this corresponds to the alternative hypothesis $|b_1 - b_2| \geq \sigma_Y/(1 - \sum a_i)$. Let $\delta = \sigma_Y/(1 - \sum a_i)$. Then the large sample results of Anderson (1971) say that the t-statistic

$$\frac{\sqrt{N}[(\hat{b}_1 - \hat{b}_2) - \delta]}{\sqrt{\hat{v}}} \tag{26.20}$$

where $\hat{v}$ is the estimate obtained using the procedure described earlier, has an approximate $N(0, 1)$ distribution if N is large. Thus, the usual test statistic

$$\frac{\sqrt{N}(\hat{b}_1 - \hat{b}_2)}{\sqrt{\hat{v}}} \tag{26.21}$$

is approximately $N(\delta\sqrt{N}/\sqrt{v}, 1)$.

We want the power, β, which is the probability that we reject the hypothesis $b_1 = b_2$. The following relation is obtained for β:

$$\begin{aligned} \beta &= 1 - P(-Z_{\alpha/2} < \sqrt{N}(\hat{b}_1 - \hat{b}_2)/\sqrt{\hat{v}} < +Z_{\alpha/2}) \\ &\approx 1 - P(-\delta\sqrt{N}/\sqrt{v} - Z_{\alpha/2} < N(0,1) < -\delta\sqrt{N}/\sqrt{v} + Z_{\alpha/2}) \\ &\approx 1 - P(N(0,1) < Z_{\alpha/2} - \delta\sqrt{N}/\sqrt{v}) \\ &= P(N(0,1) > Z_{\alpha/2} - \delta\sqrt{N}/\sqrt{v}) \end{aligned} \tag{26.22}$$

where $Z_{\alpha/2}$ is the critical point satisfying $P(\mathrm{N}(0, 1) > Z_{\alpha/2}) = \alpha/2$. The normal approximations are reasonable if N is moderately large, say greater than 25. For the last line of equation (26.22) to equal β, we need $Z_\beta = Z_{\alpha/2} - \delta\sqrt{N}/\sqrt{v}$, or $N = (Z_{\alpha/2} - Z_\beta)^2 v/\delta^2$.

Note for $\alpha = .05$ and $\beta = .90$ we have $Z_{\alpha/2} = 1.96$ and $Z_\beta = -1.28$, whence $N = (3.24)^2 v/\delta^2 = (10.50)v/\delta^2$. For one-sided tests, $Z_{\alpha/2}$ is replaced by Z_α, so that for a power $\beta = .90$ in a one-sided $\alpha = .05$, test we need $N = (1.645 + 1.28)^2 v/\delta^2 = (2.895)^2 v/\delta^2 = 8.38v/\delta^2$. For $\beta = .75$ in a one-sided $\alpha = .05$ test, we need $N = 5.39v/\delta^2$. Note also that if the size of the alternative value of $B_1 - B_2$ is doubled, δ^2 is quadrupled. Thus, N is reduced by a factor of 4. *If the size of the change in mean is halved, four times as many observations will be needed to be reasonably sure of detecting it.*

A formula for v is obtained in Appendix 26A. It can be expressed as a multiple of σ_Y^2, where the multiplier depends only on the autoregressive parameters $a_1, \ldots, a_p$, and on the proportion of observations before and after intervention. Thus, if δ is a known multiple of σ_Y, the sample size required can be specified for a particular choice of the a's and of n_1 and n_2. In terms of power, balanced designs ($n_1 = n_2$) require the fewest observations (see also Glass et al., 1975, p. 200).

Figure 26.5 is a plot of the number of observations needed either before or after intervention in a balanced design in the case where $\alpha = .05$, a two-sided test is used, $\beta = .90$, and the alternative $B_1 - B_2 = \sigma_Y$ is considered, as a function of the parameters in an AR(2) model. Only those values of a_1 and a_2 where the model is stationary (i.e., exhibits long-range stable behavior) are shown. In the worst case ($a_1 \approx -.90, a_2 \approx -.35$) 58 observations would be needed both before and after intervention. At $a_1 = 0$, $a_2 = 0$, only 31 are required. The latter number might be a better general guide; in our experience, a situation like that in the worst case where there is a large amount of negative autocorrelation is not common. Note also that the number of observations needed drops off rapidly along the upper right edge of the stationarity triangle. This is a spurious result, as from the simulation study we know that the normal approximation upon which these calculations are based is an especially poor one for large positive a_1 and a_2. The simulation does suggest a good approximation at $a_1 = a_2 = 0$ and for negative autocorrelations, so that the sample sizes above–58 in the worst case and 31 in a more reasonable one–are probably quite accurate.

For other choices of the power β and the level α, and the choice of one- or two-sided testing, these sample sizes are just scaled by a fixed proportionality

Figure 26.5 Number of observations, n_1, needed for power $\beta = .90$ when using a two-sided $\alpha = .05$ test, and alternative hypothesis $B_1 - B_2 = \sigma_Y$, for the AR(2) model.

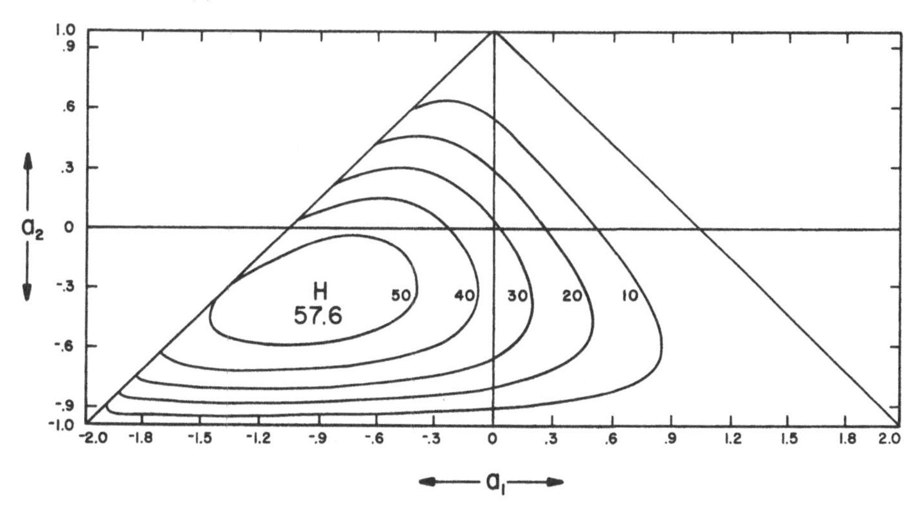

factor. Recall for $\alpha = .05$, $\beta = .90$, and two-sided tests, $N = 10.50v/\delta^2$; for $\alpha = .05$, $\beta = .90$, and a one-sided test, $N = 8.38v/\delta^2$; and for $\alpha = .05$, $\beta = .75$, and a one-sided test, $N = 5.39v/\delta^2$. Thus, in the latter two cases the largest sample-size requirement $n_1 = 58$ at $a_1 = -.95$ and $a_2 = -.35$ is reduced to $n_1 = 46$ and 30, respectively, whereas the requirement at $a_1 = a_2 = 0$ is reduced from $n_1 = 31$ to 25 or 16, respectively.

An example: the Ireland data

Following the procedure described earlier, we begin by fitting models of the form

$$Y_t = m_1 t + b_1 + \sum_{i=1}^{p} a_i Y_{t-i} + e_t \qquad \text{for } t \le 1924$$

$$Y_t = m_2 t + b_2 + \sum_{i=1}^{p} a_i Y_{t-i} + e_t \qquad \text{for } t \ge 1925 \qquad (26.23)$$

with $p = 0, 1, 2, 3, 4$, and 5. The first step was to identify the most appropriate size for the model, that is, to identify p. Table 26.11 gives the estimates a_i and the associated t statistics and significance levels for each choice of p.

Clearly, $p = 4$ is too large, the a_4 term in the AR(4) model is very small, and both the a_4 and a_5 terms in the AR(5) model are not even close to being

Table 26.11. *AR terms for model building with the Ireland data*

Number of terms	Parameter estimates	t values	Significance
$p = 1$	$a_1 = .60$	7.10	$p < .001$
$p = 2$	$a_1 = .53$	5.27	$p < .001$
	$a_2 = .16$	1.57	$p \approx .12$
$p = 3$	$a_1 = .51$	4.74	$p < .001$
	$a_2 = .092$	.77	—
	$a_3 = .14$	1.32	$p \approx .19$
$p = 4$	$a_1 = .51$	4.66	$p < .001$
	$a_2 = .13$	1.08	$p \approx .28$
	$a_3 = .096$	.79	—
	$a_4 = -.011$	−.01	—
$p = 5$	$a_1 = .50$	4.48	$p < .001$
	$a_2 = .13$	1.04	$p \approx .30$
	$a_3 = .12$	.94	$p \approx .34$
	$a_4 = -.043$	−.34	—
	$a_5 = .039$	.35	—

significant. The choice among $p = 1, 2$, or 3 is less clear cut. Although the a_2 term in the AR(2) model or the a_3 term in the AR(3) model are not significant at the $\alpha = .10$ level, they are nearly so. The model-building step requires us to include terms if there is any suggestion that they contribute to a better model. This is the exact opposite of the usual hypothesis-testing situation, where terms are assumed to be zero until proven otherwise. Strict adherence to significance levels at this stage is therefore inappropriate; the tests must be used informally. Thus, either the AR(2) or AR(3) appears reasonable. Rather than decide immediately, we shall consider the testing of the intervention effect in the context of both models.

The decision as to whether or not an intervention effect is or is not present returns us to the more usual hypothesis-testing situation. We test the null hypothesis $m_1 = m_2$ and $b_1 = b_2$ using ordinary F test. Let SS_0 denote the residual error sum of squares in the reduced model

$$Y_t = mt + b + \sum_{i=1}^{p} a_i Y_{t-i} + e_t, \qquad \text{for all } t \tag{26.24}$$

and let SS_1 denote the residual error sum of squares in the full model.[2] Then under the null hypothesis,

$$F = \frac{(SS_0 - SS_1)/2}{(SS_1)/v}$$

has an $F(2, v)$ distribution. Here v, the error degrees of freedom–equal to the number of observations (93) minus the number of "start up" observations [2 in the AR(2) model or 3 in the AR(3)] minus the number of parameters fit [6 in the AR(2), 7 in the AR(3)]–is given by the formula $v = 89 - 2p$.

In the AR(2) model $F = 5.74$ was obtained, which is significant with $p < .005$ ($v = 85$). In the AR(3) model we have $F = 5.99$, which is also significant with $p < .005$ ($v = 83$). In the context of either model we find a similar significant intervention effect.

It is also instructive to consider the lines fit to the data before and after intervention in each case. In the AR(2) model we have estimates $\hat{m}_1 = -.0630$ and $\hat{b}_1 = 137.68$ before intervention and $\hat{m}_2 = .0765$ and $\hat{b}_2 = -124.36$ after intervention. Translating these into the actual steady-state slopes and intercepts as derived in Appendix 26B, we find $\hat{M}_1 = -.208$, $\hat{B}_1 = 452.7$ before and $\hat{M}_2 = .253$, $\hat{B}_2 = 412.2$ after intervention. These lines are drawn in Figure 26.6. For the AR(3) model the estimates of m_1, m_2, b_1, and b_2 appear quite different: $\hat{m}_1 = -.0558$, $\hat{b}_1 = 121.11$, $\hat{m}_2 = .0377$, and $\hat{b}_2 = -52.211$. However, when we translate these values into the actual slopes and intercepts, $\hat{M}_1 = -.217$, $\hat{B}_1 = 472.29$, $\hat{M}_2 = .146$, and $\hat{B}_2 = -203.83$ and plot the results as is also done in Figure 26.6, we see that the lines fitted by the AR(2) and AR(3) models are similar. The AR(3) model seems to fit the data a bit better and might thus be preferred.

Note that in this example we have fit a model that incorporates a large positive autocorrelation, precisely the situation in which the level of the significance tests must be questioned. The AR(2) model we fit lies between the $a_1 = .6$, $a_2 = .3$ and $a_1 = .6$, $a_2 = 0$ models that were simulated earlier. The simulation of a simple one-degree-of-freedom test of a change in the mean should suggest the type of behavior occurring in this two-degree-of-freedom F-test. For $n_1 = n_2 = 40$ the simulation showed that a nominal $\alpha = .01$ test actually had $\alpha \approx .04$ for $a_1 = .6$ and $a_1 = .3$ and $\alpha \approx .014$ for $a_1 = .6$ and $a_2 = 0$. Thus, in our situation, where an observed significance value $p < .005$ might be claimed, the actual level might be two or three times larger. That is, $p < .01$ or $p < .015$ might be a more honest claim. This makes very little difference here, but if the usual t-significance probability was .05 instead of .005, the situation would be much less clear, and a claim of significant results could easily be questioned.

It is also interesting to observe the spurious significance if one ignored autocorrelation. If different simple linear regressions were fit to the two halves of the data, a whopping $F = 53.89$ would be found. However, as the assumptions needed to claim that this statistic can be compared to the points of an F-distribution are seriously violated, this number is not meaningful, whereas the much smaller F we computed assuming autocorrelation is meaningful. This example points out just how misleading an analysis that

Figure 26.6. Two autoregressive models fitted to the Ireland data.

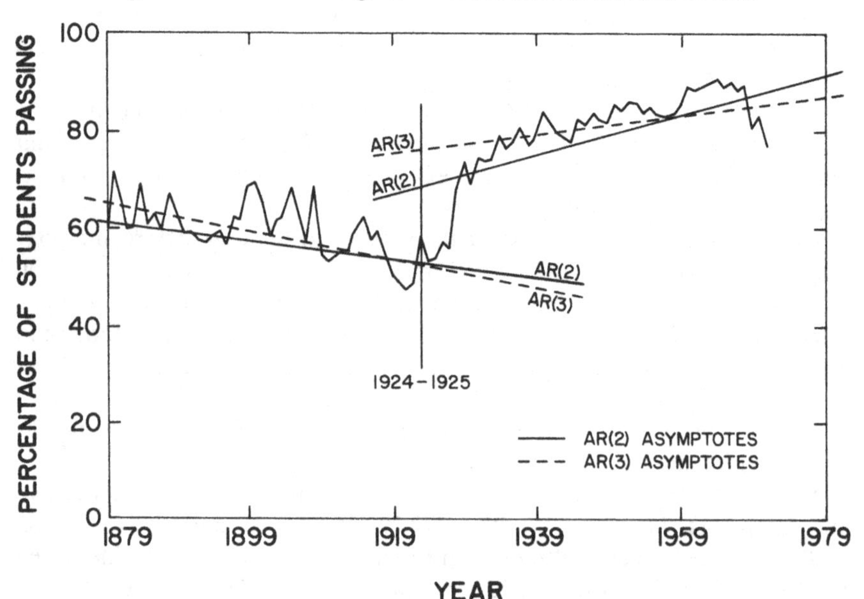

does not account for autocorrelation can be. The F statistic can be spuriously inflated tenfold.

Other approaches to the interrupted time-series experiment

It has not been intention of this chapter to dissuade the reader from employing nonlinear ARMA models in testing intervention effects. In many applications ARMA models may be best, particularly when the effect of the intervention is near the intervention point. The goal of Chapter 26 has been to propose and to explore a simple, yet uninvestigated alternative (i.e., linear AR models). This section reviews two alternative approaches that employ nonlinear ARMA models only briefly because well-written and extensive treatments of both methods are available. The reader will be referred to these sources for further reading. The two methods discussed are the approach used by Glass et al. (1975) and the Box and Tiao (1975) procedure.

Glass et al.

In the Glass et al. (1975) procedure for a simple case, suppose that the model is an AR(1):

$$x_t - L = a_1(x_{t-1} - L) + e_t \tag{26.25}$$

where e_t is $N(0, \sigma_e^2)$ and $|a_1| < 1$. In other words, after removing the mean, the time-series is an AR(1) process.

A transformation was first worked out by Box and Tiao (1965) for a few cases and then generalized to other ARIMA models by Glass et al. (1975). Rewrite equation (26.25) as follows:

$$(x_t - L) - a_1(x_{t-1} - L) = e_t \tag{26.26}$$

or

$$(x_t - a_1 x_{t-1}) = (1 - a_1)L + e_t \tag{26.27}$$

Now if we let,

$$y_1 = x_1 \text{ and } y_t = x_t - a_1 x_{t-1} \tag{26.28}$$

then, we can rewrite equations (26.26) and (26.27) in terms of the y's as

$$y_1 = L + e_1 \qquad y_t = (1 - a_1)L + e_t \tag{26.29}$$

If this is the model for the first n_1 points preceding the intervention, and we assume that the effect of the intervention is to add a constant, δ, to each point for the remaining $N - n_1$ points, then we can write all the points in terms of the y's as

$$\begin{aligned}
y_1 &= L + e_1 \\
y_2 &= (1 - a_1)L + e_2 \\
y_3 &= (1 - a_1)L + e_3 \\
\vdots &\quad \vdots \quad \vdots \\
y_{n_1} &= (1 - a_1)L + e_{n_1} \\
y_{n_1+1} &= (1 - a_1)L + \delta + e_{n_1+1} \\
y_{n_1+2} &= (1 - a_1)L + (1 - a_1)\delta + e_{n_1+2} \\
\vdots &\quad \vdots \quad \vdots \quad \vdots \\
y_N &= (1 - a_1)L + (1 - a_1)\delta + e_N
\end{aligned} \tag{26.30}$$

This transformation is useful because the only random term in each case is simply the white-noise series $\{e_i\}$. If we write equation (26.30) in matrix form we can obtain least-squares estimates for L and δ:

$$\mathbf{Y} = \mathbf{X}\boldsymbol{\beta} + \mathbf{E} \tag{26.31}$$

where

$$\begin{bmatrix} y_1 \\ y_2 \\ \vdots \\ y_{n_1} \\ \hline y_{n_1+1} \\ y_{n_1+2} \\ \vdots \\ y_N \end{bmatrix} = \begin{bmatrix} 1 & 0 \\ 1 - a_1 & 0 \\ \vdots & \vdots \\ 1 - a_1 & 0 \\ \hline 1 - a_1 & 1 \\ 1 - a_1 & 1 - a_1 \\ \vdots & \vdots \\ 1 - a_1 & 1 - a_1 \end{bmatrix} \begin{bmatrix} L \\ \delta \end{bmatrix} + \begin{bmatrix} e_1 \\ e_2 \\ \vdots \\ e_{n_1} \\ \hline e_{n_1+1} \\ e_{n_1+2} \\ \vdots \\ e_N \end{bmatrix}$$

Conditional on a_1 being known, linear least-squares estimates of L and δ are

$$\begin{bmatrix} \hat{L} \\ \hat{\delta} \end{bmatrix} = (\mathbf{X}'\mathbf{X})^{-1}\mathbf{X}'\mathbf{Y} \tag{26.32}$$

and

$$\frac{L - \hat{L}}{s_e\sqrt{c^{11}}} \sim t_{N-2} \tag{26.33}$$

$$\frac{\delta - \hat{\delta}}{s_e\sqrt{c^{22}}} \sim t_{N-2} \tag{26.34}$$

where

$$s_e^2 = (\mathbf{Y} - \mathbf{X}\hat{\boldsymbol{\beta}})'(\mathbf{Y} - \mathbf{X}\hat{\boldsymbol{\beta}})/(N - 2) = \hat{\mathbf{e}}'\hat{\mathbf{e}}/(N - 2) \tag{26.35}$$

and c^{jj} is the jth diagonal entry of $(\mathbf{X}'\mathbf{X})^{-1}$

Because, in most cases, we do not know the model parameters, a nonlinear least-squares estimation procedure is performed, selecting those values of a, L, and δ that minimize the quantity s_e^2 in equation (26.35). Once the *form* of the ARIMA model is identified, it is possible to test for the intervention effect while simultaneously estimating the model parameters. Glass et al. (1975) discussed the general transformation, and they also discussed how the design matrix can be modified for specific hypothesized intervention effects. Programs CORREL (for model identification) and TSX (for testing the intervention effect) are available from Glass for these analyses. The programs use a grid-search technique for parameter estimation, which is an inefficient procedure. Also, users have expressed difficulty in going from CORREL to TSX, but these problems can be minimized by experience with Box–Jenkins model-fitting procedures.

Box and Tiao (1975)

A second approach has recently become popular in the social sciences, primarily because of the excellent writings of Hibbs (1977) and McCleary and Hay (1980). The observed time-series can be written as

$$z(t) = y(t) + w(t)$$

where $y(t)$ is the intervention and $w(t)$ is the rest of the data, including random and deterministic effects. Now $w(t)$ can be modeled using an ARIMA scheme, which can be written in terms of the backward shift operator, B, as

$$w(t) = \Pi_1(B)e_t$$

and the intervention can be modeled as

$$y(t) = \Pi_2(B)I_t$$

where I_t is an indicator variable that is one after intervention and zero before intervention. For example, simple step intervention would have the form $y(t) = kI_t$, where k is the height of the step. A gradual intervention could be represented $y(t) = k_1 y(t-1) + k_2 I_t$; as $t \to \infty$, $y(t)$ approaches the value $y(\infty) = k_1 y(\infty) + k_2$, or $y(\infty) = k_2/(1 - k_1)$; $y(t)$ can be rewritten $(k_2/(1 - k_1 B))I_t$, where B is again the backward shift operator. The full model is then

$$z(t) = \Pi_2(B)I_t + \Pi_1(B)e(t)$$

The data prior to intervention can be used to identify $\Pi_1(B)$ (because $I_t = 0$ then) choosing a *form* of the ARIMA model, but not estimating parameters. Then the whole model would be used to estimate model parameters, either in a two-stage, least-squares procedure, or with a nonlinear search-procedure. The reader is referred to Hibbs (1977) for a lucid discussion of the approach and to McCleary and Hay (1980) for an excellent introduction with many examples. Computations in the McCleary and Hay book are unfortunately not explained; they are deferred to references to available computer software.

APPENDIX 26A

The limiting variance of $(\hat{b}_1 - \hat{b}_2)$

This section computes the variance of the least-squares estimate $(\hat{b}_1 - \hat{b}_2)$ in the model

$$Y_t = b_1 + \sum_{i=1}^{p} a_i Y_{t-i} + e_t, \qquad t = 1, 2, \ldots, n_1$$

$$Y_t = b_2 + \sum_{i=1}^{p} a_i Y_{t-i} + e_t, \qquad t = n_1 + 1, \ldots, n_1 + n_2 = N$$

in the limit as n_1 and n_2 tend to infinity. As written in equation (26.15), this model can be presented in vector notation as $\mathbf{Y} = \mathbf{XB} + \mathbf{E}$, where $\mathbf{B} = (b_1, b_2, a_1, a_2, \ldots, a_p)$. Anderson (1971) argues that $(\mathbf{X}'\mathbf{X})^{-1}\hat{\sigma}_e^2$ approximates the variance–covariance matrix of $\hat{\mathbf{B}}$, or to be more precise, he shows that $N(\hat{\mathbf{B}} - \mathbf{B})$ converges in distribution to a multivariate normal with mean $\mathbf{0}$ and covariance matrix $\mathbf{\Sigma} = \lim N(\mathbf{X}'\mathbf{X})^{-1}\hat{\sigma}_e^2$. We will compute the upper left 2×2 corner of ^{S}, which is the limiting variance-covariance matrix for $(\hat{b}_1, \hat{b}_2)$. From this the variance of $\hat{b}_1 - \hat{b}_2$ is easily obtained.

Using the expression for $\mathbf{X}$ in equation (26.15), we find that $(\mathbf{X}'\mathbf{X})$ is the symmetric matrix

$$\mathbf{X}'\mathbf{X} = \left[\begin{array}{cc|cccc}
n_1 & 0 & \sum_{0}^{n_1-1} Y_i & \sum_{-1}^{n_1-2} Y_i & \cdots & \sum_{1-p}^{n_1-p} Y_i \\
0 & n_2 & \sum_{n_1}^{N-1} Y_i & \sum_{n_1-1}^{N-2} Y_i & \cdots & \sum_{n_1-p+1}^{N-p} Y_i \\
\hline
 & & \sum_{0}^{N-1} Y_i^2 & \sum_{0}^{N-1} Y_i Y_{i-1} & \cdots & \sum_{0}^{N-1} Y_i Y_{i-p+1} \\
 & & \sum_{-1}^{N-2} Y_i Y_{i+1} & \sum_{-1}^{N-2} Y_i^2 & \cdots & \sum_{-1}^{N-2} Y_i Y_{i-p+2} \\
 & & & & \ddots & \vdots \\
 & & \sum_{1-p}^{N-p} Y_i Y_{i+p-1} & & & \sum_{1-p}^{N-p} Y_i^2
\end{array}\right]$$

Write this and its inverse as

$$\mathbf{X'X} = \left[\begin{array}{c|c} \mathbf{A} & \mathbf{B'} \\ \hline \mathbf{B} & \mathbf{C} \end{array}\right] \quad \text{and} \quad (\mathbf{X'X})^{-1} = \left[\begin{array}{c|c} \mathbf{D} & \mathbf{E'} \\ \hline \mathbf{E} & \mathbf{F} \end{array}\right]$$

where **D, E,** and **F** are of the same size as **A, B,** and **C,** respectively ($2 \times 2, 2 \times p$, and $p \times p$). Since

$$\begin{pmatrix} \mathbf{A} & \mathbf{B} \\ \mathbf{B} & \mathbf{C} \end{pmatrix}\begin{pmatrix} \mathbf{D} & \mathbf{E'} \\ \mathbf{E} & \mathbf{F} \end{pmatrix} = \begin{pmatrix} \mathbf{I}_2 & 0 \\ 0 & \mathbf{I}_p \end{pmatrix}$$

where $\mathbf{I}_2$ is the 2×2 identity matrix and $\mathbf{I}_p$ is the $p \times p$ identity, we can solve the equations $\mathbf{AD} + \mathbf{B'E} = \mathbf{I}_2$, $\mathbf{AE'} + \mathbf{B'F} = \mathbf{0}$ and $\mathbf{BE'} + \mathbf{CF} = \mathbf{I}_p$ to obtain

$$\mathbf{D} = \mathbf{A}^{-1} + (\mathbf{A}^{-1}\mathbf{B})(\mathbf{C} - \mathbf{BA}^{-1}\mathbf{B})^{-1}(\mathbf{A}^{-1}\mathbf{B})'$$

But **D** is the upper left 2×2 corner of $(\mathbf{X'X})^{-1}$, so that the limiting variance of $\sqrt{N}[(\hat{b}_1, \hat{b}_2) - (b_1, b_2)]$ will be just the limit of $(N\hat{\sigma}_e)\mathbf{D}$ as $N \to \infty$.

To compute this limit, first note that

$$\mathbf{A} = \begin{bmatrix} n_1 & 0 \\ 0 & n_2 \end{bmatrix}$$

so that if the proportions of observations before intervention, n_1/N, and after intervention, n_2/N, are fixed at, say, λ_1 and λ_2, as $N \to \infty$, then

$$N\mathbf{A}^{-1} = \begin{bmatrix} 1/\lambda_1 & 0 \\ 0 & 1/\lambda_2 \end{bmatrix}$$

Second, observe that $\mathbf{A}^{-1}\mathbf{B}$ is just

$$\begin{bmatrix} \dfrac{1}{n_1}\sum\limits_{0}^{n_1-1} Y_i & \cdots & \dfrac{1}{n_1}\sum\limits_{1-p}^{n_1-p} Y_i \\ \dfrac{1}{n_2}\sum\limits_{n_1}^{N-1} Y_i & \cdots & \dfrac{1}{n_2}\sum\limits_{n_1-p+1}^{N-p} Y_i \end{bmatrix}$$

which converges to

$$\begin{bmatrix} B_1 & B_1 & \cdots & B_1 \\ B_2 & B_2 & \cdots & B_2 \end{bmatrix}$$

Finally, the laborious but straightforward evaluation of $\mathbf{C} - \mathbf{BA}^{-1}\mathbf{B}$ yields a matrix whose entries converge to the "stripe" matrix $\mathbf{\Gamma}$, which has, in row i and j, the entry $\operatorname{cov}(Y_k, Y_{k+(i-j)}) = \gamma_{(i-j)}$. That is,

$$\mathbf{\Gamma} = \begin{bmatrix} \gamma_0 & \gamma_1 & \gamma_2 & \cdots & & \gamma_{p-1} \\ \gamma_1 & \gamma_0 & \gamma_1 & & & \gamma_{p-2} \\ \gamma_2 & \gamma_1 & \gamma_0 & \gamma_1 & & \gamma_{p-3} \\ & & \ddots & \ddots & \ddots & \vdots \\ & & & & \gamma_0 & \gamma_1 \\ \gamma_{p-1} & & & & \gamma_1 & \gamma_0 \end{bmatrix}$$

(Note $\gamma_1 = \gamma_{-1}$, $\gamma_2 = \gamma_{-2}$, and so forth.) These covariances can, in turn, be computed from the autoregressive coefficients $a_1, \ldots, a_p$ (or vice versa) by solving the Yule–Walker equations (see Anderson, 1971, p. 174):

$$\sum_{i=1}^{p} a_i \gamma_i + \sigma_e^2 = \gamma_0 = \sigma_Y^2$$

$$\sum_{i=1}^{p} a_i \gamma_{J-i} = \gamma_J, \qquad J \neq 0$$

In particular, for an AR(2) scheme the equations yield

$$\gamma_0 = \sigma_Y^2 = \sigma_e^2 \frac{1 - a_2}{(1 + a_2)(1 - 2a_2 + a_2^2 - a_1^2)}$$

$$\gamma_1 = \sigma_Y^2 \frac{a_1}{1 - a_2}$$

$$\gamma_2 = \sigma_Y^2 \frac{a_2 + a_1^2}{1 - a_2} \qquad (26A.1)$$

Putting these results together we find $(N\sigma_e^2)\mathbf{D}$, which estimates that the variance–covariance converges to

$$\begin{bmatrix} 1/\lambda_1 + B_1^2 K & B_1 B_2 K \\ B_1 B_2 K & 1/\lambda_2 + B_2^2 K \end{bmatrix} \sigma_e^2$$

where K is sum of the entries of $\mathbf{\Gamma}^{-1}$. Thus, the variance of $(\hat{b}_1 - \hat{b}_2)$ is approximated by

$$\left[\left(\frac{1}{\lambda_1} + B_1^2 K\right) - 2(B_1 B_2 K) + \left(\frac{1}{\lambda_2} + B_2^2 K\right)\right] \frac{\sigma_e^2}{N}$$

$$= \frac{1}{N} \sigma_e^2 \left[\frac{1}{\lambda_1} + \frac{1}{\lambda_2} + (B_1 - B_2)^2 K\right]$$

$$= \frac{1}{N} \{v\}, \qquad \text{say}$$

In the AR(2) case, note that

$$\mathbf{\Gamma} = \begin{bmatrix} \gamma_0 & \gamma_1 \\ \gamma_1 & \gamma_0 \end{bmatrix}$$

whence

$$K = \frac{1}{\gamma_0^2 - \gamma_1^2} (2\gamma_0 - 2\gamma_1) = \frac{2}{\gamma_0 + \gamma_1}$$

so that if

$$v = \sigma_e^2 \left[\frac{1}{\lambda_1} + \frac{1}{\lambda_2} + \frac{(B_1 - B_2)^2 2}{\gamma_0 + \gamma_1}\right]$$

and we set $\lambda_1 = \lambda_2 = \frac{1}{2}$ and $B_1 - B_2 = \sigma_Y$, as was done in the power computations of the text, we find that

$$v = \sigma_e^2\left[4 + \frac{2(1 - a_2)}{1 + a_1 - a_2}\right]$$

$$= 2\sigma_e^2 \frac{3 + 2a_1 - 3a_2}{1 + a_1 - a_2}$$

which, using the first Yule–Walker equation gives

$$v = \sigma_Y^2 \frac{2(1 + a_2)(1 - a_2 - a_1)(3 + 2a_1 - 3a_2)}{1 - a_2}$$

APPENDIX 26B
Steady-state solutions

Suppose that

$$Y_t = \sum a_i Y_{t-i} + mt + b + e_t$$

Then the expected value of Y_t is

$$E(Y_t) = \sum a_i E(Y_{t-i}) + mt + b$$

To find the steady-state solution of this difference equation in $E(Y_t)$, we assume that $E(Y_t) = Mt + B$ and solve for M and B:

$$Mt + B = \sum a_i(M(t - i) + B) + mt + b$$

Equating coefficients of t and 1, we have

$$\begin{aligned} Mt &= (\textstyle\sum a_i)Mt + mt \\ B &= \textstyle\sum a_i(-i)M + (\sum a_i)B + b \end{aligned}$$

so that

$$M = \frac{m}{1 - \sum a_i}$$

$$B = \frac{b - M(\sum ia_i)}{1 - \sum a_i}$$

27 Multivariate approaches

Time-series regression: intuitive overview

Throughout, this book has been concerned primarily with modeling the correlation among observations and has considered a nonzero mean or trend in the data a factor to be removed before analysis can continue. Chapter 26 presented some simple models in which the form of mean or trend was of more primary importance. In this chapter we consider more carefully the problem of fitting and testing a deterministic model in the presence of correlated errors. The usual least-squares procedure is based on the assumption that the errors (residuals) are not correlated. When this assumption is violated, a generalization of classical linear least-squares theory is necessary. Because regression analysis is much simpler and more elegant using matrix algebra, the reader who is not familiar with matrix notation (multiplication, transpose, and inversion) is urged to read Appendix 14A. This chapter will conclude with brief reviews of some other proposals for relating (either through regression or correlation) one series with one or more others.

Ordinary least squares

The matrix solution of the regression problem generalizes easily to multiple regression in which we try to predict Y_i with a fixed set of nonrandom variables $(x_{1i}, x_{2i}, \ldots)$:

$$Y_i = \hat{a}_0 + \hat{a}_1 x_{1i} + \hat{a}_2 x_{2i} + \cdots + \hat{a}_k x_{ki} + e_i \tag{27.1}$$

which can be written out as

$$
\begin{aligned}
Y_1 &= a_0 + a_1x_{11} + a_2x_{21} + \cdots + a_kx_{k1} + e_1 \\
Y_2 &= a_0 + a_1x_{12} + a_2x_{22} + \cdots + a_kx_{k2} + e_2 \\
&\vdots \\
Y_N &= a_0 + a_1x_{1N} + a_2x_{2N} + \cdots + a_kx_{kN} + e_N
\end{aligned} \tag{27.2}
$$

or in matrix form as

$$
\begin{bmatrix} Y_1 \\ Y_2 \\ \vdots \\ Y_N \end{bmatrix} = \begin{bmatrix} 1 & x_{11} & x_{21} & \cdots & x_{k1} \\ 1 & x_{12} & x_{22} & \cdots & x_{k2} \\ \vdots & \vdots & \vdots & & \vdots \\ 1 & x_{1N} & x_{2N} & \cdots & x_{kN} \end{bmatrix} \cdot \begin{bmatrix} a_0 \\ a_1 \\ \vdots \\ a_k \end{bmatrix} + \begin{bmatrix} e_1 \\ e_2 \\ \vdots \\ e_N \end{bmatrix} \tag{27.3}
$$

$$\mathbf{Y} = \mathbf{XB} + \mathbf{E} \tag{27.4}$$

which still has the least-squares solution (OLS)

$$\hat{\mathbf{B}} = (\mathbf{X}'\mathbf{X})^{-1}\mathbf{X}'\mathbf{Y} \tag{27.5}$$

This is also the solution if the variables, X, are random variables, *conditional* on the set of values observed in any particular set of data. A basic assumption of classical linear regression is that the errors are uncorrelated; that is, we assume that the expected value of $\mathbf{EE}'$ is equal to σ^2 times the identity matrix.

Generalized least squares

The generalization of classical linear regression (GLS) changes the assumption that the errors are uncorrelated to the assumption that

$$E(\mathbf{EE}') = \sigma^2\mathbf{\Omega} \tag{27.6}$$

where the matrix $\mathbf{\Omega}$ need not be the identity matrix, but meets some general conditions. In the classical case (OLS), $\mathbf{\Omega}$ is the identity matrix because the errors are assumed to be independently distributed. Under these conditions, Aitken (1935) generalized the Gauss–Markov theorem as follows.

According to the Cholesky decomposition theorem, if a matrix $\mathbf{D}$ is a positive definite matrix, there exists an upper triangular matrix $\mathbf{A}$ such that $\mathbf{A}$ is, in a sense, the "square root" of $\mathbf{D}$ (i.e., $\mathbf{AA}' = \mathbf{D}$). In the case of $\mathbf{\Omega}$, we can pick $\mathbf{A}$ such that $\mathbf{AA}' = \mathbf{\Omega}^{-1}$. In this case multiply the equation

$$\mathbf{Y} = \mathbf{XB} + \mathbf{E}$$

by $\mathbf{A}$:

$$\mathbf{AY} = (\mathbf{AX})\mathbf{B} + (\mathbf{AE}) \tag{27.7}$$

The covariance of these new residuals $\mathbf{AE}$ is σ^2 times the identity matrix.[1] Hence, the new covariance matrix for the transformed residuals is in the proper form for OLS. The least-squares solution for **B** is now

$$\hat{\mathbf{B}} = (\mathbf{X}'\mathbf{A}'\mathbf{A}\mathbf{X})^{-1}\mathbf{X}'\mathbf{A}'\mathbf{A}\mathbf{Y}$$

$$\boxed{\hat{\mathbf{B}} = (\mathbf{X}'\boldsymbol{\Omega}^{-1}\mathbf{X})^{-1}\mathbf{X}\boldsymbol{\Omega}^{-1}\mathbf{Y}} \tag{27.8}$$

By algebra it can also be shown that

$$\boxed{\boldsymbol{\Sigma}_{\hat{B}\hat{B}} = \sigma^2(\mathbf{X}'\boldsymbol{\Omega}^{-1}\mathbf{X})^{-1}} \tag{27.9}$$

In most cases, however, $\boldsymbol{\Omega}$ will not be known, and a two-stage procedure is suggested. First, proceed as if the residuals were independent and estimate **B** using ordinary least squares. Then, estimate $\boldsymbol{\Omega}$ from the residuals, and re-compute the estimate of **B** using the generalized least-squares method. Hibbs (1974) calls this two-stage estimation procedure "pseudo-GLS." To estimate $\boldsymbol{\Omega}$ we can use the autocovariance function. To see how, write the covariance of the residuals as follows:

$$E(\mathbf{EE}') = E\begin{bmatrix} E_1 \\ E_2 \\ \vdots \\ E_T \end{bmatrix}(E_1 E_2 \cdots E_T) = E\begin{bmatrix} E_1E_1 & E_1E_2 & \cdots & E_1E_T \\ E_2E_1 & E_2E_2 & \cdots & E_2E_T \\ \vdots & \vdots & & \vdots \\ E_TE_1 & E_TE_2 & \cdots & E_TE_T \end{bmatrix} \tag{27.10}$$

If the process E_t is assumed to be a realization of a stationary process, then

$$E(\mathbf{EE}') = \begin{bmatrix} \sigma^2 & \gamma_1 & \cdots & \gamma_{T-1} \\ \gamma_1 & \sigma^2 & \cdots & \gamma_{T-2} \\ \vdots & \vdots & & \vdots \\ \gamma_{T-1} & \gamma_{T-2} & \cdots & \sigma^2 \end{bmatrix} \tag{27.11}$$

where σ^2 is the variance of the stationary process and γ_k is the lag k auto-covariance of the process. In this case, $\boldsymbol{\Omega}$ is the striped matrix

$$\boldsymbol{\Omega} = \begin{bmatrix} 1 & \rho_1 & \cdots & \rho_{T-1} \\ \rho_1 & 1 & \cdots & \rho_{T-2} \\ \vdots & \vdots & & \vdots \\ \rho_{T-1} & \rho_{T-2} & \cdots & 1 \end{bmatrix} \tag{27.12}$$

of autocorrelations used to compute new estimates of **B**, var(**B**), and so on.

An alternative approach would be to fit an AR or ARMA model to the residuals E_t. Then the *theoretical* autocorrelation function of such a process could be used in place of the original autocorrelations. This more complicated process has the effect of smoothing the random irregularities in the original autocorrelation function and thus generally gives a better estimate of $\boldsymbol{\Omega}$.

The reader is referred to Hibbs's excellent (1974) review of the literature on the generalization of least-squares theory outlined in preceding sections of this chapter. Hibbs pointed out the bias that can be introduced in estimating the variance of parameter estimates when positive autocorrelation of the errors is ignored: that the error variance will be underestimated and "The model will appear to provide a much better fit to the empirical data than is actually the case" (p. 259). In particular, Hibbs showed that if the errors are realizations of an AR(1) process and $a_1 = .8$, classical linear regression "... would underestimate the true variance of [the regression] by 456 percent and inflate the t ratio by more than 200 percent" (p. 266).

Hibbs (1974) applied time-series regression to improve the ability of a multiple regression equation to predict Gallup polls of presidential popularity from Truman to Johnson, using 299 data points. He first demonstrated that the time dependence was fit by an AR(1) process, and then showed that his revised regression equation provided a better fit to the data than a classical equation obtained by Mueller (1970). Hibbs wrote:

> The GLS [time-series] estimate of 84.6 percent for the starting point of Truman's first-term popularity rating ... provides a much better fit to the empirical data then does the OLS {classical} estimates of 72.4 percent. Indeed Mueller's analysis of the OLS residuals led him to make special note of the poor performance of the model in accounting for Truman's extremely high initial popularity in the aftermath of Roosevelt's death. [P. 289]

It is interesting to contrast the generalized least-squares procedure with the regression used in the ITSE in Chapter 26. In Chapter 26, parameters corresponding to the slope and intercept [b_1, b_2, m_1, and m_2 in equation (26.15)] and parameters describing the correlation of successive observations were estimated simultaneously. Generalized least-squares estimates the deterministic parameters first, then separately examines the correlation of the residuals, then reestimates the deterministic parameters. Computationally, the former is easier. However, the models involved are different, and in general the parameters in a model such as equation (26.15) are rather difficult to interpret directly. The parameters in a GLS model have the same interpretations as they would in an OLS model. The model in the ITSE automatically allows for gradual cumulative changes in level; however, a simple GLS formulation might not.

Dynamic models

Hibbs (1974) reviewed dynamic models in which the past of the Y or X variables enters the model. An example of a dynamic model is

$$y_t = ay_{t-1} + bx_t + u_t \tag{27.13}$$

where u_t is a stationary time series such as

$$u_t = cu_{t-1} + v_t$$

in which v_t is white noise. We must also assume, as usual, that

$$E(u_t) = E(v_t) = E(u_{t-k}v_t) = E(y_{t-k}v_t) = 0$$

for all $k > 0$ and all t. Note that y_{t-1} is not independent of u_t. However, the equation can be rewritten as follows:

$$y_t = (a + c)y_{t-1} + bx_t - cay_{t-2} - cbx_{t-1} + v_t \tag{27.14}$$

This equation contains only three independent parameters, and all the terms are independent of v_t. The parameters can be estimated by nonlinear techniques or two-stage least squares (pseudo-GLS) can be applied directly to equation (27.13).

Of course, we could just ignore the structure in equation (27.13) and fit for parameters in (27.14), with the obvious loss of efficiency. It is true in general that an equation of the form (27.13), with a non-white-noise residual u_t, can be rewritten as a nonlinear autoregression on the past of each of the variables plus a white-noise residual v_t. The reader is referred to Hibbs's (1974) paper for a review of issues related to the estimation of these and other models.

Further remarks on transfer function models

The transfer functions employed in the Gottman–Ringland solution to the mother–infant interaction problem are an alternative to these regression models. However, here the x series and the y series are both treated as random variables, instead of just the y series. Moreover, it is possible to work with more than two series in transfer function models by regressing the current value of each series on the past of some or all of the others. Which model one prefers depends on the substantive problem, but if one is building models without the benefit of theory, multivariate autoregressive models and ITSE type models are considerably easier to use than GLS and dynamic ones.

These transfer function models can be approached alternatively by prewhitening each of the variables using autoregressive models, and then correlating residuals. This procedure was recommended by Pierce (1977), and it has distinct advantages when the predictor variables x_{ij} suffer from multicolinearity problems.

Problems of multicolinearity can be severe, and tend to occur when the time series have been generated by systems that influence one another in feedback loops. For example, the set of physiological data obtained from a subject in Robert Levenson's laboratory were heart rate (interbeat interval), respiration rate, and blood velocities from the heart to the finger and ear. The data are plotted in Figure 27.1. Note that there were four missing data points for the respiration data for this subject. The four missing respiration values (observations 34, 71, 81, and 100) of the data set were filled in using the best linear prediction of their values based on a fifth-order AR. The following stepwise procedure was used. First, the four observations were filled in with the process mean and ARs of varying order were fit. An AR(5)

Figure 27.1. Heart rate, respiration, and blood velocities to the finger and the ear from one subject.

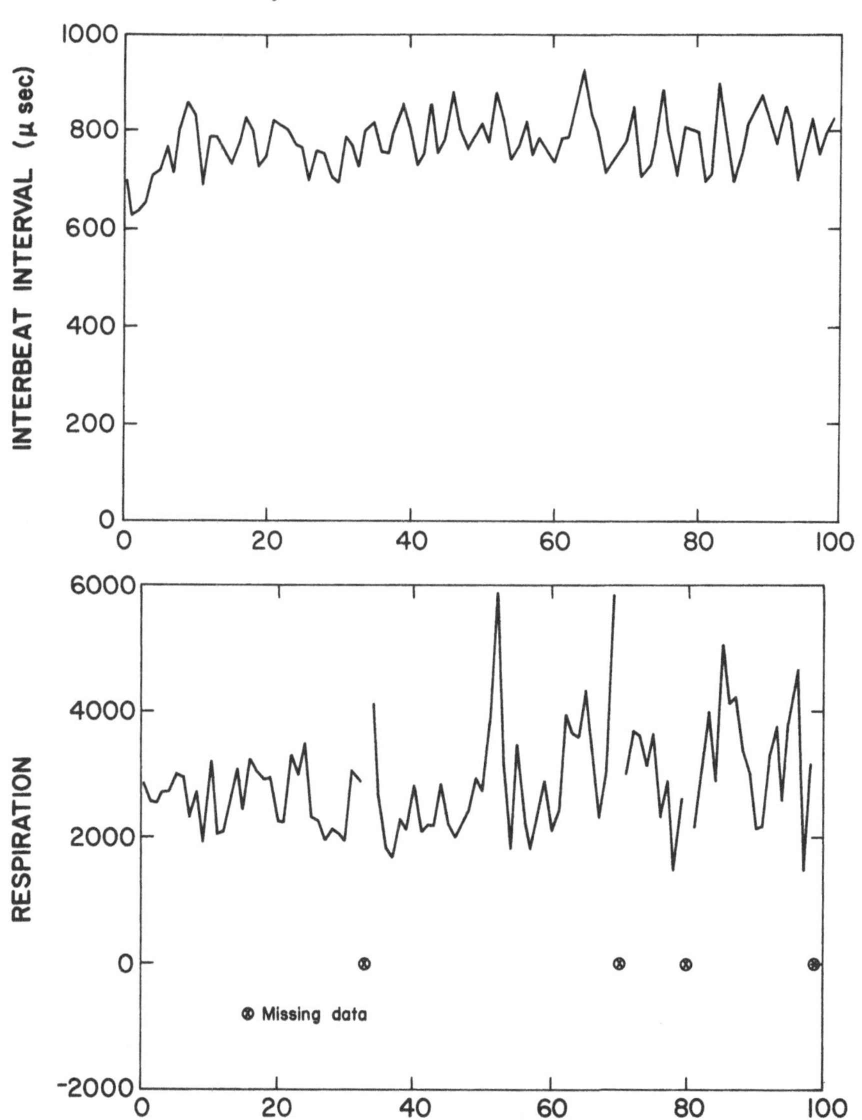

Figure 27.1. (*continued*)

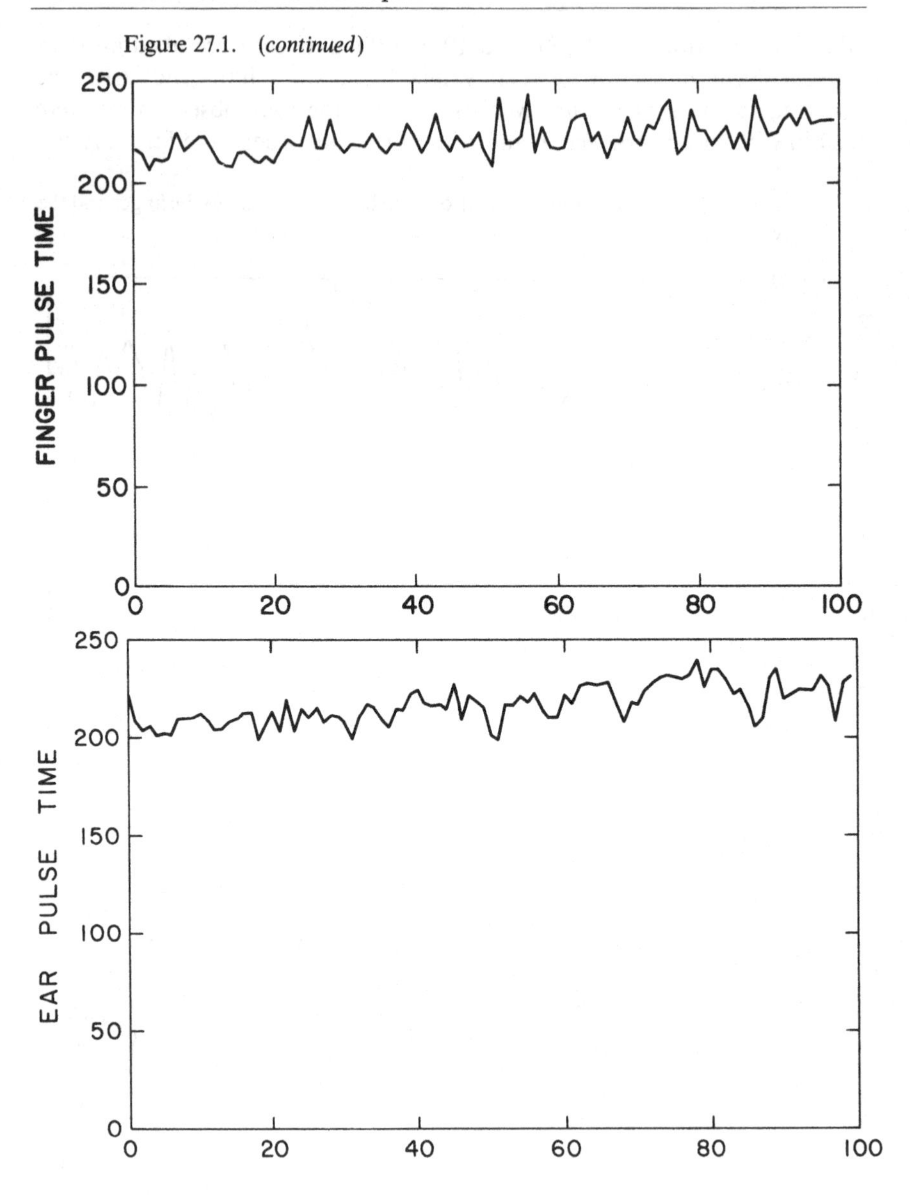

was found to be appropriate. Next, these coefficients were used to give a better prediction of what observations 34, 71, 81, and 100 should have been. The updated data were now refit and the missing values were repredicted. This procedure was iterated until the AR coefficients changed by less than .001 from one reprediction to the next.

The residual of this AR model was then correlated with the heart, finger, and ear residuals, giving the following pattern of autocorrelations and (residual) cross-correlations. Each variable was prewhitened and all possible cross-correlations between the whitened series were computed. The model represented by Figure 27.2 suggests that for this subject and these data collection procedures, respiration is less dependent on the heart's activity than are the blood velocities to the finger and ear. The essential difference between this analysis and the Gottman–Ringland transfer function for the mother–infant interaction problem is that in this analysis the model allows the *innovation* processes to be cross-correlated, even though they are not autocorrelated. Each series has been whitened by an AR fit. Nonetheless, it is possible for the white-noise residuals of each series to be correlated. For example, rainfall in one county may be unpredictable from its past but correlated perfectly with rainfall in the neighboring county. Note that although the model specifies linkages, these linkages are not unique. In Figure 27.1, for example, instead of the lines from A to B and from B to C, a direct link between A and C could have been drawn. Ringland suggests viewing the diagram as a bedspring model in the sense that motions anywhere will move parts of the spring connected to one another with the amount of motion transmitted indicated by the size of the correlation (Ringland, personal communication).

Figure 27.2. Bedspring multivariate model, using prewhitening of each time series.

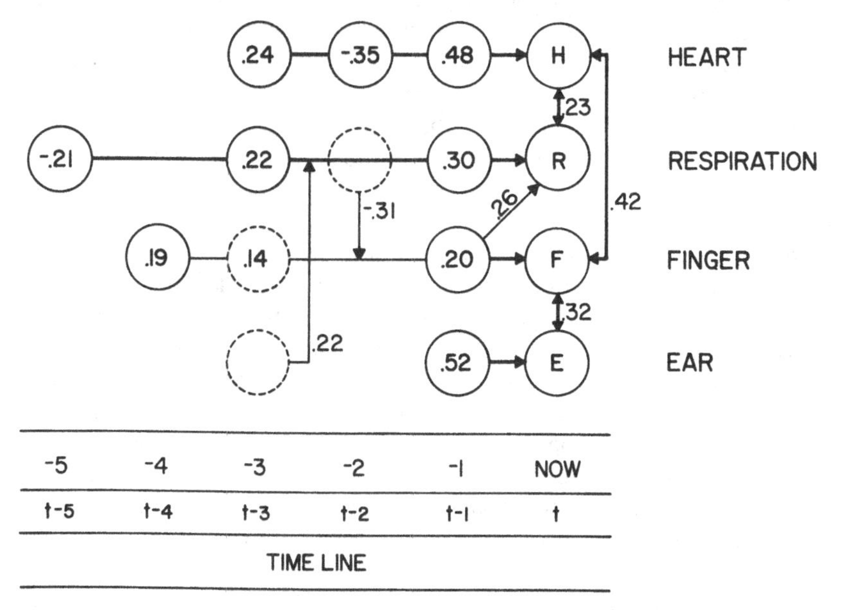

Multivariate spectral analysis

The reader is referred to Koopmans (1974) for discussions of extension from the bivariate to the multivariate case. Of course, the bivariate analysis on each pair forms a basis of multivariate analysis. But concepts of multiple and partial coherence, for instance, can be added. Koopmans (1974, pp. 157–8) reviewed a biological application of multivariate spectral analysis by Gersh and Goddard (1970), who recorded EEG patterns from six sites of an epileptic focus of the brain of a cat and found that the time series obtained from one focus was driving the other five. This analysis involved the use of partial coherence.

An additional multivariate frequency-domain approach was suggested by Brillinger (1975). If $x(t)$ is a vector of different time series $[x_1(t), x_2(t), \ldots, x_r(t)]$, an autocovariance matrix $\boldsymbol{C}_{xx}(u)$ and a spectral density matrix $\boldsymbol{p}_{xx}(f)$ can be computed. Brillinger suggests a principal-components analysis of this spectral density matrix, which will have a set of principal axes for each frequency.

What to read next

The time-series literature is sizeable and much of it is highly technical; for the beginner it can be bewildering to know what to read. This section is my recommendation about a few useful books.

Kendall (1973) is an excellent presentation of selected methods. It is a readable book and contains many figures and worked examples. One of its major advantages is that it is full of practical advice from a working statistician. It includes selected topics in both the time domain and frequency domains. I would recommend it as the next book to read.

Chatfield (1975) is less practical than Kendall (1973), and at times it requires a knowledge of calculus. Its major advantages are that it is a reasonably good review of both time-domain and frequency-domain methods and that it is well written.

Hannan's (1960) little book (152 pocket-sized pages) must be included in any list of what to read in time-series analysis. This book also includes topics in both the time and the frequency domains. It is amazing how succinctly so much valuable ground is covered by Hannan in this book. It should, however, be read slowly and carefully; the reader will probably repeatedly return to this book. A knowledge of calculus is necessary. The remaining list of recommended readings contains works that are more specialized than Kendall, Chatfield, or Hannan.

For writings about time-series analysis in the frequency domain, I highly recommend Jenkins and Watts (1968). Knowledge of calculus is required. This book is a useful introduction to the area from an applied standpoint. It is well written and ideas are carefully developed. On a somewhat higher mathematical level is an excellent monograph by Koopmans (1974). Together, these two books are a good introduction to spectral analysis.

For writings about time-series analysis in the time domain, Box and Jenkins's (1970) classic book is a thorough introduction to univariate and multivariate ARIMA model building and forecasting. However, the reader may find it tough going the first few times through, and I recommend skimming Anderson's (1975) little book that summarizes some of Box and Jenkins (1970). Anderson (1971) is a fine, but extremely detailed and also a tough-going supplement; it is particularly useful with respect to autoregressive models and inferential statistics.

An informative and practical book on forecasting is Montgomery and Johnson (1976). Hamming's (1977) book on digital filters is clear, readable, and practical, as are all of Hamming's books. An elementary and lucid introduction to generalized least squares (GLS) can be found in Wonnacott and Wonnacott (1970); for reading about GLS as applied to time-series problems, the reader is again referred to Hibbs (1974).

The Gottman–Williams computer programs

In response to what I see as the need for a flexible set of time-series programs that are primarily written for the *beginner*, a set of ten computer programs (in FORTRAN IV) and a manual were written. These programs provide the kind of output that social scientists require, such as statistical tests and confidence intervals. The manual, titled *Time series analysis computer programs: a manual for social scientists*, is available from Cambridge University Press. There are three packages of programs. The order form for the programs is provided in the manual.

Package 1 is for *univariate model fitting*. This package includes the following time domain programs: (1) *DETRND* removes trend, tests the significance of the trend-line parameters, and outputs the deviations from the trend line; (2) *LINFIL* transforms a time series in a variety of ways, including differencing, smoothing, and removal of a deterministic curve (as in the Porges example in Chapter 9); (3) *DESINE* removes a cycle of any period (specified by the user), performing a least-squares fit that finds the best amplitude and phase parameters and outputting a test of the significance of the cycle removed and a residual time series; (4) *ARFIT* performs an autoregressive model fit *up to* the order specified by the user, outputting the variance of the series, its autocovariances, autocorrelations, partial autocorrelation coefficients, the autoregressive model coefficients, tests of the significance of the coefficients, the residual time series from the AR model fit, and the Box-Pierce test of whether the residual is white noise. In the time domain, program *SPEC* computes the autocovariances, autocorrelations, spectral density estimates (using a Tukey-Hanning window), confidence intervals for each frequency, and a test of significance that compares the spectral density estimate to the theoretical density of white noise. The program also computes the periodogram. The density functions and confidence intervals are plotted as well as printed. The user specifies only the maximum lag for the calculation. Recall that larger maximum lags divide the frequency range into a finer grid than smaller maximum lags.

Package 2 is for the *application* of univariate model fitting. It includes two programs: (1) *FORCST* computes the best one-step ahead prediction, a confidence interval and the prediction variance using the autoregressive model fitting procedure; (2) *ITSE* stands for interrupted time-series analysis and performs an analysis for change in level and slope parameters following an intervention, using the linear autoregressive model procedure described in chapter 26. The user supplies only the order of the AR model; the output includes pre- and post-intervention slope and intercept parameters, significance tests of the parameters, an overall F-test, and t-tests for change in level and slope.

Package 3 is for *multivariate time series analysis*. In the time domain this package includes two parts. The first is *BIVAR*, which performs the computations of the Gottman-Ringland procedure (Chapter 25). The output is extensive and will not be described here. It includes a test of the adequacy of the stepwise procedure as well as tests of asymmetrical and symmetrical association between the two series, controlling for autocorrelation in each series. The user specifies only the maximum lags to be used [A, B, C, and D in Equations (25.15) and (25.16)]. The second part is *TSREG*, which performs a two-stage generalized least-squares regression. The program predicts one criterion time series from a set of other predictor time series; if used in combination with *LINFIL, lagged and bedspring models are possible.* This program outputs the correlogram, autoregressive coefficients, t-tests of these coefficients, and PACF for the first-stage residuals, the variance accounted for by the regression, t-tests for the generalized least-square beta weights, and the residual time series. In the frequency domain, Package 3 consists of *CRSPEC*, which performs a bivariate frequency domain time-series analyses. The user supplies only the maximum lag for the analysis. The program outputs the spectral density functions for each series, the autocovariances, the cross-covariances, the coherence spectrum–with a significance test for zero coherence–and the phase spectrum, with confidence intervals. All the spectra are plotted. The user can specify a shift parameter to reduce bias in estimating the coherence (see Appendix 23B).

The manual includes sample data and worked problems. Detailed information about the computational procedures employed in each program, the significance tests computed, and specific steps for modifying the capability of each program are given in the manual. The portability of the programs has also been tested.

APPENDIX I

Proof of the fundamental theorem of linear filtering

This proof refers to Chapter 17. Let $w = 2\pi f$, and let the covariance of the y_t process be $\gamma_{y,L}$, and the covariance of the x_t process be $\gamma_{x,L}$, where $L = $ lag, which are related by the equation $Y_t = \sum a_u X_{t-u}$. Also, let the spectral density of Y_t be denoted as $p_u(w)$ and the spectral density of x_t be denoted as $p_x(w)$. Then

$$\gamma_{y,L} = E(Y_s Y_{s+L})$$

$$= E\left(\sum_u a_u x_{s-u} \sum_v a_v x_{s+L-v}\right)$$

$$= \sum_u \sum_v a_u a_v \gamma_{x,L+u-v}$$

$$= \sum_u \sum_v a_u a_v \int_{-\pi}^{-\pi} e^{-iw(L+u-v)} p_x(w)\, dw$$

This last result is possible by the symmetry of the Weiner–Khintchine theorem–the theorem that the autocovariance is the Fourier transform of the spectral density function:

$$\gamma_{y,L} = \int_{-\pi}^{-\pi} \left(\sum_u a_u e^{-iwu}\right)\left(\sum_v a_v e^{iwv}\right) e^{-iwL} p_x(w)\, dw$$

$$= \int_{-\pi}^{-\pi} A(e^{-iw}) A(e^{iw})\, e^{-iwL} p_x(w)\, dw$$

$$= \int_{-\pi}^{-\pi} e^{iwL} [|A(e^{iw})|^2 p_x(w)]\, dw$$

The quantity in brackets in the preceding equation must be $p_y(w)$. Hence,

$$p_y(w) = |A(e^{iw})|^2 p_x(w)$$

APPENDIX II

Transfer-function weights from frequency-domain statistics

This appendix refers to Chapter 24. The linear filter in the time domain can be expressed.

$$y_t = h_0 x_t + h_1 x_{t-1} + \cdots + h_M x_{t-M} + e_t$$

and this can be expressed in operator notation (see Chapter 14) as

$$y_t = h(B)x_t + e_t$$

where

$$h(B) = \sum_{k=0}^{M} h_k B^k$$

If we replace the operator B by $e^{-2\pi f}$, it can be shown that

$$h(e^{-2\pi f}) = G_{yx}(f)e^{i\phi_{yx}(f)} = \sum h_k e^{-i2\pi fk}$$

where $G_{yx}(f)$ is the gain and $\phi_{yx}(f)$ the phase at frequency f. This function $h(e^{-i2\pi f})$ is called the frequency-response function. We can solve for h_k and obtain

$$h_k = \text{Re}\left[\int_0^{1/2} \frac{p_{yx}(f)}{p_x(f)} e^{i2\pi fk}\, df\right]$$

Jenkins and Watts (1968, p. 440) claim that this is often computationally a more tractable method for obtaining the impulse-response function than is the method described in Chapter 24.

Notes

Chapter 1. The search for hidden structures

1. The reader may also notice that there is a high-powered peak at $f = .5$. For the purposes of our discussion, this peak will be ignored.

Chapter 7. Models and the problem of correlated data

1. The computations that will be introduced in Chapter 8 were used to compute the autocorrelations. This footnote is for the reader who cycles back to this chapter after reading Chapter 8, or who already knows what "autocorrelation" and "lag" are. The autocorrelations for the 20 baseline days (prior to the intervention of using praise plus participation in a favorite activity contingent on not talking out) were: lag 1, .28; lag 2, .29; lag 3, .09; lag 4, .10; lag 5, $-.05$; lag 6, $-.01$. For $N = 20$, the Bartlett 2SD confidence bound is $2/\sqrt{N} = 2/\sqrt{20} = .45$, and so we could assume that these data are white noise. However, it might be noted here that just because the autocorrelation coefficients are not significant, this does not imply that autocorrelation does not exist, but that there is no power with this N to detect it.
2. The reader who makes a second pass through these chapters may be interested in noting that the estimates for one series need not be biased. For example, if we fit $y = \mu + e$, where e is normally, independently distributed with zero mean and variance σ^2, when either (1) and AR(1) model is correct (i.e., $y_t = ay_{t-1} + \mu^* + e$) or (2) an MA(1) model is correct, [i.e., $y_t = \mu^{**} + (be_{t-1} + w_t)$]. The $\bar{y}$ from the original model is unbiased in the AR case but biased for the MA case.
3. $2/\sqrt{N} = 2/\sqrt{5}$ is the approximate band around the autocorrelation of lag-1.
4. Scheffé (1959, p. 338) showed that if a time series $(y_1, y_2, \ldots)$ is distributed normally with zero mean and variance σ^2, and if the data have only lag 1 autocorrelation, ρ, the variance of the sample means of consecutive observations of size n will be

$$\frac{\sigma^2}{n}\left[1 + 2\rho\left(1 - \frac{1}{n}\right)\right]$$

Thus, if ρ is positive, then σ^2/n will underestimate the variance in sample means, and if ρ is negative, σ^2/n will overestimate it. The proof of this fact is not difficult. Simply expand the sample estimate of the variance in means from K consecutive samples, and take expectations.

Chapter 8. An introduction to time-series models: stationarity

1. The condition expressed in equation (8.8) is necessary for the two definitions–the ensemble definition and the series definition–of stationarity to be equivalent, that is, for the ergodic theorem for stationary processes to hold (see Koopmans, 1974, chap. 2).

It is not difficult to show that a sine-wave *process* is stationary using the first definition of stationarity. That is, if the process defines a family of curves of the form

$$x_t = A \sin(\theta t + \phi)$$

and ϕ is a uniformly distributed random variable on $(0, 2\pi)$ independent of the amplitude, A, which is also a random variable, then, whatever distribution of the random variable A has, the process will be stationary if $E(A) = 0$ and $\text{var}(A) = \sigma^2 =$ a constant. The autocovariance can be shown to be

$$\gamma(k) = \frac{\sigma^2}{2\pi} \cos \theta k$$

which is independent of historical time, t. Nonetheless, each *realization* of the process will be a deterministic sine wave. It is thus essential to keep the two definitions of stationarity distinct, because

$$\sum r_k^2 = \sum_k \cos^2 \theta k = \infty$$

so that the process is not stationary by the Wiener definition.

Chapter 9. What if the data are not stationary?

1. It can also be shown that for linear trend, the spectral density function will be an inverse square of the frequency (i.e., $1/f^2$) (see Kendall, 1973 p. 99 for a derivation).
2. Once again a note for the reader making a second pass through the book. This periodogram test will work only if the same bandwidth is employed for one-half the data as is employed for all the data.
3. One potential *advantage* of a differencing transformation is that it can be used to eliminate nonstationarity other than trend. For example, if the two halves of the time series are dominated by two different high-amplitude, low-frequency oscillations, they will both be eliminated by a differencing transformation. Conversely, a moving-average transformation has the potential for eliminating nonstationarity in the high-frequency end. Gottman and Ringland (in press) used this to eliminate nonstationarity in mother–infant play data from Tronick et al. (1977). The play data showed that in some cases, as the mother and infant played their cyclicities increased with time (i.e., their tempos became synchronously faster). This was an interesting result, and it reveals the value of *modeling* the nonstationarity instead of just eliminating it.

Chapter 14. The duality of MA and AR processes

1. *Proof:* $S = 1 + x + x^2 + x^3 + \cdots$; $xS = x + x^2 + \cdots$; $S - Sx = 1$; $S(1 - x) = 1$; $S = 1/1 - x$ (although this is true only if $|x| < 1$, it is used heuristically in Chapter 14.)

Appendix 14A. Matrix algebra primer

1. For example, the transpose of matrix **A** in Chapter 14 is the matrix

$$\mathbf{A}' = \begin{bmatrix} a_{11} & a_{21} \\ a_{12} & a_{2} \end{bmatrix}$$

Chapter 15. The spectral density function

1. Wold (1938) credits the arrangement in Table 15.1 to Buys-Ballot (1847); he points out that the arrangement was used before Buys-Ballot, and that the method was developed in detail by Stewart and Dodgson (1879) and sharpened by Whittaker (1911).
2. Wold (1938) referred to a course he took with H. Cramer in 1933 in which Cramer derived an "expectance theory of the Whittaker periodogram. He showed, i.e., that if the given series is a random sample of a normally distributed variables, then $\bar{D}^2(M_i)$ [the variance of the means, M, in Table 15.1] follows the well-known Student's distribution" (p. 25). James Ringland (personal communication) noted that this is basic analysis of variance. If $Y = XB + E$, where

$$X = \left[\begin{array}{cccccc} 1 & & & & & \\ & 1 & & & \bigcirc & \\ & & 1 & & & \\ & & & 1 & & \\ & \bigcirc & & & 1 & \\ & & & & & 1 \\ \hline 1 & & & & & \\ & 1 & & & \bigcirc & \\ & & 1 & & & \\ & & & 1 & & \\ & \bigcirc & & & 1 & \\ & & & & & 1 \\ \hline & & & \vdots & & \end{array}\right]$$

that is, $X_t = X_{t+6}$ repeating, then the $B = (B_1, B_2, \ldots, B_6)$ and the F test for the hypothesis H_0: $B_1 = B_2 = \cdots = B_6$ is

$$F_{5,N-6} \sim \frac{\sum_1^6 (\hat{\beta}_i - \beta_i)^2/5}{\sum_1^N (Y_i - \hat{Y})^2/(N-6)}$$

The denominator is similar to (but not quite equal to) σ_U^2. This test is just like one-way analysis of variance.

3. Wold (1938) referred to the stochastic property of a number of series, including terrestrial magnetism, air pressure in Potsdam, temperature in Batavia, and economic data. To make a case for the stochastic nature of these data, he employed Bartels's (1935) extension of the Buys-Ballot table. Bartels defined a "persistency characteristic" that would aid in an effort to distinguish deterministic from stochastic sine waves. The persistency characteristic is based on the well-known result that the periodogram of a deterministic sine wave increases proportionally with the number of observations. See also Chapter 16.

Chapter 16. The periodogram

1. There are other expressions for the periodogram. As Chatfield (1975) pointed out, the advantage of this definition "is that the total area under the periodogram is equal to the variance of the time series" (p. 135). This holds *not* for the simple model (16.3), but it will hold if we consider all available frequencies.

2. If we have a continuous spectrum, as T increases, we include more frequencies, so $I(f)$ remains bounded. Note now the mathematical difference between line and continuous underlying spectra. For a line spectrum, the periodogram increases without bound as T increases at the target cycle, f_0 [i.e., $I(f_0) \to O(T)$ as $T \to \infty$; this statement is read "$I(f)$ increases with the order of T as T goes to infinity"]; $I(f)$ goes to zero elsewhere, for $f \neq f_0$. To interpret this, recall that the total energy or variability is proportional to $T\sigma^2$ and all shows up at f_0. However, the frequency band around f_0 shrinks on the order of $1/T$, so that the total area under the periodogram is of the order of 1.0. For continuous spectra, on the other hand, the periodogram is bounded, approaches the spectral density function, and the increasing total energy is accounted for by including more frequencies.

3. The proof of the equation follows.

$$\hat{A} = \frac{2}{T} \sum x_t \cos 2\pi f t,$$

with $E(\hat{A}) = 0$; then

$$\operatorname{var} A = \frac{4\sigma^2}{T^2} \sum \cos^2 2\pi f t = \frac{4\sigma^2}{T^2} \frac{T}{2} = \frac{2\sigma^2}{T}$$

Thus,

$$\hat{A}^2 \sim \frac{2\sigma^2}{T} \chi^2(1)$$

and

$$I(f) \sim \frac{T}{4\pi} \left[\frac{2\sigma^2 \chi^2(1)}{2T} + \frac{2\sigma^2 \chi^2(1)}{2T} \right] = \frac{\sigma^2}{4\pi} [\chi^2(1) + \chi^2(1)] = \frac{\sigma^2}{2\pi} \frac{\chi^2(2)}{2}$$

4. Note that throughout this book, the autocovariance is estimated from the (zero-mean) data using the expression

$$c_k = \frac{1}{T} \sum_{1}^{T-k} x_t x_{t+k}$$

This estimate is biased for large lags, and

$$c_k = \frac{1}{T-k} \sum_{1}^{T-k} x_t x_{t+k}$$

is used in many other time-series texts.

Chapter 17. Spectral windows and window carpentry

1. The general expression for the Fourier transform that takes a time-domain window $w(k)$ into a frequency-domain window $W(f)$ is

$$W(f) = \frac{1}{2\pi} \sum_{k=-\infty}^{\infty} e^{-i2\pi fk} w(k)$$

2. In practice the window is truncated so that we have

$$\hat{p}(f) = \frac{1}{2\pi} \sum_{k=-\text{MAXLAG}}^{k=\text{MAXLAG}} w_k c_k \cos 2\pi fk$$

The user specifies MAXLAG for the program. The equivalent degrees of freedom for chi-square is $2N/\sum w_k^2$, which in the case of the Tukey–Hanning window is $2.67T/\text{MAXLAG}$. See Chatfield (1975), pages 149–50.

Chapter 18. Explanation of the Slutzky effect

1. For white noise, the autocovariance function is zero for lags greater than zero. Substitute these values into equation (16.19) and note that the following result is unaffected by windowed estimates to prove that the theoretical spectral density of white noise is

$$p(f) = \sigma_e^2/2\pi$$

2. This can be shown by using calculus, setting the derivative of equation (18.19) to zero and solving for f_0.

Chapter 19. AR model fitting and estimation: Mann–Wald procedure

1. The general expression can be written as follows. Let $a(k, k)$ be the kth coefficient of the kth-order AR model that is fit, and let $a(k, i)$ be the ith coefficient of the kth-order

AR model that is fit. Then the coefficients are written recursively as

$$a(k,k) = \frac{r_k - \sum_{j=1}^{k-1} r_j a(k-1, k-j)}{1 - \sum_{j=1}^{k-1} r_j a(k-1, j)}$$

and

$$a(k,i) = a(k-1,i) - a(k-1,k-i)a(k,k)$$

where r_j are the autocorrelations.

2. In this last set of numbers, the order p would be retained if $\sqrt{Tb_p}$ were < 1.96; that is, $\sqrt{Tb_p}$ is compared to a standard normal distribution.

3. The parameter d is the number of times the data had to be differenced to obtain a stationary series (see Chapter 20).

Chapter 20. Box–Jenkins model fitting: the ARIMA models

1. Aigner (1971) concluded that on the basis of a series of Monte Carlo experiments by Pagan (1970), there is little reason to prefer any other estimation method to setting the initial e_i equal to their expected value of zero. Note that if the e_i are assumed to be independent, identically *normal*, their least squares is a maximum likelihood approach, and a_i and b_i are the maximum likelihood estimators.

2. Once again, note that a grid-search procedure is sensible only if there are one or two model parameters. More general methods of nonlinear optimization exist for other cases that are more powerful and less clumsy (such as the algorithm of Marquardt) commonly available for computer computation.

3. One limitation of the criterion of minimizing residual variance in model fitting may be illustrated by the fact that it is possible to have a process that is generated by a pseudoperiodic AR(2) model with a_1 large and a_2 negative but small, so that the peak frequency is near zero. We may fit this to an AR(1) model because a_2 is small and accounts for little variance; however, this term is vital to understanding the process and its cyclicity. In general, it is wise to examine the spectral density estimate of the residual series.

Chapter 21. Forecasting

1. The forecast error must also include the error of the parameter estimates, a point not considered by Box and Jenkins (1970).

Chapter 23. Bivariate frequency-domain analysis

1. In fact, the estimate of the coherence using the periodogram and cross-periodogram is identically equal to 1.0, so smoothing is absolutely essential.

Proof. The periodogram for series $x(t)$ is

$$I_{xx}(\lambda) = \left| \frac{1}{(2\pi T)^{1/2}} \sum e^{i\lambda t} x(t) \right|^2$$

$$= z_x^*(\lambda) z_x(\lambda)$$

Similarly the cross-periodogram is

$$I_{xy}(\lambda) = z_x^*(\lambda) z_y(\lambda)$$

and the periodogram for the series $y(t)$ is

$$I_{yy}(\lambda) = z_y^*(\lambda) z_y(\lambda)$$

Hence, the unsmoothed estimate of the periodogram is

$$\frac{|I_{xy}|^2}{I_{xx} I_{yy}} = \frac{|z_x^* z_y|^2}{z_x^* z_x z_y^* z_y} \equiv 1$$

Chapter 24. Bivariate frequency example: mother–infant play

1. Equation (23.9) can be used to determine that the critical $\hat{\rho}^2$ is .198; for this computation, we need to know that the equivalent degrees of freedom for the analysis (which is a function of the number of lags and the spectral window selected) was 9.0.

Chapter 25. Bivariate time–domain analysis

1. Differencing can be suggested by the least-squares autoregressive parameters; for example, if $a_1 = 1.0$ or is close to 1.0, first differencing may be useful.
2. For the Tronick et al. (1977) data, it was also necessary to truncate the data, tossing out the last observation of each series. The last observation is an outlier, uncharacteristic of the pattern in the data, which represents the end of the interaction (see Figure 8.4, dyads 1 and 3).

Chapter 26. The interrupted time-series experiment

1. If the matrix is

$$W = \begin{bmatrix} a & b \\ c & d \end{bmatrix}$$

its inverse is

$$W^{-1} = \frac{1}{ad - bc} \begin{bmatrix} d & -b \\ -c & a \end{bmatrix}$$

2. Some explanation may be necessary for being able to compute SS_0 and SS_1. First, SS_1 is the residual sum of squares, $\mathbf{E}'\mathbf{E}$, computed from equation (26.15) or equation

(26.22). Second, SS_0 is $\mathbf{E'E}$, from the reduced model in equation (26.23). In matrix form, equation (26.23) is again $\mathbf{Y} = \mathbf{XB} + \mathbf{E}$, where

$$\mathbf{Y} = \begin{bmatrix} Y_{p+1} \\ Y_{p+2} \\ \vdots \\ Y_n \end{bmatrix}, \qquad \mathbf{B} = \begin{bmatrix} b \\ m \\ a_1 \\ a_2 \\ \vdots \\ a_p \end{bmatrix},$$

$$\mathbf{X} = \begin{bmatrix} 1 & p+1 & Y_p & Y_{p-1} & \cdots & Y_1 \\ 1 & p+2 & Y_{p+1} & Y_p & \cdots & Y_2 \\ 1 & p+3 & Y_{p+2} & Y_{p+1} & \cdots & Y_3 \\ \vdots & \vdots & \vdots & & & \\ 1 & N & Y_{N-1} & Y_{N-2} & \cdots & Y_{N-p} \end{bmatrix}$$

and

$$\mathbf{E} = \begin{bmatrix} e_{p+1} \\ e_{p+2} \\ \vdots \\ e_N \end{bmatrix}$$

In this model SS_1 is again $\mathbf{E'E}$.

Chapter 27. Multivariate approaches

1. By definition of the covariance,

$$\begin{aligned} \mathrm{cov}(\mathbf{AE}) &= E[\mathbf{AE}(\mathbf{AE})'] \\ &= E(\mathbf{AEE'A'}) \\ &= \mathbf{A}E(\mathbf{EE'})\mathbf{A'} \\ &= \mathbf{A}\sigma^2\mathbf{\Omega}\mathbf{A'} \\ &= \mathbf{A}\sigma^2\mathbf{\Omega}\mathbf{A'}(\mathbf{AA}^{-1}) \\ &= \mathbf{A}\sigma^2\mathbf{\Omega}(\mathbf{A'A})\mathbf{A}^{-1} \\ &= \mathbf{A}\sigma^2\mathbf{\Omega}\mathbf{\Omega}^{-1}\mathbf{A}^{-1} \\ &= \mathbf{A}\sigma^2\mathbf{I}\mathbf{A}^{-1} \\ &= \sigma^2\mathbf{I} \end{aligned}$$

References

Aigner, D. J. (1971) A compendium on estimation of the autoregressive moving average model from time-series data. *International Economic Review, 12*:348–71.

Aitken, A. C. (1935) On least squares and linear combination of observations. *Proceedings of the Royal Society of Edinburgh, 55*:42–8.

Anderson, O. D. (1975) *Time-series analysis and forecasting: the Box-Jenkins approach.* London: Butterworths.

Anderson, T. W. (1971) *The statistical analysis of time-series.* New York: Wiley.

Bartels, J. (1935) Zur Morphologie geophysikalischer Zeitfunktionen. *Sitzungsberichte der Berliner Wissenschaften, 139*:1–72.

Bartlett, M. S. (1946) On the theoretical specification of sampling properties of autocorrelated time-series. *Journal of the Royal Statistical Society, B8*:27–41.

(1955) *An introduction to stochastic processes.* Cambridge: Cambridge University Press.

Beveridge, W. H. (1921) Weather and harvest cycles. *Economics Journal, 31*:429–52.

Blackman, R. B., and Tukey, J. W. (1958) *The measurement of power spectra.* New York: Dover.

Bloomfield, P. (1976) *Fourier analysis of time-series: an introduction.* New York: Wiley.

Box, G. E. P., and Jenkins, G. M. (1970) *Time-series analysis: forecasting and control.* San Francisco: Holden-Day.

Box, G. E. P., and Pierce, D. S. (1970) Distribution of residual autocorrelations in autoregressive-integrated moving average time-series models. *Journal of the American Statistical Association, 65*:1509–26.

Box, G. E. P., and Tiao, G. C. (1965) A change in level of a nonstationary time-series. *Biometrika, 52*:181–92.

(1975) Intervention analysis with applications to economic and environmental problems. *Journal of the American Statistical Association, 70*:70–9.

Brazelton, T. B., Koslowski, B., and Main, M. (1974) The origins of reciprocity: the early mother–infant interaction. In *The effect of the infant on its caregiver* (ed. by M. Lewis and L. A. Rosenblum). New York: Wiley.

Brigham, E. O. (1976) *The fast Fourier transform.* Englewood Cliffs, N.J.: Prentice-Hall.

Brillinger, D. R. (1975) *Time-series: data analysis and theory.* New York: Holt, Rinehart and Winston.

Buys-Ballot, C. H. D. (1847) *Les Changements périodiques de température*. Utrecht.

Campbell, D. T., and Stanley, J. C. (1963) *Experimental and quasi-experimental designs for research*. Chicago: Rand McNally.

Chatfield, C. (1975) *The analysis of time-series: theory and practice*. London: Chapman & Hall.

Cox, D. R., and Miller, H. D. (1968) *The theory of stochastic processes*. New York: Wiley.

Durbin, J. (1970) Testing for serial correlation in least-squares regression when some of the regressors are lagged dependent variables. *Econometrica*, *38*:410–21.

Einstein, A. (1906) Zur Theorie der Brownschen Bewegung. *Annalen der Physik*, *19*: 372–81.

Fisher, R. A. (1970) *Statistical methods for research workers*. New York: Hafner Press.

Fourier, J. (1822) *La Théorie analytique de la chaleur*. Paris: Didot. (Translated by Alexander Freeman as *The analytical theory of heat*. Cambridge, 1878.)

Gersh, W., and Goddard, G. V. (1970) Epileptic focus location: spectral analysis method. *Science*, *169*:701–2.

Glass, G. V (1968) Analysis of data on the Connecticut speeding crackdown as a time-series quasi-experiment. *Law and Society Review*, *3*:55–76.

Glass, G. V, Willson, V. L., and Gottman, J. M. (1975) *Design and analysis of time-series experiments*. Boulder, Colo.: Colorado University Associated Press.

Gottman, J. M., and Glass, G. V (1978) Time-series analysis of interrupted time-series experiments. In *Single subject research* (ed. by T. Kratochwill). New York: Academic Press.

Gottman, J. M., and Parkhurst, J. (1980) A developmental theory of friendship and acquaintanceship processes. In *Minnesota symposia on child psychology*, vol. 13 (ed. by A. Collins). Norwood, N.J.: Lawrence Erlbaum.

Gottman, J. M., and Ringland, J. T. The analysis of dominance and bidirectionality in social development. *Child Development*, in press.

Gottman, J. M., Markman, H., and Notarius, C. (1977) The topography of marital conflict: a sequential analysis of verbal and nonverbal behavior. *Journal of Marriage and the Family*, *39*:461–77.

Granger, C. W. J., and Hatanaka, M. (1964) *Spectral analysis of economic time-series*. Princeton, N.J.: Princeton University Press.

Hall, R. V., Fox, R., Willard, D., Goldsmith, L., Emerson, M., Owen, M., Davis, F., and Procia, E. (1971) The teacher as observer and experimenter in modification of disputing and talking-out behaviors. *Journal of Applied Behavior Analysis*, *4*:141–9.

Hamming, R. W. (1977) *Digital filters*. Englewood Cliffs, N.J.: Prentice-Hall.

Hannan, E. J. (1960) *Time-series analysis*. London: Methuen.

(1970) *Multiple time-series*. New York: Wiley.

Hawkins, T. (1970) *Lebesque's theory of integration: its origin and development*. New York: Chelsea.

Hersen, M., and Barlow, D. (1976) *Single-case experimental designs: strategies for studying behavior change*. New York: Pergamon Press.

Hibbs, D. A. (1974) Problems of statistical estimation and causal inference in time-series regression models. In *Sociological methodology* (ed. by H. L. Costner). San Francisco: Jossey-Bass.

Holtzman, W. (1963) Statistical models for the study of change in the single case. In *Problems in measuring change* (ed. by C. W. Harris). Madison, Wis.: University of Wisconsin Press.

Huntington, E. (1945) *Mainsprings of civilization*. New York: Wiley.

Jenkins, G. M., and Watts, D. G. (1968) *Spectral analysis and its applications*. San Francisco: Holden-Day.

Johnston, J. (1972) *Econometric methods*. New York: McGraw-Hill.

Jones, R. H. (1965) A reappraisal of the periodogram in spectral analysis. *Technometrics*, *7*:531–42.

Jones, R. R., Vaught, R. S., and Weinrott, M. (1977) Time-series analysis in operant research. *Journal of Applied Behavior Analysis*, *10*:151–66.

Kendall, Sir M. (1973) *Time-series*. New York: Hafner Press.

Koopmans, L. H. (1974) *The spectral analysis of time-series*. New York: Academic Press.

Mann, H. B., and Wald, A. (1943) On the statistical treatment of linear stochastic difference equations. *Econometrica*, *11*:173–220.

McCleary, R., and Hay, R. A., Jr. (1980) *Applied time series analysis for the social sciences*. Beverley Hills: Sage.

Montgomery, D. C., and Johnson, L. A. (1976) *Forecasting and time-series analysis*. New York: McGraw-Hill.

Mueller, J. E. (1970) Presidential popularity from Truman to Johnson. *American Political Science Review*, *64*:18–34.

Ostrom, C. W., Jr. (1978) *Time-series analysis: regression techniques*. Beverly Hills, Calif.: Sage.

Padia, W. L. (1975) The consequence of model misidentification in the interrupted time-series experiment. Unpublished doctoral dissertation, University of Colorado.

Pagan, A. (1971) The estimation of models with moving average disturbance terms. Unpublished mimeo, Australian National University, cited in Aigner (1971).

Pierce, D. A. (1977) Relationships–and the lack thereof–between economic time-series, with special reference to money and interest rates. *Journal of the American Statistical Association*, *73*:11–26.

Quenouille, M. H. A large-sample of autoregressive schemes. *Journal of the Royal Statistical Society*, *110*:123–9.

Reiner, I. (1971) *Introduction to matrix theory and linear algebra*. New York: Holt, Rinehart and Winston.

Sackett, G. P. (1980) Lag sequential analysis as a data reduction technique in social interaction research. In *Psychosocial risks in infant–environment transactions* (ed. by D. B. Sawin, R. C. Hawkins, L. O. Walker, and J. H. Penticuff). New York: Bruner-Mazel.

Scheffe, H. (1959) *The analysis of variance*. New York: Wiley.

Schuster, A. (1898) On the investigation of hidden periodicities with application to a supposed 26-day period of meteorological phenomena. *Terrestrial Magnetism*, *3*:13–41.

Shewart, W. A. (1931) *The economic control of the quality of manufactured product*. New York: Macmillan.

Slutzky, E. (1937) The summation of random causes as the source of cyclic processes.

Econometrica, *5*:105–46. (This was the English translation of the 1927 Russian paper.)
Steidinger, D. (1979) Unpublished lab memo. University of Illinois.
Stern, D. (1977) *The first relationship: infant and mother*. Cambridge, Mass.: Harvard University Press.
Stewart, B., and Dodgson, W. (1879) Preliminary report to the committee on solar physics on a method of detecting the unknown inequalities of a series of observations. *Proceedings of the Royal Society*, *29*:15–37.
Tronick, E. E., Als, H., and Brazelton, T. B. (1977) Mutuality in mother–infant interaction. *Journal of Communication*, *27*:74–9.
Tukey, J. W. (1977) *Exploratory data analysis*. New York: Addison-Wesley.
Walker, G. (1931) On periodicity in series of related terms. *Proceedings of the Royal Society*, *A131*:195–215.
Whittaker, E. T. (1911) On the law which governs the variations of SS cygni. *Monthly Notes of the Royal Astronomical Society*, *71*:12–24.
Whittaker, E. T., and Robinson, G. (1924) *The calculus of observations*. New York: D. Van Nostrand.
Wiener, N. (1930) Generalized harmonic analysis. *Acta Mathematica*, *55*:117–258.
(1933) *The Fourier integral and certain of its applications*. Cambridge: Cambridge University Press.
(1948) *Cybernetics: or control and communication in the animal and the machine*. Cambridge, Mass.: MIT Press.
(1949) *The extrapolation, interpolation and smoothing of stationary time-series with engineering applications*. New York: Wiley.
Wold, H. (1938) A study in the analysis of the stationary time-series. Uppsala: Almquist & Wiksell.
Wonnacott, R. J., and Wonnacott, T. H. (1970) *Econometrics*. New York: Wiley.
Wu, S. M., and Pandit, S. M. Time-series and systems analysis: modeling and applications. In preparation.
Yule, G. U. (1971a) On a method of investigating periodicities in disturbed series with special reference to Wolfer's sunspot numbers. (Published in 1927.) In *Statistical papers of George Udny Yule* (ed. by A. Stuart and M. Kendall). New York: Hafner Press.
(1971b) On the time-correlation problems with special reference to the variate-difference correlation method. (Paper read in 1921.) In *Statistical papers of George Udny Yule* (ed. by A. Stuart and M. Kendall). New York: Hafner Press.
(1971c) Why do we sometimes get nonsense correlations between time-series? – A study in sampling and the nature of time-series. (Published in 1927.) In *Statistical papers of George Udny Yule* (ed. by A. Stuart and M. Kendall). New York: Hafner Press.
(1971d) On a method of studying time-series based on their internal correlations. In *Statistical papers of George Udny Yule* (ed. by A. Stuart and M. Kendall). New York: Hafner Press.
Zimring, F. E. (1975) Firearms and federal law: the gun control act of 1968. *The Journal of Legal Studies*, *6*:133–91.

Index